CISM International Centre for Mechanical Sciences

Courses and Lectures

Volume 574

The series presents lecture notes, monographs, edited works and proceedings in the field of Mechanics, Engineering, Computer Science and Applied Mathematics. Purpose of the series is to make known in the international scientific and technical community results obtained in some of the activities organized by CISM, the International Centre for Mechanical Sciences.

More information about this series at http://www.springer.com/series/76

Marta Boaro · Antonino Salvatore Aricò
Editors

Advances in Medium and High Temperature Solid Oxide Fuel Cell Technology

Springer

Editors
Marta Boaro
Department Polytechnic of Engineering and Architecture
University of Udine
Udine
Italy

Antonino Salvatore Aricò
Institute for Advances Energy Technologies (ITAE)
National Research Council (CNR)
Messina
Italy

ISSN 0254-1971 ISSN 2309-3706 (electronic)
CISM International Centre for Mechanical Sciences
ISBN 978-3-319-46145-8 ISBN 978-3-319-46146-5 (eBook)
DOI 10.1007/978-3-319-46146-5

Library of Congress Control Number: 2016952903

Printed on acid-free paper

This Springer imprint is published by Springer Nature
The registered company is Springer International Publishing AG
The registered company address is: Gewerbestrasse 11, 6330 Cham, Switzerland

Preface

Modern societies are starving of energy to support their activities and growth. Today, the energy consumption of a country represents a reliable measure of its wealth and of its industrial development. The most industrialized countries consume about 20 MWh per capita per year while the world average energy consumption is only of 2.4 MWh/year. Today, 81 % of energy production comes from the exploitation and transformation of fossil fuels. These evidences suggest that we need to rethink both the way how energy is produced and its conversion into electricity with the aim to achieve a sustainable and balanced development for every nation. Valorization of renewable resources and distributed energy generation are very appropriate strategies to strike a balance between the growing energy demand and the need to reduce the environmental impact on human activities.

This requires the development of green technologies for the production of energy and for a rational and more efficient use of resources. Fuel cells, which convert chemical energy directly to electricity, offer higher efficiencies and significantly lower emissions than conventional technologies. The modular configuration of fuel cells makes these devices suitable for a wide range of potential applications, including combined heat and power (CHP) production, distributed power generation and transport, which reduces reliance on imports of primary energy carriers and encourages local productivity.

In a transitional phase from an economy based on the predominant use of fossil fuels to one capable of efficiently exploiting renewable sources, medium/high temperature fuel cells based on ceramic oxides (solid oxide fuel cells, SOFCs) can play a fundamental role in accelerating the transformation. Thanks to their ability of using hydrocarbon-based fuels, the increased durability, and higher tolerance to contaminants, these type of cells could be immediately implemented in the established grid infrastructure and therefore have a rapid market penetration on a large scale.

Considering this potential, in recent decades, both the public and private sectors have invested resources to bring SOFCs to the commercial and residential markets, albeit with limited success. Progress in price reduction and performance increase has remained incremental, and a real launch of the technology has long resided "just around the corner." Insufficient longevity, reliability, and especially prohibitive costs are the main issues to be addressed.

The challenge of making fuel cells reliable, lasting, and efficient stemmed from the complexity of how they work, which needs to be truly understood, requires an interdisciplinary approach integrating scientific and technical knowledge. Several examples suggest that this type of methodology could achieve ground-breaking discoveries that would ultimately lead to viable commercial products.

For this purpose, the education of a class of scientists and engineers aware of how the technology is progressing and able to tackle the current challenges is crucial. In view of such request, this book gathers and updates a series of lectures given at the "Advances in Medium and High temperature Solid Oxide Fuel Cell Technology" international school organized by the editors in cooperation with the International Centre for Mechanical Sciences (CISM) in July 2014 in Udine (Italy). The book presents an overview of the recent advances in the field of SOFC technology and is intended as an introduction to the challenging issues that need to be addressed.

The chapters are written by internationally renowned scientists currently working at the leading edge of fuel cell research and development, and cover several themes such as the fundamentals of fuel cell thermodynamics, materials and components properties, electrodic processes, and the modeling principles for SOFC systems.

In chapter "Introduction to Fuel Cell Basics," fundamental electrochemistry of medium- and high-temperature fuel cells (solid oxide fuel cell, SOFCs) is explained with emphasis addressed to different aspects of high-temperature solid-state electrochemistry compared to low-temperature electrochemistry.

Chapter "Testing of Electrodes, Cells and Short Stacks" illustrates how to obtain reliable, accurate and reproducible electrochemical measurements through a proper selection of cell geometries and set-up. Principles, benefits, and drawbacks of different characterization techniques are discussed. Moreover, the concept of area-specific resistance (ASR) and how direct and alternate current methods can be optimized to provide not only the total ASR but also an electrochemical characterization of specific components (electrolyte and electrodes) of a fuel cell are described. The authors conclude by introducing the readers to some of the approaches used to study the effects of impurities on cell performance and to the problem of gas leakage in high-temperature fuel cells.

Chapters "Proton-Conducting Electrolytes for Solid Oxide Fuel Cell Application" and "Interconnects for Solid Oxide Fuel Cells" deal with new categories of materials to achieve the target of making SOFCs efficiently operative at temperatures as low as 550–600 °C. To this end, proton-conducting oxides have attracted widespread interest as electrolyte materials, alternative to traditional oxygen ion conductors. Chapter "Proton-Conducting Electrolytes for Solid Oxide Fuel Cell Application" presents an overview of main advances in the field of solid oxide proton-conducting materials describing several classes of materials such as perovskite-based materials (e.g., doped $BaCeO_3$, $BaZrO_3$, $BaCeO_3$-$BaZrO_3$ $SrCeO_3$, $LaScO_3$) or fluorite- or pyrochlore-based materials (e.g., doped $Ba_2In_2O_5$, CeO_2, $LaNbO_4$). Composition, transport, thermal and structural properties of the materials have been correlated with their conductivity and stability with the aim of indicating the most suitable materials for SOFC applications. Chapter "Interconnects for Solid Oxide Fuel Cells" deals with materials and strategies to prepare suitable interconnects. Authors illustrate the two main categories of interconnects: ceramic and metallic-type, pointing out advantages and disadvantages of both. The recent strategy of coating metallic interconnects with redox active oxide is also discussed.

In chapter "Catalysts and Processes in Solid Oxide Fuel Cells," physical and physicochemical properties of electrode materials are examined. It is pointed out that operation of anodes with fuels other than hydrogen (fossil and renewable fuels) is commercially necessary and challenging because of carbon deposition. The complex properties required for SOFC anodes are described, and issues related to nickel anode degradation are specifically addressed. The ultimate approach of producing stable direct oxidation in the anode is compared to the classical fuel reforming processes (internal and external to the anode), discussing advantages and disadvantages of the different methods.

The final chapters of the book discuss the SOFC technology in terms of an integrated system whose efficiency depends on several other subsystems. In this light, modeling is a key tool to design and optimize such systems avoiding extensive experimental investigations. Chapters "Energy System Analysis of SOFC Systems," "DOE Methodologies for Analysis of Large SOFC Systems," and "Solid Oxide Fuel Cells Modeling" describe the principle of mathematical modeling addressed to FC systems and processes optimization. The aim of Chapter "DOE Methodologies for Analysis of Large SOFC Systems" is to design experiment methodologies for analyzing large SOFC systems and the presentation of a case history using the CHP-100 kWe SOFC Field Unit (Siemens Power Generation-Stationary Fuel Cells) installed at TurboCare (Torino, Italy). Chapter "Solid Oxide Fuel Cells Modeling" provides an introduction to SOFC modeling using a macroscopic, physically based approach for the description of the chemical and electrochemical processes occurring at the electrodes.

We would like to thank all the authors for their valuable contribution to this book, safe in the knowledge that their work will provide graduate students, young researchers, and engineers with the scientific and technical know-hows to uncover the secrets of solid.

Udine, Italy

Marta Boaro
Antonino Salvatore Aricò

Contents

Introduction to Fuel Cell Basics 1
Robert Steinberger-Wilckens

Testing of Electrodes, Cells, and Short Stacks 31
Anne Hauch and Mogens B. Mogensen

Proton-Conducting Electrolytes for Solid Oxide Fuel Cell Applications 77
Dmitry Medvedev, Angeliki Brouzgou, Anatoly Demin and Panagiotis Tsiakaras

Interconnects for Solid Oxide Fuel Cells 119
Angeliki Brouzgou, Anatoly Demin and Panagiotis Tsiakaras

Catalysts and Processes in Solid Oxide Fuel Cells 155
Alfonsina Pappacena, Luca Bardini and Marta Boaro

Energy System Analysis of SOFC Systems 223
Andrea Lanzini, Domenico Ferrero and Massimo Santarelli

DOE Methodologies for Analysis of Large SOFC Systems 265
Domenico Ferrero, Andrea Lanzini, Pierlugi Leone and Massimo Santarelli

Solid Oxide Fuel Cells Modeling 291
Domenico Ferrero, Andrea Lanzini and Massimo Santarelli

Introduction to Fuel Cell Basics

Robert Steinberger-Wilckens

Abstract Basic principles of catalysis, thermodynamics, and reaction kinetics are very similar across the different types of fuel cells. Nevertheless, the small differences bring decisive variations in performance, function, limiting factors, sensitivity to impurity poisoning, and finally applications. This chapter explains the basic function of fuel cells, concentrating on the ceramic solid oxide fuel cell. Understanding the scientific basis of fuel cell operation helps in designing and optimising fuel cells, fuel cell stacks, and fuel cell systems.

1 World Energy Needs

In 2014, the average world energy use per capita was 1.8 tons of oil equivalent (toe, 1.8 toe/a*cap approx. being 25.2 MWh/a*cap) (BP 2015). This encompasses all primary energy use of the world economies divided by world population and includes residential (personal) energy use as well as transport, industry, military, commercial, administrative, infrastructure, and finally non-energetic use of oil and natural gas for chemical products.

Although this number may seem moderate, the regional distribution gives rise to concerns. Whilst Bangladesh forms bottom of the list with 0.2 toe/a*cap (2.8 MWh/a*cap, less than the average European household electricity bill), the USA is the first runner with 7.3 toe/a*cap (102.2 MWh/a*cap), followed by a number of other industrialised countries, including Japan and the EU with 3.7 and 3.3 toe/a*cap, respectively (51.8 and 46.2 MWh/a*cap). Table 1 shows primary energy use in some selected regions of the world together with the related CO_2 emissions.

R. Steinberger-Wilckens (✉)
School of Chemical Engineering, Centre for Fuel Cell and Hydrogen Research, University of Birmingham, Birmingham, UK
e-mail: r.steinbergerwilckens@bham.ac.uk

M. Boaro and A.S. Aricò (eds.), *Advances in Medium and High Temperature Solid Oxide Fuel Cell Technology*, CISM International Centre for Mechanical Sciences 574,
DOI 10.1007/978-3-319-46146-5_1

Table 1 Primary energy use, CO_2 emissions, and GDP per capita across prominent countries worldwide (based on (BP 2015) data for 2014, Worldbank, and UN world population data, Sept. 2015)

Country	PE use toe/ capita year	PE use MWh/ capita year	$t\,CO_2$/ capita year	GDP in US$/ capita year
Canada	9.3	111.3	17.31	50,235
USA	7.1	85.7	18.62	54,629
Australia	5.1	61.7	15.68	61,925
Russian Fed	4.7	55.8	11.30	12,735
Germany	3.8	46.0	9.84	47,821
Japan	3.6	43.2	10.59	36,194
EU 27	3.1	37.1	7.26	27,400
UK	2.9	34.7	7.25	46,331
PR China	2.2	26.0	7.11	7590
Brazil	1.4	17.3	2.84	11,384
India	0.5	6.0	1.63	1581
Bangladesh	0.2	2.1	0.45	1086

Judging from standards of living and gross domestic product (GDP) values in comparing the USA, the EU and Japan, for example, no major differences can be identified and the question arises for what reason and due to what circumstances this reft has persisted over now many decades.

Concepts of 'energy justice' (Sovacool and Dworkin 2014) would require that all citizens of the world—in principal—have equal and unbiased access to natural resources, of course within the limits of regional availability, and amongst these especially energy. 'Energy poverty' can be the cause of continued low economical development when economies cannot afford the purchase of enough energy to fuel their growth, thus perpetuating their backlog of income.

The simple mind experiment of having everyone worldwide move up to the same level of energy use as the USA or even Japan or the EU would completely shatter the world energy market and deplete the natural fossil reserves within no time at all. So obviously this is not the path forward and a balanced and more just world energy economy has to be built on other solutions than providing plentiful and cheap (subsidised) fuel for everyone. Haldi and Favrat (2006) have sketched out the '2 kW society', where the specific allocation of 2 kW average power (corresponding to 17.5 MWh or 1.25 toe annual energy use—by coincidence—or not—close to the current world average) to every world citizen allows sustainable and both environmentally as well as socially benign world economic growth.

The primary answer to excessive energy use, though, is increased energy efficiency (EU 2012). In the years since the first and second oil crisis, the linear dependency of economic growth on energy consumption has been completely overturned, with GDP growth nowadays being independent of energy use in the OECD economies (Worldbank 2015), even including the USA, albeit at a higher level.

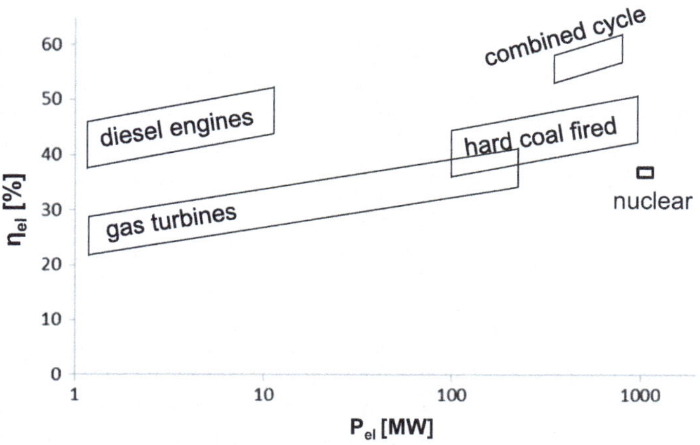

Fig. 1 Typical electrical efficiency of large-scale power stations as taken from data commonly published by utilities and manufacturers. Nuclear power stations by definition have an electrical efficiency of 33 %

And here you can now enter the fuel cell. It bears the promise of substantially increased efficiencies and of 'green' fuels thus allowing high exploitation levels of fuels and at the same time low to zero emissions. Since—in contrast to thermal machines which are based on a volumetric process—fuel cells and electrochemistry are a surface- and interface-based technology, they are easily scaled without changing the basics of the processes. Therefore, Fig. 1 shows the electrical efficiencies (in principle) achievable from conventional power generation equipment at different power ratings. Fuel cells can achieve efficiencies of up to 65 % (LHV) at any rating which puts them way above any of the competing technologies, especially since this is possible from size 1 kW upwards already.

The increased efficiency, though, is a promise. And surely, enough can be done to disappoint by bad engineering and ignorance to basic thermodynamical and electrochemical facts. In the following, we will further inspect these to form the base for a better understanding and help avoid the development of underperforming technology.

1.1 A Little Fuel Cell History

The basic principles and definitions of the terms 'electrolyte', 'electrode', and various electrochemical process descriptions were established by Michael Faraday (1791–1867). The electrolysis of water was first described by William Nicholson and Anthony Carlisle (1800). Reverse electrolysis was first realised by Sir William Grove (1811–1896) in 1839 on the basis of his own and Friedrich Schönbein's (1799–1868) research. Grove and Schönbein, a Swiss scholar, had been in close contact about this discovery.

Nevertheless, further progress was slow and major contributions to electrochemistry theory were only supplied by Wilhelm Ostwald (1853–1932) at the turn of the nineteenth to twentieth century. First technical developments were made and patents submitted between 1902 and 1913, when the company VARTA was assessing whether to invest in battery or rather fuel cell R&D. Armenian physicist Oganes Davtyan (1911–1990), who taught in Moscow and Odessa, published the first fuel cell monograph (his habilitation thesis) in 1947 (Davtyan 1947). He achieved major breakthroughs in quantum chemistry in the late 1960s, which did not gain him unconditional affection from the Soviet government and won him an exile in Armenia until his death.

The main reason for this slow progress was the insufficient understanding of reaction kinetics and materials' issues which play a significant role in improving the performance of fuel cells. Having today progressed considerably in the way of materials research and manipulation, we might say that if the eighteenth and nineteenth centuries were those of coal and steam, and the twentieth century that of oil, the twenty-first century might be the one of materials and electrochemistry?

As far as solid oxide fuel cells (SOFCs) are concerned, essential discoveries arrived even later: in the 1890s, the first description of ionic conductivity in ZrO_2 was delivered by German chemist Walter Nernst in the quest to replace gas lighting by electric filaments (Nernst 2003). In 1937 E. Baur and H. Preis built the first SOFC prototype (Baur and Preis 1937). It took practically a century more than for the platinum electrode-based Grove/Schönbein model to build an operational SOFC.

From the mid-1960s, the development of high-temperature fuel cells took off in the USA driven by Westinghouse and in Japan by Tokyo Gas and Mitsubishi Heavy Industries (MHI). In the 1990s, Siemens acquired Westinghouse (becoming Siemens Westinghouse Power Corporation, SWPC) and consequentially shut down its own SOFC development in Germany in 1996. In 1998, SWPC started their first tests with a 100-kW-size system in Arnhem, The Netherlands. At about the same time, 1999, Swiss technology company Sulzer had formed its subsidiary SulzerHexis and started first field trials with a 1 kW residential system.

Comparing the dates of discovery of the basic principles of electrochemistry with those of thermal energy conversion techniques, we note the century between the advent of the steam engine and the discovery of the fuel cell effect, and another century to the first SOFC prototype. Nevertheless, this should not be used as an excuse for the seemingly late arrival of the SOFC as a commercial product. Seeing the developments today, little still needs to be done to catch up with the developments of the low-temperature fuel cells.

1.2 The Very Basics of Fuel Cell Principles

The basic principle of fuel cells and electrolysis is to decompose a chemical reaction into several steps by prohibiting the free exchange of electrons in an oxidation reaction. Instead, the electrical transfer is divided into an electronic and an ionic charge transfer. As a result, the electron transfer can be utilised as an 'external' electrical current and the useful energy (see below) of the reaction transferred into the phenomenon known as 'electricity' whilst the 'internal' transfer is done by ionic transport.

In order to close the electric circuit and separate the electron transfer from the ionic charge transfer, a material is required that will have to satisfy the following conditions:

- electrically insulating,
- conductive for ions, and
- impenetrable to gases (fuels).

The electrochemical component that delivers this function is called the 'electrolyte'. It provides the separation of fuel and oxidant and serves as electronic insulator and ionic conductor. Technically, this is often called a 'membrane', a term we do not really use in the SOFC world. The interfaces at which the electronic/ionic separation takes place are the 'electrodes'.

Integrating all three properties in one material is not necessarily straightforward and explains some of the initial difficulties in developing operational fuel cell prototypes. Figure 2 gives an impression of the basic functions and elements of a fuel cell for a type with hydrogen ion (proton) conductivity and one with oxygen ion conductivity. The whole assembly of membrane and electrodes as depicted in Fig. 2 is generally called a 'membrane electrode' assembly (MEA)—alas, not with SOFC, where we speak of an 'SOFC cell' for the same combination of electrolyte and electrodes. This is mainly due to the fact that contrary to the low-temperature fuel cell world, electrolyte and electrodes of SOFCs can hardly be manufactured singly and highly build on each other.

The use of the labels 'cathode' and 'anode' requires caution. The electrochemical definition of the 'cathode' is the electrode at which:

- the reduction reaction takes place, or
- the transferred electrons enter the electrolyte (or where they combine with a positively charged ion).

Both definitions essentially say the same. Common sense, though, tells us that the 'cathode' will be the negatively charged electrode. This is correct if we inspect an electrolyser (a reverse fuel cell that will split water into hydrogen and oxygen). For a fuel cell, though, we find that the cathode is positively charged and the anode negatively. This is due to the reversal of the current flow in reversing the water-splitting reaction. Therefore, in speaking with scientists working in electrolysis, it is better to speak of the 'oxygen' and 'hydrogen' (or air and fuel) electrodes.

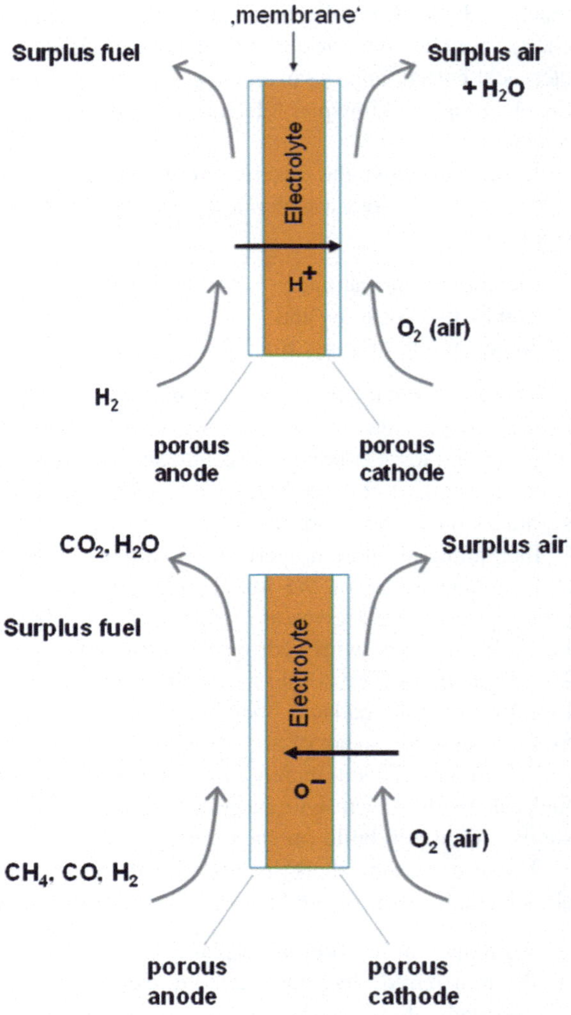

Fig. 2 Basic principle of a fuel cell: (*top*) with a proton-conducting electrolyte (membrane), as for instance in polymer electrolyte fuel cells (*PEFC*) operating at temperatures below 200 °C; (*bottom*) with an oxygen ion-conducting electrolyte, as in SOFCs operating between 500 and 950 °C

In both cases shown in Fig. 2, the direction of current flow is identical, as are the reaction products as long as the same fuel and oxidant are offered.

A fuel cell with a proton-conducting membrane, the typical polymer electrolyte fuel cell (PEFC) with a modified polymer membrane electrolyte (upper graph in Fig. 2), will display the electrode reactions (half-cell reactions)

$$\text{cathode side} \quad O_2 + 4\,H^+ + 4\,e^- \rightarrow 2H_2O \tag{1}$$

$$\text{anode side} \quad 2H_2 \rightarrow 4H^+ + 4e^- \tag{2}$$

giving an overall reaction of

$$2H_2 + O_2 \rightarrow 2H_2O \tag{3}$$

The oxygen ion-conducting SOFC (lower graph in Fig. 2), for example, delivers

$$\text{cathode side} \quad O_2 + 4e^- \rightarrow 2O^{2-} \tag{4}$$

$$\text{anode side} \quad 2H_2 + 2O^{2-} \rightarrow 2H_2O + 4e^- \tag{5}$$

giving the same overall reaction of

$$2H_2 + O_2 \rightarrow 2H_2O$$

Fuel cells are categorised according to two different systems: temperature and electrolyte material. Let us first look at the electrolyte classes (Fig. 3).

The PEFC is the most common proton-conducting electrolyte fuel cell. Its membrane consists of a modified Teflon that is available under the trade name Nafion®. It is often also called proton exchange membrane fuel cell (PEMFC), which is inconvenient since other fuel cells also work with protonic charge carriers. Therefore, the strict labelling according to electrolyte material should be preferred.

Historically, the phosphoric acid and alkaline fuel cells (PAFC and AFC, respectively) were the first types technically realised. They use liquid sulphuric

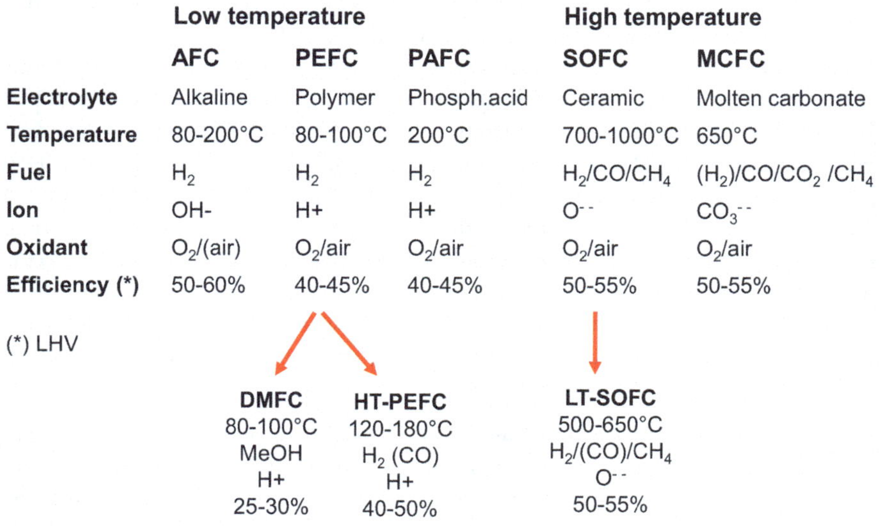

Fig. 3 Categories of fuel cell types sorted by temperature, stating various other characteristics. AFC alkaline fuel cell, PEFC polymer electrolyte fuel cell, PAFC phosphoric acid fuel cell, DMFC direct methanol fuel cell, HT-PEFC high-temperature PEFC, SOFC solid oxide fuel cell, LT-SOFC low-temperature SOFC, MCFC molten carbonate fuel cell

acidic or caustic potash electrolytes, respectively. The PAFC builds on a proton-conducting phosphoric acid electrolyte, whereas the AFC uses hydroxide ions (OH^-) as charge carriers.

All three types operate in the range of 40–100 °C with the PAFC running at temperatures as high as 200 °C. They require very pure hydrogen (5 point, 99.999 % pure). The electrolyte of the AFC will react with the carbon dioxide in air and form an insoluble precipitate of potassium carbonate that will eventually destroy the fuel cell. Operation on pure oxygen or at least reduced levels of CO_2 is therefore required. The PEFC strongly relies on platinum catalysts which will react with carbon monoxide impurities in the fuel—this deactivates the catalyst irreversibly.

In recent years, the high-temperature PEFC has been developed that operates at temperatures up to as high as 180 °C. This allows the HT-PEFC to cope with up to 3 % of CO in the fuel gas. Nevertheless, the performance is very much lower than with a standard PEFC.

One of the only two fuel cells to be labelled according to their fuel is the direct methanol fuel cell, which is a variant of the PEFC running on liquid fuel, namely methanol.

The molten carbonate fuel cell (MCFC) uses a salt as electrolyte that is liquid at the elevated operating temperature of 550–650 °C. The salt is contained in a ceramic structure and the charge carrier is a double negatively charged carbonate ion ($CO_3^{=}$), in essence a compound of a CO_2 molecule with a 'piggybacking' oxygen ion.

Finally, the solid oxide fuel cell (SOFC) is based on thin ceramic layers. It also uses oxygen ions as charge carriers and relies on high temperatures (550–950 °C) for the electrolyte to achieve sufficient ionic conductivity. Both MCFC and SOFC often use nickel at the anode and ceramic oxides as cathodes.

Due to the many possibilities of shaping ceramics, the SOFC has many forms and geometric variants. This will be discussed in the later chapter. It also has three subvariants. The intermediate-temperature SOFC (IT-SOFC) operates at temperatures below 600 °C. The proton-conducting fuel cell (PCFC) is a confusing label given to SOFC with hydrogen ion instead of oxygen ion conductivity of the electrolyte. And finally, the direct carbon fuel cell (DCFC) is the other variant named according to its fuel and integrates a solid carbon layer on the anode into the SOFC or MCFC structure.

From the above, two distinct realms of fuel cells can be determined: the low- and the high-temperature world. Very briefly, these can be summarised with the following properties:

- Low-temperature fuel cells (PEFC, AFC, PAFC) use hydrogen fuel with very high purity; they are typically found in mobile and portable applications; stationary applications are possible when combined with a reformer unit (on-site hydrogen production); predominantly they consist of lightweight materials (plastics, graphite, etc.).
- High-temperature fuel cells (MCFC, SOFC) can operate on hydrogen fuels but due to their elevated operating temperature can also directly convert natural gas

Type	CO	CO_2	N_2	H_2S	Cl	Particles
PEFC	<10 ppm	tol.	tol.	0	<0,05 ppm	0
PAFC	<2%	tol.	<2%	<50 ppm	<1 ppm	<1 mg/m³
MCFC	tol.	tol.	tol.	<0,1 ppm	<1 ppm	<1 μm
SOFC	tol.	tol.	tol.	<1 ppm	<1 ppm	<1 mg/m³

Fig. 4 Fuel impurity tolerance of the different fuel cell types. Apart from corrosive agents and dust which are not taken well by any fuel cell type, the low-temperature fuel cells require higher fuel purity

and hydrocarbons; their typical application is stationary power, but they are also found in auxiliary power units on vehicles; predominantly they consist of heavyweight materials, e.g. steels and ceramics.

Due to their much higher operating temperatures, the high-temperature fuel cells have the advantage of lower sensitivity to fuel impurities. Carbon monoxide can be used in both MCFC and SOFC as a fuel (Fig. 4).

1.3 Basic Thermodynamics for Fuel Cells

A thermodynamic 'system' is contained within a system boundary and is characterised by a number of primary variables, namely temperature T, pressure p, volume V, and mass m. Such a system could be simply imagined as a balloon that contains a specific mass of gas in a volume V at temperature T and pressure p.

The ideal gas law links the four variables as follows:

$$pV = mRT \tag{6}$$

This simple system will, at any time t, possess a certain state of energy described for instance by its temperature and pressure. This is quantified as its inner energy state, U. The energy state can be altered by exchanging activities with the surroundings, for instance by transporting energy across the balloon skin (boundary). This could happen either in the form of heating or cooling, or in the form of compression energy (work), either by reducing the volume, thus according to Eq. 6 increasing the pressure, or by expansion and cooling.

In an adiabatic system that is insulated from its surroundings, thus not transferring any heat or mass, Eq. 6 can be written as follows:

$$pV/T = \text{const} \tag{7}$$

indicating that if work is exerted on the system by compression/expansion, not only the pressure will change but also the temperature. The variables p and V could be considered as independent variables with T as the dependent variable.

Any change in inner energy state U has to be balanced by the exchange of heat Q and work W according to

$$U = Q + W \tag{8}$$

This is the first law of thermodynamics and the formulation of the conservation of energy. No energy can be 'created' or 'destroyed' ('consumed'). Within a complete system description that represents all processes within the system boundary and across it, the total amount of energy is a constant. This represents a 'closed' system. If the mass within the system was to be changed (and therefore again U), the system would have to be 'open' and allow an increase or decrease of m. This would for instance be the case if we blew gas into the balloon or let gas escape.

An arbitrary system might be subjected to a potential in a field of some kind (gravitational, electric, magnetic, etc.). Therefore, U will change, if we for instance pick up the balloon from ground level to some height h. We add energy to the system, which it will lose again when we let go of the balloon—it will fall to the ground and regain its initial state of energy. Nevertheless, for anything happening on the inside of the balloon, this event will be rather meaningless. Neglecting any subtle gradients in atmosphere, neither T, p, nor V will be affected by our lifting of the system.

In the following, we will concentrate on chemical and electrochemical processes and reactions. These will always occur between a 'before' (the reaction) and 'after'. Therefore, we are only interested in the change in U between the start of a process or reaction at time t_1 and its termination at t_2. We thus prefer writing Eq. 8 in the form

$$\mathrm{d}U = \mathrm{d}Q + \mathrm{d}W \tag{9}$$

It follows that in an isolated (adiabatic) system,

$$\mathrm{d}U = 0$$

Strictly speaking, all thermodynamic laws are only valid in the steady state (equilibrium). A transition from one state to another must therefore be calculated in infinitesimal steps. These are individually assumed to be in balance. Thermodynamics as such is about the balance between energy flows and the equilibrium states a system reaches after a change, not about speed or rate of reactions, which is what we call kinetics. This approach seems a bit formalistic but we will later see that it has some essential consequences.

The process occurring between two states at times t_1 and t_2 going through a transition according to Eq. 9 is called 'reversible' if it can be 'undone' (reversed) without loss of energy and the system returns to the state at t_1.

The second law of thermodynamics postulates the existence of a property S (called 'entropy') that can be calculated from

$$S = Q/T \quad \text{or} \quad \mathrm{d}S = \mathrm{d}Q/T \tag{10}$$

for reversible processes. Q in this case would also be called 'reversible heat'. Entropy in this definition can be considered as a measure of quality of heat, connecting thermal energy with the temperature that it is delivered or represented at.

If a process is irreversible, then

$$\mathrm{d}S > \mathrm{d}Q/T \tag{11}$$

since more heat is lost ('dissipated') in the process than can be reclaimed and converted back to work.

Combining Eqs. 9 and 10, we get

$$\mathrm{d}U = \mathrm{d}W + \mathrm{d}S \cdot T \quad \text{or} \quad \mathrm{d}W = \mathrm{d}U - \mathrm{d}S \cdot T \tag{12}$$

which is the equation most will recognise as the second law of thermodynamics. In a fully reversible process, work W (useful energy) can be converted from heat Q as per Eq. 9, and vice versa. If the processes are not reversible, losses will occur and U is reduced by more than the heat necessary for performing the work W. Thus, Eq. 11 cuts in.

In chemistry, we prefer to write 'enthalpy' H instead of U, and 'Gibb's free energy' G instead of work W. Using Δ instead of d to denote differences, we obtain

$$\Delta H = \Delta G + T\,\Delta S \quad \text{or} \quad \Delta G = \Delta H - T\,\Delta \mathrm{S} \tag{13}$$

neglecting any changes in pressure (expansion work) as is regularly done when analysing chemical reactions.

1.4 Efficiency

A simple thermodynamic system converting heat Q to work W can be constructed as shown in Fig. 5. This is a representation of a 'Carnot machine', the basic and most efficient thermal machine.

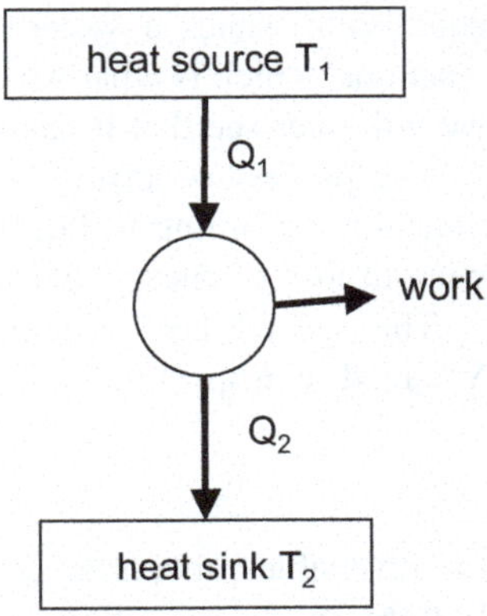

Fig. 5 A simple representation of a Carnot machine working between two temperature reservoirs at temperatures T_1 and T_2 and performing work W

Obviously, we would consider the efficiency of the process to be the ratio of the useful outcome, in this case the work W, to the energy input, in this case Q_1. The losses are represented by Q_2.

$$\eta_c = W/Q_1 \tag{14}$$

From the first law of thermodynamics, it follows

$$Q_1 - Q_2 - W = 0 \quad \text{or} \quad W = Q_1 - Q_2$$

and substituting Eq. 10 we obtain

$$S = Q_1/T_1 \quad \text{and} \quad S = Q_2/T_2$$

thus, it follows that

$$\begin{aligned}\eta_c &= (T_1 - T_2)/T_1 \\ &= 1 - T_2/T_1\end{aligned} \tag{15}$$

which is the maximum (idealised) efficiency a thermal machine can achieve and is called the 'Carnot' efficiency. Figure 6 shows the Carnot efficiency as a function of the input temperature T_1. In order to maximise the Carnot efficiency, either T_1 has to be considerably increased, or T_2 decreased, or both. For 100 % efficiency, T_1 would need to be infinite, or T_2 zero Kelvin, both not possible on the earth.

1.5 Chemical Reactions

Consider a chemical reaction between substances A and B, forming products C and D:

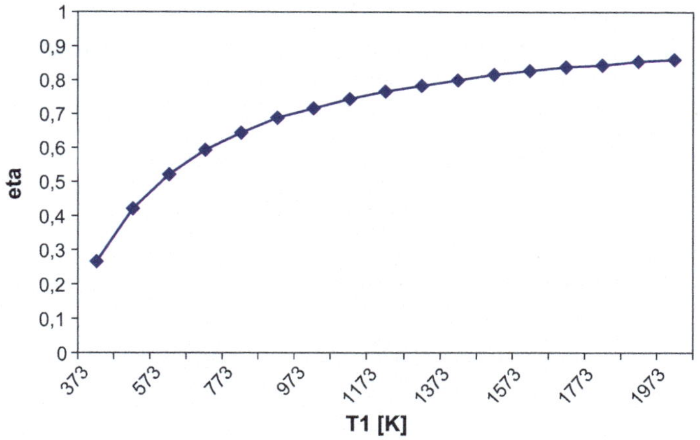

Fig. 6 Temperature dependency of the Carnot efficiency. As can be seen from Eq. 16, η_C approaches 1 asymptotically with growing process temperature T_1 when reject temperature T_2 is kept constant (0 °C in this case)

$$\alpha A + \beta B \leftrightarrow \gamma C + \delta D \tag{16}$$

where the α, β, γ, and δ give the relative concentrations of the single reactants and products, the '↔' indicating the reaction is in equilibrium (balance). This equilibrium of reaction is temperature dependent and implies that the forward and backward reactions are in balance (i.e. have the same rate).

We can then define a temperature-dependent equilibrium constant

$$K(T) = \frac{[C]^{\gamma}[D]^{\delta}}{[A]^{\alpha}[B]^{\beta}} \tag{17}$$

where $[n]$ is the 'activity' (expressed as concentrations in liquids or as partial pressures in gases) of reactant or reaction product n and K is used as K_c for the molarity or concentration-related equilibrium constant, K_p when partial pressures are used.

If $K \sim 1$, both reactants and products coexist at similar shares. Whatever the value of $K(T)$, thermodynamics tells us that the composition is in balance, i.e. in equilibrium. What pathway the reaction takes, how fast it processes, and how long it takes to reach equilibrium, this approach does not tell us. Thus, any gas composition including the (in this case) four components will eventually progress to this specific equilibrium.

If
$K > 1$, products will prevail,
$K \gg 1$, the reaction will be complete (with only very marginal amounts of reactants left),

$K < 1$, reactants prevail,
$K \ll 1$, no reaction takes place at all (with only very marginal reaction products present).

For the reaction

$$H_2 + \frac{1}{2}O_2 \leftrightarrow H_2O$$

at $T > 823$ K, the equilibrium coefficient is

$$K = \frac{[H_2O]^1}{[H_2]^1[O]^{1/2}} = 4.64\ \text{E26}$$

which underpins that at temperatures above ~550 °C hydrogen will spontaneously react with any oxygen present and form water.

Let us inspect two reactions, one of which is endothermic (hydrogen iodide synthesis), the other exothermic (ammonia synthesis). Figure 7 shows the temperature dependence of K_p in an Arrhenius plot (over *1/T*).

We can clearly see the linear relationship between $\ln(K_p)$ and *1/T* and can write Eq. 18 for the energy of formation as

$$\Delta G^\circ = -\text{RT}\ln K_p \tag{18}$$

Inserting in the second law of thermodynamics and rearranging gives us

$$\ln K_p = -(\Delta H^\circ/RT) + \Delta S^\circ/R \tag{19}$$

for the case that ΔH° and ΔG° are themselves not strongly temperature dependent.

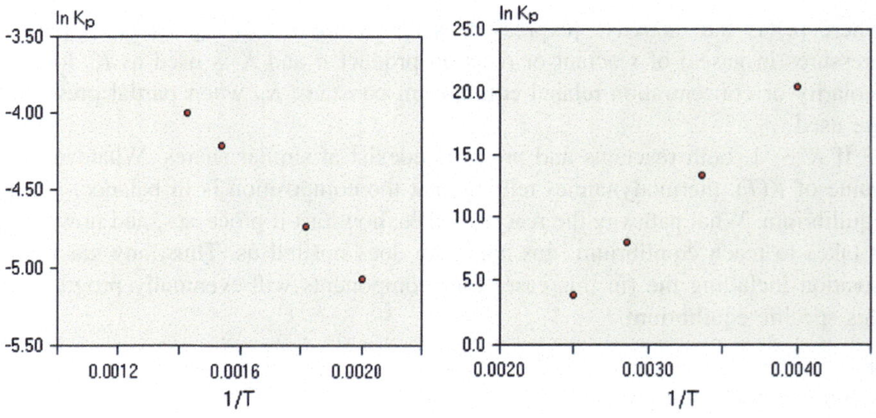

Fig. 7 Temperature dependence of K_p for an endothermic (hydrogen iodide synthesis, HI, *left*) and an exothermic reaction (ammonia synthesis, NH_3, *right*)

Table 2 HI and ammonia synthesis values for K_p

HI synthesis (endothermic)		Ammonia synthesis (exothermic)	
T (K)	K_p (−)	T (K)	K_p (−)
500	6.25×10^{-3}	250	7×10^{8}
550	8.81×10^{-3}	298	6×10^{5}
650	1.49×10^{-2}	350	2×10^{3}
700	1.84×10^{-2}	400	36
720	1.98×10^{-2}		

A catalyst does not influence the chemical equilibrium—it merely influences the reaction rates (in both directions!) by reducing the intermediary energy transitions (activation energy).

The rate of reaction can also be shifted to one side of the reaction equation by increasing or decreasing the concentration of reactants or products, respectively. This is called *Le Chatelier's* principle. It describes the property of reactions to attempt to reach thermodynamical equilibrium. If the balance is artificially disturbed by removing or adding reactants or products, the equilibrium tendency will try to counterbalance. If the product is removed (its partial pressure is reduced), the reaction rate will increase in order to re-establish the balance. The same effect will be achieved when increasing the concentration (partial pressure) of the reactants.

The ammonia synthesis reaction cited in Table 2 and Fig. 7 can be written as follows

$$N_2 + 3H_2 \rightleftharpoons 2NH_3 + Q, \quad \Delta H = -92\,\mathrm{kJ\,mol^{-1}} \tag{20}$$

where the heat released from the exothermic reaction is shown as Q. Applying *Le Chatelier's* principle, it can be concluded that temperature and heat can also influence the reaction balance. And indeed, increasing the temperature (or synonymously: the heat input) will reduce K in exothermic reactions. This is precisely what Fig. 7 shows. On the other hand, lowering T will reduce the reaction rate (independent of the balance as expressed by K). This will obviously lead to an optimum temperature for a given reaction at which a maximum of product is achieved.

In endothermic reactions, K will increase with T and heat input, as Fig. 7 shows for the HI synthesis.

1.6 Basic Electrochemistry

A well-known and simple way of producing hydrogen is water electrolysis. Two electrodes are placed in a container with water, a voltage is applied, and bubbles of gas will form at the two electrodes. These can be identified as hydrogen at the negative terminal and oxygen at the positive terminal. Transferring this mind

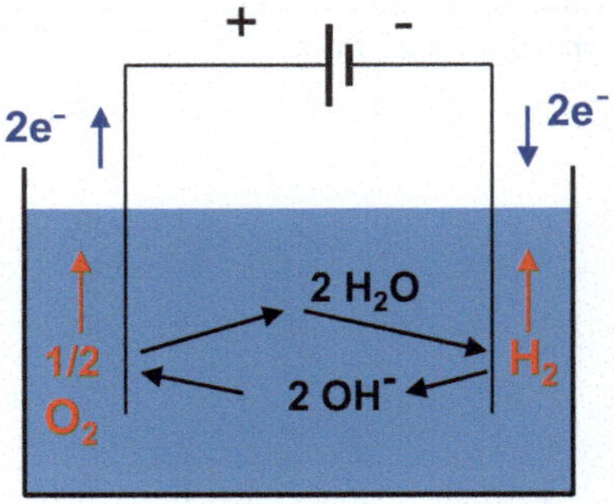

Fig. 8 Charge transport and half-cell reactions in alkaline water electrolysis

experiment into practice is not quite as trivial as it seems, but essentially this is how high-purity hydrogen is produced.

The following reaction equations show one of the several possible water-splitting reaction pathways:

$$\text{cathode half–reaction} \quad 2H_2O + 2e^- \rightarrow H_2 + 2OH^- \tag{21}$$

$$\text{anode half – reaction} \quad 2OH^- \rightarrow \frac{1}{2}O_2 + H_2O + 2e^- \tag{22}$$

$$\text{overall reaction} \quad H_2O \rightarrow H_2 + \frac{1}{2}O_2 \tag{23}$$

Figure 8 shows a schematic of these reactions. At the negative terminal (in this case the cathode), electrons are received from the feeding electrical circuit and used to form two negatively charged hydroxyl ions from two water molecules, thus shedding a hydrogen molecule. The negatively charged ions diffuse through the water towards the positively charged other electrode, the anode, and these are reshuffled into a water molecule, transferring two electrons to the electrode and releasing an oxygen radical. This will eventually recombine with another oxygen radical to form an oxygen molecule.

The takeaway insights are as follows:

- we observe a feed of electrical current (charge) into the water that will eventually lead to splitting the water molecule,
- the amount of charge transferred is equal to two electrons per molecule of water,
- at the electrodes, electrons are transferred; the transport of charge between the electrodes is performed not by electrical current but by ionic charge transfer (diffusion of ions),
- obviously, an electrical current through the liquid would have a detrimental impact (internal short circuit),
- mixing of the two gases will be undesirable,
- it will be of advantage, if the liquid contains hydroxyl ions, which will support the ionic conductivity.

We call the liquid in the container an 'electrolyte'. It separates the two electrodes by

- not allowing electronic conductivity (electrically isolating),
- preventing mixing of the two gas streams (gas tightness),
- supporting ionic conductivity (high conductivity for the charge-carrying ions).

The 'cathode' is the negatively charged terminal. The definition of a 'cathode' is that it is the terminal at which electrons are transferred to the ionic charge carriers. This is synonymous with the location of the 'reduction reaction', where the half-reaction product, the hydroxyl ion, is reduced (made more negative in charge) by receiving the electrons. The other terminal is obviously the 'anode', where the complementary oxidising reaction takes place in which electrons are transferred away from the ions to form the oxygen molecule. This all sounds very familiar in light of what was explained with Fig. 2 and, indeed, the electrolysis and fuel cell reactions are the exact reverse of each other.

The example shown in Eqs. 21 and 22 is a so-called alkaline electrolyser that uses potassium hydroxide as the electrolyte. In practical application, water will continuously have to be fed into the liquid electrolyte in balance with the gases formed.

There are a number of further things we know about the water-splitting reaction:

- the total charge transferred per mol of water split is twice the Faraday constant F, which is the amount of charge a 'mol' of electrons carries, since two electrons are required per water molecule; F is equal to 96.587 C/mol;
- the energy required to split water is equal to the energy of formation (to be precise: with the inverse sign).

From calorimetry, the latter can be distinguished as follows:

$$\Delta G_o = 237\,\mathrm{kJ/mol} \tag{24}$$

at equilibrium conditions at STP (see below). We are therefore considering a process that is in balance and thus neither produces a net balance of hydrogen or oxygen, nor consumes any energy. This condition is also called open-circuit voltage (OCV) since it describes a situation with no (net) current flow.

The work performed in splitting the water molecule is identical with the electric energy fed into the electrolyte. This is equal to the voltage between the terminals times the charge flow, in this case represented by the electrons transferred. From the first law of thermodynamics, it then follows that

$$\Delta G_o = 2FU_o^o \tag{25}$$

at 'standard state' (standard temperature and pressure, STP, superscript '0', mostly defined in tabulated data as 25 °C and 1 bar). We can insert Eq. 24 into Eq. 25 and receive

$$U^{o}{}_{o} = 1.23\,\mathrm{V} \tag{26}$$

which is the exact voltage necessary for water splitting at STP under equilibrium conditions.

In practice, of course, the process of water splitting will not be performed without losses (second law of thermodynamics!). The entropy induced will express itself in heat produced by the current flow (ohmic heat), kinetic processes that do not perform according to the theoretical ideal ('overpotential' necessary to create a current flow), and competition of ionic reactions at the electrodes that lead to 'parasitic' currents. As a result, the voltage required for electrolysis will always be higher than given in Eq. 26, once a current is fed to the device and the equilibrium condition (OCV) is left. Generally, this will be in the order of 1.7–1.9 V, instead of the ideal 1.23 V at STP.

The theoretical amount of energy needed per $Nm^3$of H_2 (a cubic metre of hydrogen at normal temperature and pressure conditions, NTP, 20 °C and 1 bar) is

$$\Delta G_o / V_N = 237\,\mathrm{kJ\,mol^{-1}} / 22.41\,\mathrm{l\,mol^{-1}} \approx 3\,\mathrm{kWh/Nm^3} \tag{27}$$

where V_N is the volume of a mol of an ideal gas at NTP. State-of-the-art electrolysers will require about 4.2–4.8 kWh/Nm^3, giving the process an efficiency of typically ~65 %.

Due to the second law of thermodynamics

$$\Delta G = \Delta H - T\,\Delta S \tag{28}$$

we can deduct that the energy for splitting water may not only be supplied by electricity (which is pure Gibb's free energy) but can also be supplied by heat ($T\ \Delta\ S$). Figure 9 shows the relationship of ΔG and ΔH with temperature. Due to Eq. (28), the amount of energy supplied as ΔG declines with rising temperatures, letting an increasing part of the water splitting be done by heat input $Q = T\ \Delta\ S$.

Assuming electrolysis and ΔG taking the form of electricity, this means we need increasingly less electricity and substitute this by heat. From Eq. (28), we can derive that the voltage at higher temperatures is reduced—which implies less electrical energy at a given hydrogen production rate. The production rate will depend on the rate at which water is split, which again will depend on the current flow, which finally depends on the amount of charge transferred. Therefore, the current drawn will remain constant whilst we thermodynamically require a lower voltage.

Table 3 shows how ΔG (and thus voltage) decreases with temperature. For an electrolyser (by definition running on electricity), this means the output (in the shape of hydrogen) per unit of electrical energy (for instance kWh) increases with temperature—as long as we do not have to account for the heat input. Using a rather conventional definition of 'energy input' as being equal to the electricity input alone (which thermodynamically is incorrect), we receive efficiencies above 100 % which is of course physically not possible. Nevertheless, we can quite rightly claim that up to 100 % of the electrical input is converted to chemical energy in the form of hydrogen, the remainder of the work being done by heat.

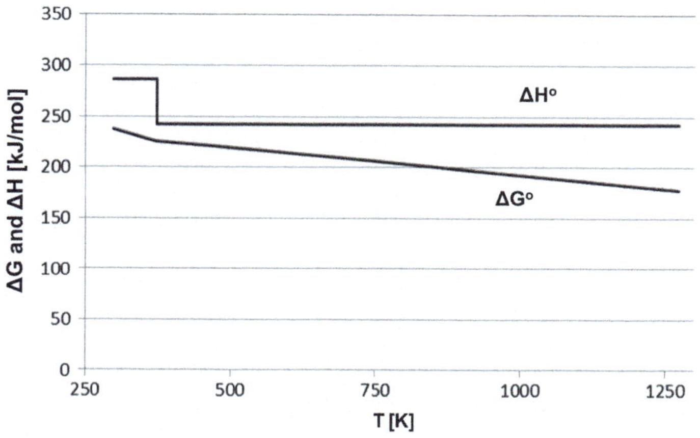

Fig. 9 Dependence of ΔH and ΔG on temperature at ambient pressure for the reaction according to Eqs. (21–23). The step change in ΔH corresponds to the evaporation enthalpy at 100 °C

Table 3 Dependence of ΔG on temperature at ambient pressure for the reaction according to Eqs. (21–23) and Eqs. (29–31)

Water phase	Temperature (°C)	ΔG (kJ/mol)	Equilibrium voltage (V)
Liquid	25	237.2	1.23
Liquid	80	228.2	1.18
Gaseous	100	225.2	1.17
Gaseous	200	220.4	1.14
Gaseous	400	210.3	1.09
Gaseous	1000	177.4	0.92

The table also gives the resulting open-circuit voltage (OCV) according to Eq. (25)

As was mentioned several times, the reactions Eqs. (21–23) can be assumed to be reversible under equilibrium conditions. Therefore, the water-splitting reactions of electrolysis would be in balance with the water-forming reactions of the fuel cell reactions. We can write the equations

$$\text{oxygen half-reaction: } \frac{1}{2}O_2 + H_2O + 2e^- \rightarrow 2OH^- \tag{29}$$

$$\text{fuel half-reaction: } H_2 + 2OH^- \rightarrow 2H_2O + 2e^- \tag{30}$$

$$\text{overall reaction: } H_2 + \frac{1}{2}O_2 \rightarrow H_2O \tag{31}$$

that correspond to the overall reaction process shown in Fig. 10, again with an alkaline (KOH) electrolyte. The system Eqs. (29–31) is quite obviously Eqs. (21–23) flipped around, as can be expected from a reversible reaction.

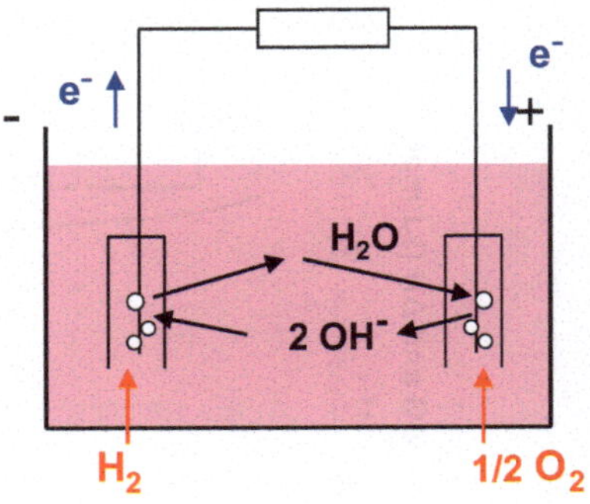

Fig. 10 Charge transport and half-cell reactions in alkaline fuel cell

It should be considered here, though, which electrode is the cathode and which the anode. By the definition presented above, the electrode at which the reduction reaction takes place (or the negative charge enters the electrolyte) will be labelled 'cathode'. Eq. (29) fulfils these conditions, but comparison with Eq. (21) shows that the electrolyte 'cathode' is the hydrogen electrode whereas in the fuel cell it is the oxygen electrode. On second thought, this is logical since the current has been reversed and therefore the entry point of the electronic charge into the electrolyte has to switch around. This thought train, though, goes to show that caution has to be applied when electrolyser and fuel cell researchers talk with each other about electrodes. What is more, is the change in polarity. Whereas the electrolyser cathode is the 'fuel' electrode where the water is supplied and is the negative terminal, in a fuel cell the fuel electrode is the positive terminal.

Since the reaction is the same, the open-circuit equilibrium voltage U_o is identical to that of electrolysis (but negative!) and

$$\Delta G^{o}{}_{o} = 2FU^{o}{}_{o} = -237\,\mathrm{kJ/mol}$$

at standard conditions, 20 °C, for a H_2–O_2 cell and

$$U^{o}{}_{o} = 1.23\,\mathrm{V}$$

for convenience, U is not marked as negative.

Now, it seems a little pointless to consider equilibrium conditions if we are talking about devices that produce hydrogen or, vice versa, convert hydrogen into water, simultaneously consuming or generating electrical and thermal energy, respectively. According to Le Chatelier's principle, the removal of the reaction product or the consumption (conversion) of reactants will influence the reaction equilibrium coefficients.

For any reaction that is not in equilibrium, the following equation holds:

$$\Delta G = \Delta G_{o} - \mathrm{RT}\ln\left(P_{A}P_{B}\right)/\left(P_{C}P_{D}\right)$$

with A, B being the reactants and C, D the reaction products. P_i is the (absolute) pressure of the individual component i, which can be separated into the partial pressure and the absolute (overall) pressure components p_i and P, respectively, where P will be the same across the entire reaction and at standard state (superscript '0') will be 1 bar.

Specifically written here for the hydrogen–oxygen reaction, this gives:

$$\Delta G = \Delta G_o - \mathrm{RT}\ln\left(P_{H_2} P_{O_2}{}^{1/2}\right)/P_{H_2O} \tag{32}$$

Using Eq. (32), we can express this as the Nernst equation where the actual voltage at a given condition of reactant and reaction product concentrations ('activities') is modified to

$$\begin{aligned} U &= U^o{}_o + \mathrm{RT}/2F\ln\left([H_2][O_2]^{1/2}\right)/[H_2O] \\ &= U^o{}_o + \mathrm{RT}/2\mathrm{F}\ln\left(p_{H_2} * p_{O_2}{}^{1/2}\right)/p_{H_2O} + \mathrm{RT}/2F\ln P/P_o \end{aligned} \tag{33}$$

where P is an arbitrary pressure at which the whole reaction takes place, $[A]$ is the 'activity' of a reactant or reaction product. In the case of gases (as in the cases analysed here), this is equivalent to the partial pressure of the gas as shown in Eq. 33.

1.7 Electrolysis and Fuel Cell Operation

It has been pointed out repeatedly that thermodynamics is only strictly valid in the steady state or equilibrium. An electrochemical device, though, will always be operating off-equilibrium, because it is actively converting water to hydrogen or hydrogen to water. The product of this reaction is the desired output and will be removed instantly. This leads to an imbalance that the reaction attempts to compensate for through Le Chatelier's rule. So both electrolysis units and fuel cells will by definition not be operating in equilibrium.

Therefore, the rules discussed above are only valid at OCV when no current flows. This gives us a baseline of what performance to expect in the ideal case. If we now assume that the process of converting hydrogen to water, generating electricity and heat, was without any losses, we will retain the OCV at any rate of hydrogen conversion since $\Delta G = \Delta H$. The same would be true for electrolysis and the conversion of water to hydrogen (and oxygen).

In a real system, though, we will have losses. The hydrogen–oxygen reaction involves electron transfer, which we now use in an external circuit in our electrochemical device as the current 'produced' or 'consumed' (fuel cell and electrolysis case, respectively). A loss in this system will therefore express itself as a voltage drop (in the fuel cell case), where we obtain less voltage than theoretically (ideally) expected. We can consider these losses as 'internal resistance' and according to Ohm's law

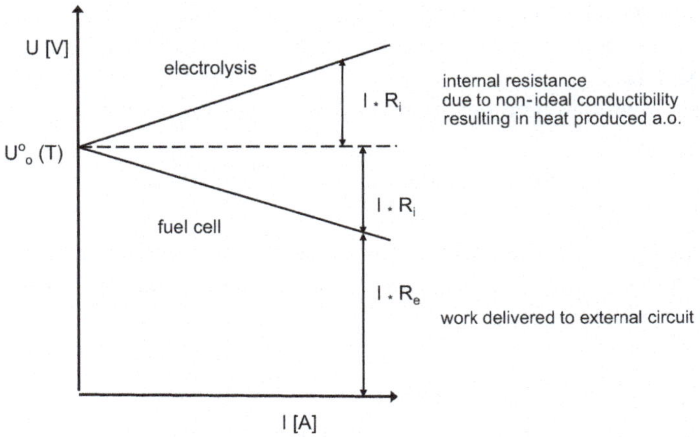

Fig. 11 Schematic I–V curve of electrolysis (*upper*) and fuel cell (*lower*) operation of the same electrochemical cell. The *dashed line* gives the equilibrium voltage (which is temperature dependent according to Table 3). The internal ohmic losses result in a voltage increase with growing current for electrolysis and an equivalent voltage drop for fuel cell operation

$$\Delta U = U^{\mathrm{o}}{}_{\mathrm{o}} - U = I * R_{\mathrm{i}} \tag{34}$$

The same is true for the electrolysis case; the difference being that in this case, we are 'driving' the reaction by the current and we can expect an increased voltage, termed 'overpotential', which is higher than the theoretically expected but with the same value as given by Eq. (34).

Figure 11 shows a schematic of the real voltage of electrolysis (top) and fuel cell (bottom) cases as a function of current assuming a purely ohmic behaviour (i.e. a linear relationship between current and voltage).

Let us now concentrate on fuel cell behaviour and assume that electrolysers will behave more or less symmetrically (which they almost do). Figure 12 shows a 'typical' I–V curve for a fuel cell device. The curve is only linear in its central section, indicating that here Ohm's law is valid, although only in the formulation according to Eq. (34) for voltage differentials

$$\Delta U = \Delta I * R_{\mathrm{i}} \tag{35}$$

Both the region close to $I = 0$ and towards large currents require more explanation, though.

Figure 13 shows a 'Tafel plot' which is essentially a modified I–V diagram. It plots the ΔU created by applying a current across an electrochemically active electrode. Obviously, there is a linear relationship between current and voltage, starting from a minimum current i_{o}. Further, different reactions have differences in the slope of the plot.

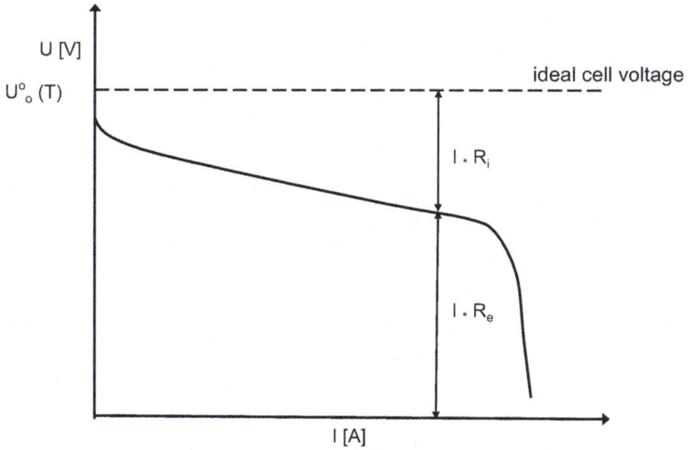

Fig. 12 I–V curve for a fuel cell device including the nonlinear segments not represented in Fig. 11. 'Activation' voltage losses shown close to zero current; 'transport limitation' losses shown at high currents

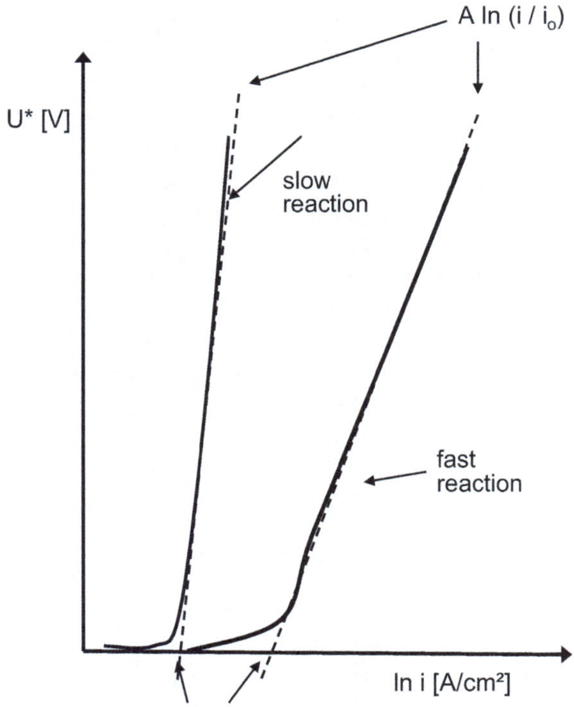

Fig. 13 'Tafel' plot representing the correlation of exchange current and overvoltage; the exchange current i_0 offers a measure of the catalytic activity according to Table 4

The explanation is as follows: if we consider a reaction in equilibrium, such as (gradually beginning to bore you with this reaction):

$$H_2 + \frac{1}{2}O_2 \leftrightarrow H_2O \tag{36}$$

In equilibrium, we have learned, there is no net production rate, since the equation is balanced according to its equilibrium coefficient. Nevertheless, a reaction activity continuously takes place from reactants to product and back again. In this reaction, charge is transferred. Therefore, it can be assumed that an associated current flows, since charge transfer is synonymous with current. The current is circular (or rather undirected) and therefore does not create a potential difference, so that the OCV is undisturbed. It is called the 'exchange current'. The lower case '*i*' indicates that we talk of current 'density' here in the unit of A/cm^2, rather than absolute currents. This gives us a measure of the specific properties of an electrode and makes comparisons easier without needing to know anything about the geometric dimensions of the samples, etc.

If we want to get an effective current flowing, we have to overcome the circular current and give it the desired direction. This is the minimum current we have to apply until we see a net flow that expresses itself in a voltage (difference) according to Ohm's law.

We can safely assume that the higher the exchange current i_o, the higher the effective reaction rate and therefore the activity of the catalyst. We have to apply a higher current to overcome this, but the ensuing slope of the I–V curve is lower. This can be taken as a measure for the rate of reaction. A very steep line indicates a large change in voltage with a small change in current. If we apply this to the section very close to the OCV in Fig. 12, we can see that a high increase in overvoltage (voltage drop from the ideal voltage) will lead to a rapid decline of the voltage and a high ohmic resistance. A small voltage drop with increasing current will result in higher performance and a lower slope of the I–V curve.

Table 4 gives some values for i_o of a variety of catalysts. These are given for a hydrogen electrode. The oxygen reduction reaction has a factor of 10^{-5} lower values, which explains why the fuel cell cathode is generally considered as the rate-limiting component in the fuel cell. The overvoltage U^* (Fig. 13) associated with the exchange current is expressed as

$$U^* = A \ln (i/i_o) \tag{37}$$

where A is the slope of the i–V* curve and given as

$$A = RT/2\alpha F \tag{38}$$

with α defined as a 'charge transfer coefficient' that will take on values between $0 < \alpha < 1$. α increases the factor $RT/2F$ [cf. Eq. (33)], the less efficient the transfer of charge at the electrode is. The actual voltage will, in amendment to Eq. (33), therefore be

$$U = U_o - U^* = U_o - A \ln(i/i_o) \tag{39}$$

The voltage drop created by an insufficient charge transfer is labelled the 'activation loss'. It is due to the limitations in the catalyst performance. This will be influenced by temperature, available surface area (electrochemically active surface area, ECAS), and partial pressure of reactants.

The other large voltage drop occurs at high currents. Here, the main driver is the balance of electrical current flow, the ionic current within the electrolyte (and parts of the electrodes), and the fuel flow. In order to maintain an overall current flow, all these factors have to match and be fully balanced. At the end of the curve where the voltage approaches 'zero', it can be also assumed that all reactants have been consumed and there is nothing left to run the fuel cell on.

The voltage drop can be then expressed as follows: assume the situation where only one reactant (oxygen or hydrogen) varies and the other is constant. The constant can then be eliminated from the voltage drop equation when using the Nernst Eq. (33):

$$\Delta U = \mathrm{RT}/2F \ln(P_2/P_1) \tag{40}$$

We now label the current density at a given fuel flow; all chemical energy in the fuel is converted (the 'fuel utilisation' u_F is equal to 'one' or 100 %) as limiting current density, i_{lim}. At i_{lim}, the fuel (reactant 2) is fully consumed and $P_2 = 0$. We can then write

$$\begin{aligned} P_2/P_1 &= (i_{lim} - i)/i_{lim}, \quad \text{or} \\ P_2 &= P_1(1 - i/i_{lim}) \end{aligned} \tag{41}$$

The performance of the fuel cell at high currents again is determined by the processes explained for the activation losses. In addition, the actual transport (diffusion) of species to and from the reaction centres is now in focus. If the fuel and oxidant cannot be transported to the reaction sites rapidly enough, or, vice

Table 4 Exchange current density i_0 for a variety of metal catalysts

Metal	i_o (A/cm^2)
Pb	2.5×10^{-13}
Zn	3×10^{-11}
Ag	4×10^{-7}
Ni	6×10^{-6}
Pt	5×10^{-4}
Pd	4×10^{-3}

versa, the water cannot be removed quickly enough, the circuit of current flow and charge transport tips into imbalance. The flow of reactants cannot cope with the desired current flow anymore and the voltage subsequently collapses.

Strategies to overcome this limitation are improved transport properties of the electrode for gases (porosity and tortuosity) and sufficient removal of reaction products.

1.8 Fuel Cell Efficiency

This section does not necessarily discuss electrochemistry, but it follows on from a number of discussion points raised earlier and is an important aspect of fuel cell evaluation. So this chapter ends with considerations about fuel cell efficiency.

First, let us consider what the baseline for an efficiency calculation could be. The total energy 'stored' in a molecule or in a chemical compound is the 'energy of formation'. It is the balance of all energy flows that led to the production of this molecule or compound. This is the enthalpy H_o or rather ΔH_o, since we are interested in what happens during the reaction (i.e. the difference between 'before' and 'after' the reaction), not in any other possible contributions to the total energy state of the compound. We use the '0' superscript to denote that we are talking of 'standard' conditions, in essence ambient pressure.

We know that the enthalpy is temperature independent but has a component from phase change (cf. Fig. 9). To cite our favourite reaction of oxygen and hydrogen to water, the enthalpy will be 285.6 kJ/mol at 20 °C if the product water is liquid, and 241.8 kJ/mol if the product water is in the vapour phase. This enthalpy value is the highest we can possibly gain from producing water or, alternatively, the maximum work we will need to split the molecule by electrolysis. We should note that the issue whether the product water is in the liquid or vapour form is the same as that of the higher and lower heating value (HHV and LHV) reference in efficiency calculations of regular thermal machines or combustion processes.

The efficiency of a process (or reaction in our case) will be the 'useful' output divided by the energy input. In this case, this is the Gibb's free energy versus the total enthalpy:

$$\eta = \Delta G / \Delta H \tag{42}$$

Care has to be taken to reference the correct ΔH. One would be tempted to take water vapour as the reference point. This would clearly result in higher efficiency values. Nevertheless, in a fuel cell, water can very well (and will) also manifest itself in the liquid phase, which is why we here choose the liquid phase value of 285.8 kJ/mol as the reference throughout.

The efficiency of the water producing process can now be written as

$$\eta_{th} = \Delta G/\Delta H = \Delta G/285.8\,\text{kJ mol}^{-1} \tag{43}$$

using the subscript 'th' for 'thermodynamical'. From Table 3, we can see that ΔG drops with temperature. Therefore, the thermodynamical efficiency of fuel cells decreases with temperature. At 25 °C, ΔG = 237 kJ/mol if the water product is liquid, and 218 kJ/mol if the water is in the vapour phase. This gives us efficiencies of 83 and 77 % at 25 °C. If we move to 700 °C, we only have a thermodynamical efficiency of 68 %.

This tells us two things: firstly, the thermodynamical efficiency of high-temperature fuel cells is definitely lower than that of low-temperature fuel cells due to the additional entropy the higher temperature introduces; secondly, in exploiting the full potential of the fuel cell, the higher heating value (HHV) should be the point of reference and efforts should be made to fully use the latent heat stored in the water vapour. This is something often not considered in fuel cell system engineering.

A second useful reference for calculating a performance indicator (efficiency) is the OCV. Reviewing Fig. 12, we can write the voltage efficiency as

$$\eta_V = U/E_o = U/\text{OCV} \tag{44}$$

with E_o = OCV for standard pressure. This is a measure for the voltage losses within the fuel cell and therefore the lack or dominance of activation, ohmic, and concentration losses.

The 'fuel utilisation' u_F is the percentage of fuel fed to a fuel cell that is actually converted. The remainder will leave the device and be lost, if no further measures (e.g. recycling) are taken. u_F is the ratio of utilised fuel to fuel feed

$$u_F = ((n_{\text{fuel in}} - n_{\text{fuel out}})/\Delta t * \Delta H_o)/(n_{\text{fuel in}}/\Delta t * \Delta H_o) \tag{45}$$

with $n/\Delta t$ being the fuel flow in mol per unit time. u_F is not so accessible to measurement as the voltage efficiency. The measurement of unused fuel flow is not simple. Alternatively, the total current can be balanced against the potential current, which can be evaluated from the fuel flow, and the theoretically available number of electrons. This can provide

$$u_F = I/(N * F * n_{\text{fuel in}}/\Delta t) \tag{46}$$

with N the number of electrons one mol of fuel can provide (2 in the case of hydrogen) and F the Faraday constant.

Finally, we can define a 'current efficiency' (also called Faradaic efficiency) that gauges the amount of current delivered to the external circuit against the ionic flow through the electrolyte. The loss that is quantified by this efficiency is that of internal leak currents that cause electrical currents within the fuel cell, for instance by nonzero electronic conductivity of the electrolyte or any form of internal short circuit, either in the fuel cell, or in the fuel cell mounting

$$\eta_{I} = I/I_{ionic} \tag{47}$$

The overall fuel cell efficiency η_c will be the product of all these efficiency values. Nevertheless, by convention, the thermodynamical efficiency is generally left out of the equation since it is given for a specific device running at a set temperature. Maybe also because it is a little frustrating for high-temperature fuel cell developers? It should be pointed out, though, that in the process of internal reforming, high-temperature fuel cells operating on methane can in effect use part of the entropy produced and convert it back into chemical energy.

We therefore find

$$\eta_c = \eta_V * \eta_I * u_F \tag{48}$$

as the equation generally used for computing the efficiency of a single fuel cell or a fuel cell stack in producing a direct current. η_I is in many cases even neglected and assumed to equal 100 %, so that Eq. (48) simplifies to

$$\eta_c = \eta_V * \mathrm{u_F} \tag{49}$$

The fuel utilisation will never reach 100 %, because this means p_{fuel} becomes 'zero' and we reach the limiting current (density). At this point, the fuel cell is completely 'starved', i.e. no fuel is available, at which the fuel cell voltage becomes 'zero', too. The ensuing distribution of potential across the cell can cause damage to the fuel cell layers and lead to re-oxidation of the anode since both the lack of reducing agent (i.e. fuel) and the desire to uphold the current flow and the further transport of oxygen ions through the electrolyte will lead to higher than desired oxygen partial pressures at the anode side. The obvious loss of fuel by retaining a nonzero flow of fuel out of the cell can be managed by recycling the anode off-gas to the anode inlet and thus reusing the excess fuel.

Returning to Eq. (49) and using Eq. (44), though, at a given $\mathrm{u_F}$, the efficiency of the cell will only depend on the value of the actual cell voltage. The higher the voltage, the higher the efficiency. This is synonymous with low activation and concentration losses and calls for high-performing electrodes and electrolytes with low ohmic, ionic, and transfer resistances. A plot of efficiency versus cell voltage shows a linear dependency, as expected (Fig. 14). This indicates that in contrast to the Carnot efficiency (Fig. 6), a fuel cell can potentially reach extremely high efficiencies as long as materials can be employed that produce extremely low overpotentials. In essence even, any fuel cell operating at low current densities will by principle have very high efficiencies. Technically this is meaningless, because the power delivered will be low. But it does go to show the valuable potential dormant in fuel cell technology.

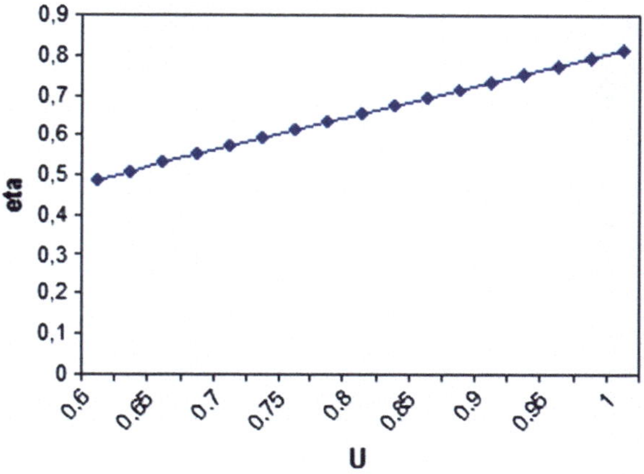

Fig. 14 Increase of fuel cell voltage efficiency (Eq. 44) with open-cell voltage for a PEFC at 80 °C operating temperature

References

Baur, E., & Preis, H. (1937). Über Brennstoff-Ketten mit Festleitern. *Zeitscrift für Elektrochemie, 43*, 727.

BP. (2015). *Statistical review of world energy*. To be found at http://www.bp.com/en/global/corporate/energy-economics/statistical-review-of-world-energy.html. Last accessed September 28, 2015.

Davtyan, O. K. (1947). *The problem of direct transformation of chemical energy of fuel into electrical one*. Academy of Science of the USSR, Moscow, 145 p. (in Russian).

EU. (2012). *Directive 2012/27/EU of the European Parliament and of the Council on energy efficiency*. Brussels October 25, 2012 (published November 14, 2012).

Haldi, P.-A., & Favrat, D. (2006). Methodological aspects of the definition of a 2 kW society. *Energy, 31*, 3159–3170.

Nernst, W. (2003) *Reasoning of theoretical chemistry: Nine papers (1889–1921)*. (Begründung der Theoretischen Chemie: Neun Abhandlungen, 1889–1921). Frankfurt am Main: Verlag Harri Deutsch.

Sovacool, B. K., & Dworkin, M. H. (2014). *Global energy justice—Problems, principles, and practices*. Cambridge: Cambridge University Press.

Worldbank. (2015). Data on GDP and energy demand per capita, to be found at data worldbank. org. Last accessed September 28, 2015.

Testing of Electrodes, Cells, and Short Stacks

Anne Hauch and Mogens B. Mogensen

Abstract The present contribution describes the electrochemical testing and characterization of electrodes, cells, and short stacks. To achieve the maximum insight and results from testing of electrodes and cells, it is obviously necessary to have a good understanding of the fundamental principles of electrochemistry, but it also requires proper test geometries and set up, well-chosen operating conditions for different test purposes, correct probing of voltages and temperatures, and solid knowledge on benefits and drawbacks of different characterization techniques to obtain reliable, accurate, and reproducible electrochemical measurements, and this will be the focus of this chapter. First, the important issue of understanding potential differences and measurements of potentials, which is linked to the choice of proper electrode geometries and test set up configurations for electrode and cell testing, is presented. Then probing of voltages and temperatures, choice of sealing and contacting, as well as considerations regarding the choice of operating conditions for different purposes mainly for single cell testing are outlined. Having considered optimization of test set up, geometries, and the selection of optimal operating conditions, the details of measurement of the electrochemical performance of the electrode, cell, or stack are explained. As part of this, the concept of area specific resistance (ASR) and how DC and AC methods can be used and optimized to provide not only the total ASR, but also the electrochemical characterization of specific parts (electrolyte, each electrode) in a full cell are described. Some experimental results are provided including illustrative examples of breakdown of losses in full cells and determination of their temperature and gas composition dependencies, and finally, challenging issues, such as the effects of impurities and the problem of leakage in cell testing, are discussed as well.

A. Hauch (✉) · M.B. Mogensen
Department of Energy Storage and Conversion, Technical University of Denmark, Risø Campus, Roskilde, Denmark
e-mail: hauc@dtu.dk

M.B. Mogensen
e-mail: momo@dtu.dk

M. Boaro and A.S. Aricò (eds.), *Advances in Medium and High Temperature Solid Oxide Fuel Cell Technology*, CISM International Centre for Mechanical Sciences 574,
DOI 10.1007/978-3-319-46146-5_2

1 Introduction

This chapter describes the electrochemical testing and characterization of electrodes and cells and provides a brief discussion of short stack testing as well. To achieve the maximum obtainable knowledge from testing of electrodes and cells, it is obviously necessary to have a good understanding of the fundamental principles of electrochemistry (Bockris 1970; Holze 2007; Hamann 1998; Greef 1985) and especially solid-state electrochemistry (Kharton 2009, 2011). Furthermore, a good and sound basic knowledge in materials science, especially in high-temperature materials (Ibach 2009; Smart and More 1996; Carter 2013), will inevitably increase the understanding of the results from electrode and single cell solid oxide cell (SOC) tests.

There are good reasons to pursue all three levels of tests: electrodes, cells, and short stacks. Testing of electrodes can provide detailed insight into the electrochemical performance and its correlation with parameters such as gas composition, exact chemical composition of electrode materials, and the effect of specific micro- or nano-structures on the electrochemical performance and durability; all electrode-specific information as elegantly shown in literature (Ramos et al. 2010; Nielsen et al. 2011; Kromp et al. 2012; Shearing et al. 2010a, b; Utz et al. 2011; Vels Jensen et al. 2001). Testing of full cells complicates the setup, test procedures, and analyses of the results, but on the other hand, testing of full cells provides insight into the effect of parameters such as high current density, i.e., overpotential gradients and current density gradient along the fuel flow direction due to fuel utilization as shown many times in recent years (Rasmussen and Hagen 2010; Nielsen et al. 2010; Hauch et al. 2014; Tietz et al. 2013). Testing of short stacks, sometimes named single repeating units (SRU), includes interconnects for current collection. Such test enables studies of, e.g., effects of corrosion of interconnect material, poisoning electrodes due to interconnects, and contacting issues. Detailed probing of voltages, gas compositions, temperature, and impedance measurements can still be obtained for SRU. If stacks are to be tested as a part of development of a commercial product, it is desirable to test a complete system including balance-of-plant components. Such tests are expensive and complex, and it is difficult to obtained detailed electrochemical information from such test and to interpret the results in detail. Stack testing will not be treated further in this chapter on electrochemical testing of SOC; however, recent work elegantly shows the possibilities of characterization of individual single cells in a stack (Mosbaek et al. 2013).

Thousands of reports can be found in literature on electrochemical testing of SOC electrodes, cells, and short stacks. As there is no general agreement on test protocols even inside regions such as USA and/or EU, it is particularly important that the testers carefully report the obtained fundamental parameters, such as area specific resistance (ASR), activation energy (E_a), and polarization losses, and provide the exact test conditions for the study to minimize misleading interpretations and inappropriate comparison of electrochemical test results.

This chapter will focus on the proper testing of these electrodes and cells—specifically the use and position of reference electrodes (Sect. 2), test set up and proper probing for cell tests and choice of test conditions and specific test sequences and conditions (Sect. 3). This section also describes the issue of impurities in gasses and raw materials which often complicates the interpretation of the results from electrochemical testing. As the majority of testing of electrodes and cells is performed to obtain a measure for the performance of the cell, e.g., given as the ASR of the cell, the concept of ASR, and a thorough description of its calculation is given here (Sect. 4) together with a review of key characterization methods for electrodes and cells (Sect. 5). This provides a more practical and test-related description of measurements to obtain results on contributions to the ohmic and polarization losses compared to descriptions found in literature (Boukamp 2004; Macdonald and Barsoukov 2005; Orazem 2008). Section 5 also provides examples of complementary non-electrochemical characterization techniques for SOC and finally describes the issue of gas leakage in cell testing (Sect. 6).

2 Electrodes

2.1 Measurement of Potentials and Understanding of Potential Gradients

The main problem with characterizing the individual electrodes in single cells and short stacks is the insertion of the reference electrodes, which are used to judge the performance of the individual components and interfaces in the cell. Since a single cell stack is made up of five components, an electrolyte, two electrodes, and two interconnects, there are four interfaces at which reference electrodes can be inserted. Unfortunately, such reference electrodes will not work in the geometries, which are normally employed.

Results of tests on cells with thin electrolyte layers using one or more reference electrodes have been reported on many occasions (Gödickemeier et al. 1995; Erning et al. 1995; van Heuveln et al. 1993; Kleinlogel et al. 1997), but the electrodes under investigation appeared not to be polarized; the short explanation is that the reference electrodes were not working. The geometric requirements for the position of electrodes in three-electrode set-ups have been treated in detail by several researchers (Nagata et al. 1994; Jacobsen and Skou 1997; Winkler et al.1998; Reinhardt and Göpel 1998; Adler 2002; Mogensen 2002), and there is a general agreement among these researchers. In spite of this, there still seems to be a great need in the SOC community for basic information on how to measure electrode potentials properly; some of these details are given below.

Figures 1 and 2 illustrate the problem. An electrode-supported cell with a "reference" electrode is often sketched as shown in Fig. 1a. However, such a sketch is very deceiving when it is used for an assessment of the current distribution. For this purpose, the sketch should be drawn to scale, i.e., the electrolyte thickness

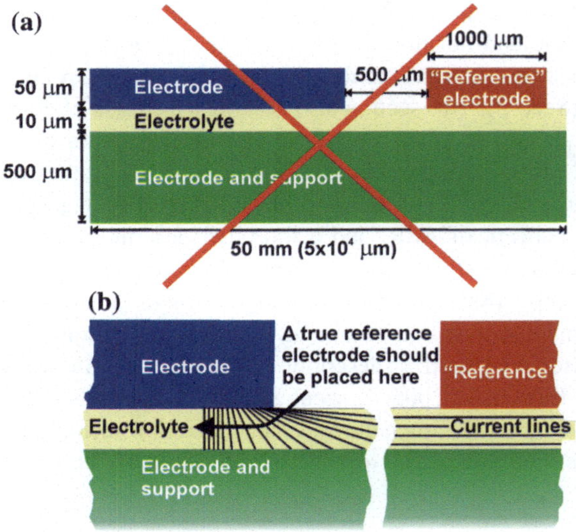

Fig. 1 **a** A typical sketch of an electrode-supported cell which is out of scale, because the gap between the top electrodes should be 50 times greater than the electrolyte thickness and **b** expanded view of electrode corners showing the current distribution indicated by schematic current lines. The current density, apart from being approximately parallel to the electrolyte plane, is very small at the position of the "reference" electrode, at least 50 (500 μm/10 μm) times smaller than the current density of the cell.

should be the relevant unit of length. When the correct length scale is used, as in Fig. 1b, it is evident that the gap between the upper working electrode and the "reference" electrode is huge. This means that the current distribution around the right-hand edge of the working electrode becomes very different from the even current distribution in the main part of the cell. Furthermore, the current in the vicinity of the "reference" electrode becomes parallel to the electrolyte plane, i.e., there is only a minute voltage difference across the electrolyte at the "reference" electrode position. The correct position of the reference electrode would be inside the electrolyte of the cell and some distance away from the corner, but this is difficult in a 10-μm-thick electrolyte.

Figure 2a illustrates the electrostatic potential across a cell at open circuit voltage (OCV) condition. Note that for a good electrolyte (ionic conduction only), there will be no potential gradient inside the electrolyte at zero current. The whole potential change across the cell is localized at the interfaces between the electrolyte and the electrodes. These regions are the so-called electrochemical double layers with thickness in the nano-meter range and with high space charge concentrations as a result of the very high potential gradients. Figure 2b and c show the potential across a cell when it is loaded, i.e., a current flows through it, in fuel cell, and electrolysis mode, respectively. Note that there is a potential loss across the electrolyte due to the electrolyte resistance. The potential steps at the interfaces are now smaller compared to the OCV condition in the fuel cell case and larger in the electrolysis case due to the

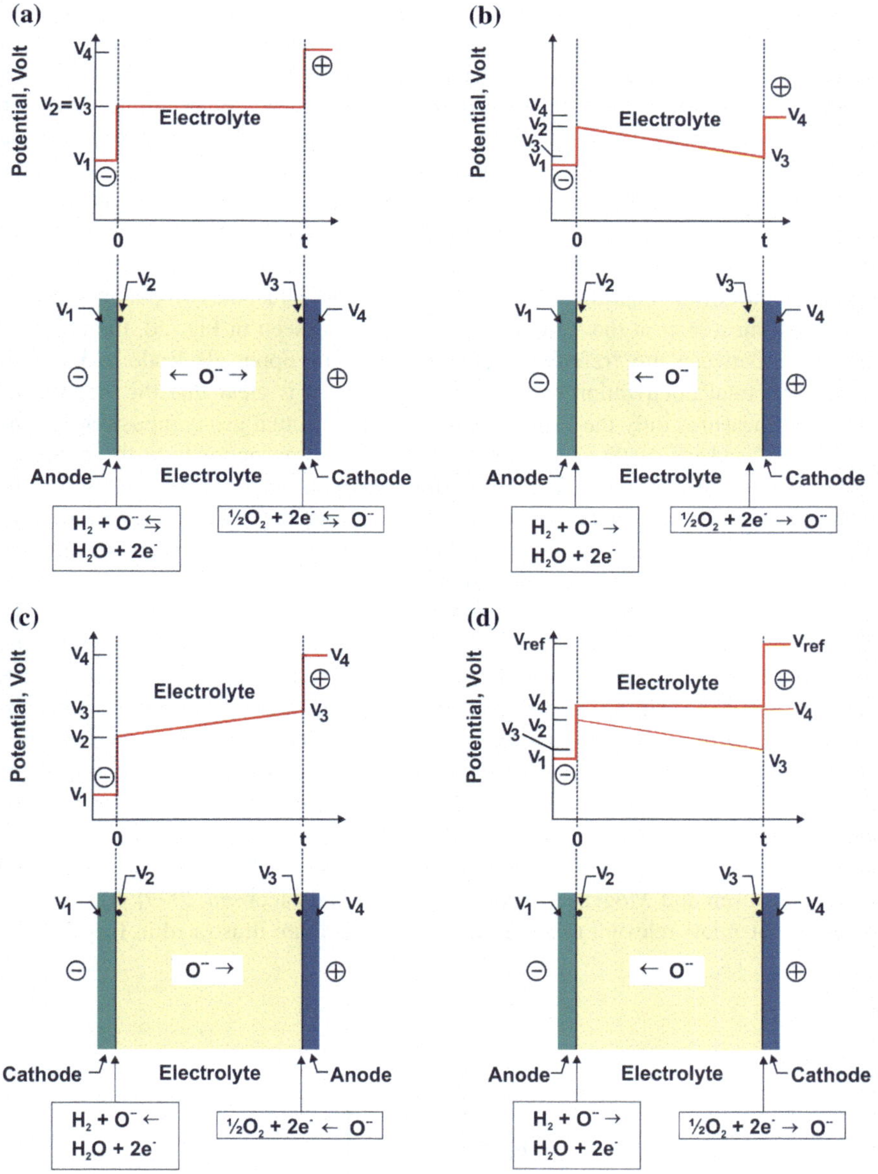

Fig. 2 Electrostatic potential through the electrode-supported cell with **a** no current, **b** current load in fuel cell mode, **c** current load in electrolysis mode, and **d** the potential across the electrolyte at the "reference" electrode position (*thick line*) and through the cell part with the current load (*thin line*). It is seen that: $(V_{ref} - V_4)/i = R_{p,anode} + R_{p,cathode} + R_{elyt}$ = the total polarization resistance of the cell apart from concentration polarization resistance. Note that the electrostatic potential single value cannot be measured with a voltmeter, but voltage (= potential difference between two points) can

losses originating from the polarization resistance of the electrode processes. Figure 2d illustrates the case of a fuel electrode—supported cell with an attempt of placing a reference electrode on the electrolyte surface next to the oxygen electrode. The potential is given for both the position of the "reference" electrode where no current flows across the electrolyte (i.e., the electrical potential is constant across the electrolyte) and a position far away from the electrode edges as in Fig. 2b. As the potential along the fuel electrode, which is a very good electronic conductor, is the same everywhere (the electrode constitutes an iso-potential plane), the potential of the fuel electrode at the "reference" electrode must be equal to the potential in the middle of the current bearing part of the fuel electrode. Thus, the two potential curves in Fig. 2d must start at the same point. Therefore, as seen in Fig. 2d, the potential difference between the "reference" electrode and the upper electrode in Fig. 1 is simply the total polarization of the full cell. Thus, it is clear that the "reference" electrode measures only the Emf of the cell with the actual gas compositions at the "reference" and inside the support at the lateral position opposite to the reference while the current is flowing. Thus, no information on what happens on any of the working electrodes can be derived from such measurements. If the concentrations of the reactants and products at the lateral position of the reference electrode were the same as in the active electrode/electrolyte interfaces, then it would be possible to deduce the total concentration overpotential. This is, however, in general not the case. This means that the voltage difference actually measured between the working and the "reference" electrodes cannot be assigned any clear meaning. In the present context, it is also helpful to remember that a single electrode potential cannot be measured directly; it is only possible to measure a potential difference between two electrodes. Furthermore, it may be useful to know that the Fermi potential of the electrons—also of mobile electrons inside an electrolyte—is not at all the same as the electrostatic potentials. The gradient of the Fermi potential will even have the opposite sign of the electrostatic potential gradient across the electrolyte in fuel cell mode (Jacobsen and Mogensen 2008; Mogensen and Jacobsen 2009). The various types of potentials relevant to an electrochemical cell are illustrated in Fig. 3.

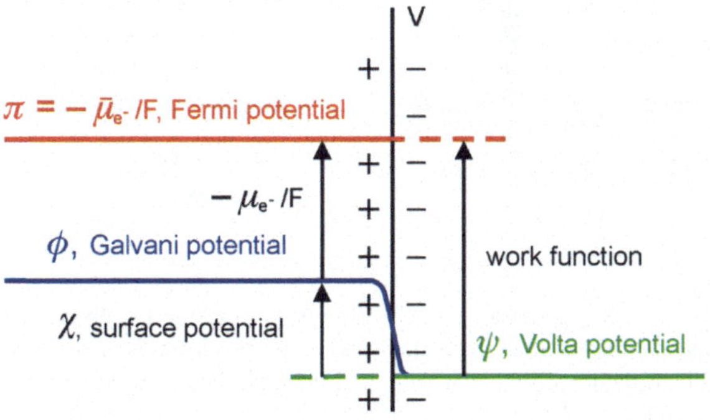

Fig. 3 Potentials in a solid conductor relative to the outer electric potential in vacuum. $\bar{\mu}_{e^-}$ is the electrochemical potential of the electrons and F is the Faraday's number. The Galvani potential is the same as the electrostatic potential, and the Fermi potential, π, is also called the electromotive potential (Jacobsen et al. 2014)

Figure 4 shows the course of the Fermi potential (π) and the electrostatic potential (ϕ) across the electrolyte in a YSZ-based SOC for the cases of OCV, fuel cell mode (SOFC), and electrolysis mode (SOEC). It is worth noting that the *S*-shaped π-curve moves to the right when polarized in fuel cell mode and to the left in electrolysis mode. The ϕ-curves are linear, reflecting the simple ohmic loss in the electrolyte, and change the sign in its slope when going form fuel cell mode through OCV to electrolysis mode. The Fermi potential (π) and the electrostatic potential (ϕ) determine together the oxygen potential inside the electrolyte. If the oxygen potential is expressed as oxygen partial pressure (pO_2), then in case of electrolyte materials with constant stoichiometry like YSZ, the electrode equilibrium potential, ($\pi - \phi$), is related to pO_2 as:

$$(\pi - \phi) = (\pi - \phi)^{\ominus} + \frac{RT}{4F} \ln \frac{p_{O_2}}{p^{\ominus}} \tag{1}$$

where $p^{\ominus}$ is taken at standard pressure of 1 bar (Jacobsen et al. 2014).

Apart from the demand that the reference electrode must sense a relevant electrostatic potential inside the electrolyte near the interface of the electrolyte–working-electrode at a position where the current density is uniform and representative of the electrode condition, the reference electrode must also have a well-defined electrode potential of its own. This means that it may be the electrode potential of the half-cell $Pt|YSZ|O^{2-}|O_2$ (at 1 atm. and given temperature), which is very suitable for SOC purposes. The general theory of reference electrodes in three-electrode set-ups may be found in most electrochemical textbooks. The three-electrode concept as such implies that it is not enough to imbed a wire of Pt or other metal inside the electrolyte. The "wire" inside the electrolyte must be an electrically insulated "wire" consisting of electrolyte in similarity to the Luggin capillary (in liquid electrochemistry). That provides the ionic contact to the half-cell that contains the reference electrode, which usually is placed outside the cell with the working and counter electrodes. As this is very difficult in solid-state electrochemistry, special attention has to be given to the geometry of three-electrode cells or to use the electrode test strategies, in which the electrode response can be measured or calculated without using a three-electrode set-up. This is the subject of the next section.

2.2 Electrode Testing

To avoid such problems, it is necessary to test the electrodes using a suitable three-electrode set-up or a symmetrical two-electrode cell, which gives interpretable results at OCV and in the overpotential region, where the current density is a linear function of overpotential (Jørgensen et al.1999), even though all methods have their

Fig. 4 Calculated potentials versus fractional distance across the electrolyte of a Ni|YSZ|LSM SOFC. The *symbols* in Fig. 3 are used for the potentials. The three cases are: **a** OCV, **b** fuel cell mode, and **c** electrolysis cell mode. In **b** and **c**, the OCV potentials are also plotted as *dotted curves* (Jacobsen et al. 2014)

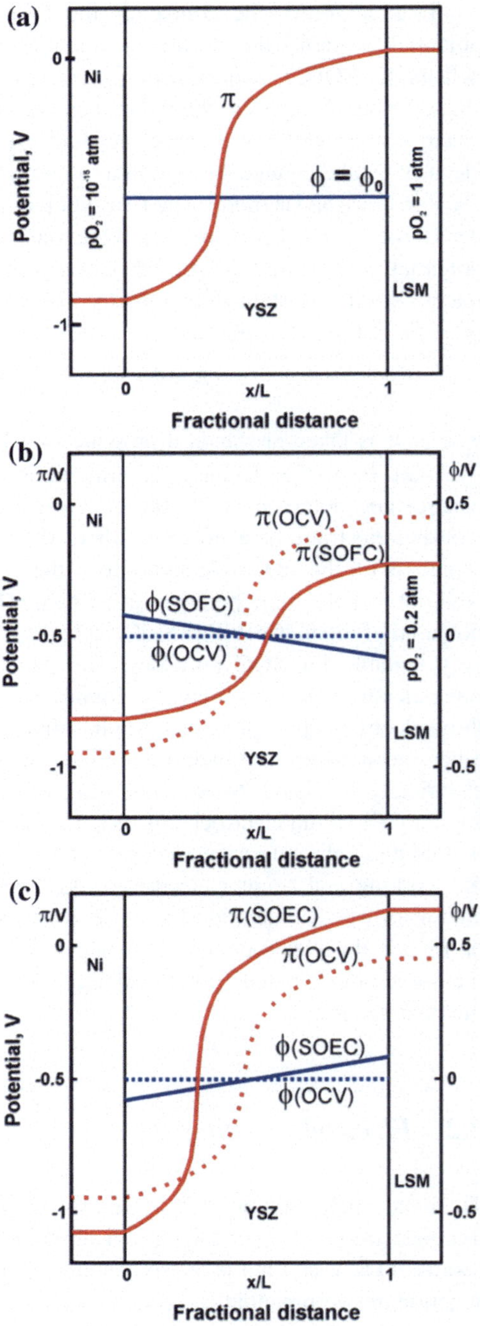

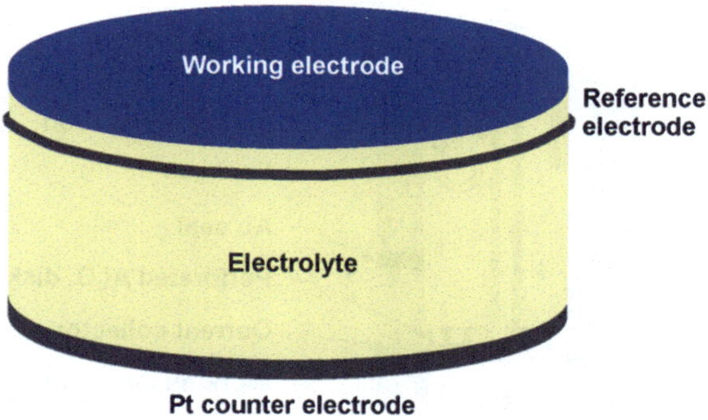

Fig. 5 Illustration of a simple three-electrode set-up. This can naturally only be used in a one atmosphere compartment, and therefore, it is important to continuously monitor the pO_2 in this compartment. The reference electrode may be a Pt-wire, which must be strictly parallel to the edge of the electrode

shortcomings. Some examples of useful set-ups for studying electrode performance are briefly presented here. The set-ups are based on zirconia pellets with an electrode arrangement suitable for three-electrode studies. The simplest type—"pellet with a belt"—is illustrated in Fig. 5. This can only be used in one atmosphere compartment set-up, and it is therefore absolutely important to monitor pO_2 continuously using a pO_2 sensor placed in the test compartment.

Another more sophisticated three-electrode set-up with two atmospheres is sketched in Fig. 6. If pure oxygen (1 atm.) is used in the counter and reference electrode compartment, then this set-up is a genuine three-electrode set-up. It is a correct geometry, and the gas composition is constant during testing. Such pellet-like geometries, where the reference electrode can be suitably placed (in a bore as in Fig. 6 or as a ring around the pellet in Fig. 5), are suitable for fundamental studies of electrode kinetics. The pellet-like test cell geometries suffer from two disadvantages; it is difficult to ensure that the fabrication process for the electrodes used is identical to the one used for the technological larger scale full cells, and the ohmic resistance between the working and the reference electrodes is quite substantial which may result in a "signal-to-noise" problem, when very high performing electrodes are studied.

To measure a particular electrode performance in detail, a symmetrical cell with identical electrodes on each side of the electrolyte can be used as shown in Fig. 7. This has a platinum mesh to ensure good contact with the electrodes and two platinum wires coming out from each side of the cell, one to measure current and the other for potential determination. Such a test cell is well suited for electrode development work, because there is no ambiguity about the source of electrode properties and performance in this case. However, its use is limited to the investigations close to OCV, where the electrode loss does not depend on whether it is anodic or cathodic polarized.

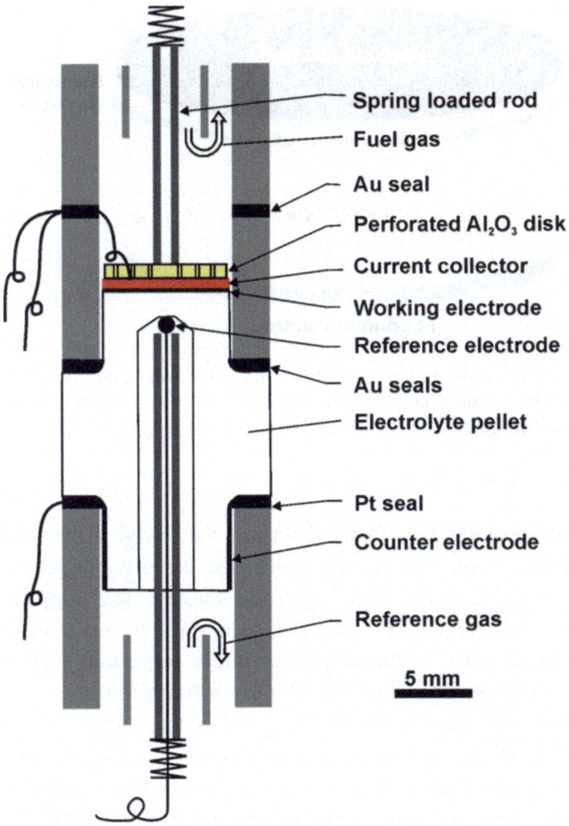

Fig. 6 Sketch of a three-electrode set-up based on a cylindrical YSZ electrolyte pellet, which is held between two alumina tubes that are supplying each their gas to the three-electrode cell. In order to make the composition of the reference and counter electrodes gas independent of the gas conversion on the counter electrode, it is recommended to use pure oxygen. In principle, other materials and gases may be used in such a set-up

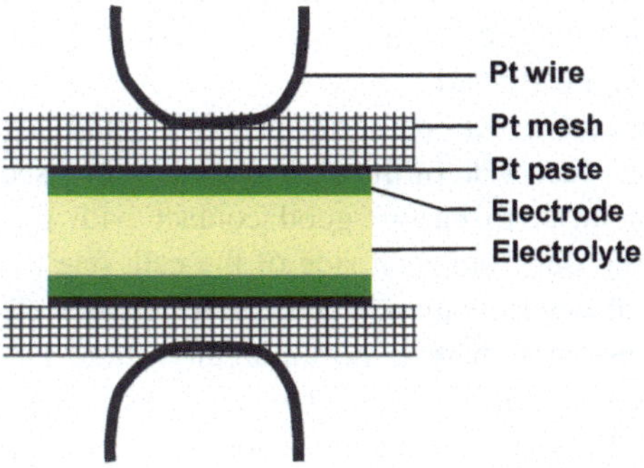

Fig. 7 A symmetrical two-electrode cell arrangement for measurements near OCV, i.e., inside the potential regime of polarization voltage in which the current density—overvoltage curve is linear for the tested electrode type

3 Cells and "Short" Stacks—Geometries, Test Setup, and Conditions

This section describes possible test set-up and cell assembly. Furthermore, focus will be on the probing of voltages and temperature, sealing and contacting of the cell; all important aspects of proper cell testing for obtaining reliable results; also including the aspect of impurities.

3.1 Test House and Cell Assembly for Full Cell Test

For testing of planar SOC, a test house and cell assemblies as illustrated in Fig. 8 can be used. As illustrated in Fig. 8a the cell will be sandwiched in between a fuel and air distributor (grooved) plates or meshes, and these are contacted with platinum or gold foil and Ni foil for the fuel side. The cell is sealed using glass seals (gas tightness) along the edges between two alumina blocks that constitute the cell house. This ensures that fuel and air are led to the electrodes and not simply mixed in the test house. The alumina blocks have built in gas channels for air (or oxygen) inlet and outlet and for fuel inlet and outlet. Different designs for test housing and cell assembly can be found in literature (Jensen et al. 2009a, b; Ebbesen et al. 2010; Kornely et al. 2011) but based on similar concept, i.e., the cell sandwiched between gas distributors and current collecting foils and applying sealing to avoid mixing of fuel and air. The whole assembly shown in Fig. 8 is subsequently enclosed in a furnace within a ventilated hood into which gases are fed from a manifold system. This allows a range of gasses to be led to the fuel side, e.g., H_2, O_2, N_2, CO, CO_2 and CH_4 and to the oxygen electrode side, e.g., air, O_2 and N_2. For humidification, H_2 can be led through a thermo-stated water bubbler, hydrogen and oxygen can be mixed in the fuel inlet tubing, or an external evaporator can be connected. For safety reasons, the ventilated hood should be equipped with sensors to monitor possible H_2 and CO leakages and monitor proper ventilation.

For stack testing, there are several examples on results—experimental and modeling—that illustrate the importance of awareness of the choice of flow design and geometries, e.g., cross-flow versus co-flow versus counter-flow, when considering temperature profiles (Lawlor 2013; Zhang 2012; Kromp et al. 2011). The effect of chosen flow geometry will be much smaller when considering single cell test compared to testing of large stacks. Besides considering air and fuel flow directions, it is also worthwhile to consider the applied flow field design given by the chosen gas distributor plates, meshes, or interconnects. As illustrated in Fig. 9 via experimental data and results from modeling (Kornely et al. 2011), the flow field design obtained by different design of gas distributor plates for single cell test can have a significant influence of obtained cell performance.

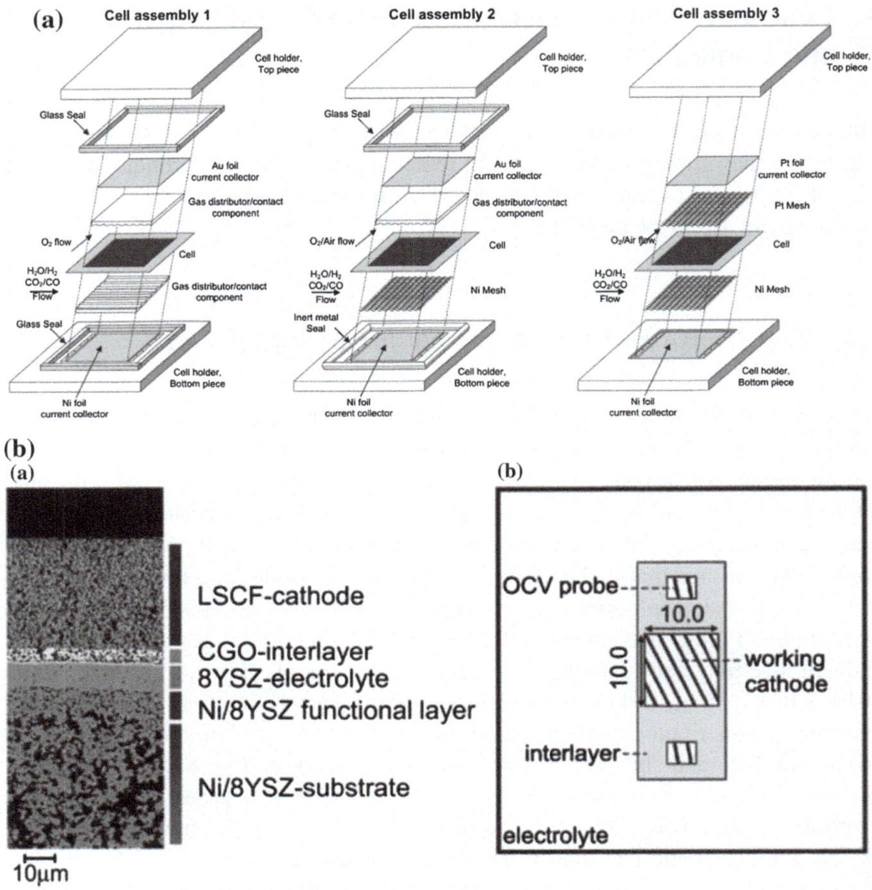

Fig. 8 **a** The assembly of cell and gas distribution plates in a cross-flow pattern with varying choice of gas distributors and solutions for sealing for 53 × 53 mm^2 single cells (Ebbesen et al. 2010) and **b** for a design-wise different test set-up/geometry for a 10 × 10 mm^2 single cell (Kromp et al. 2011)

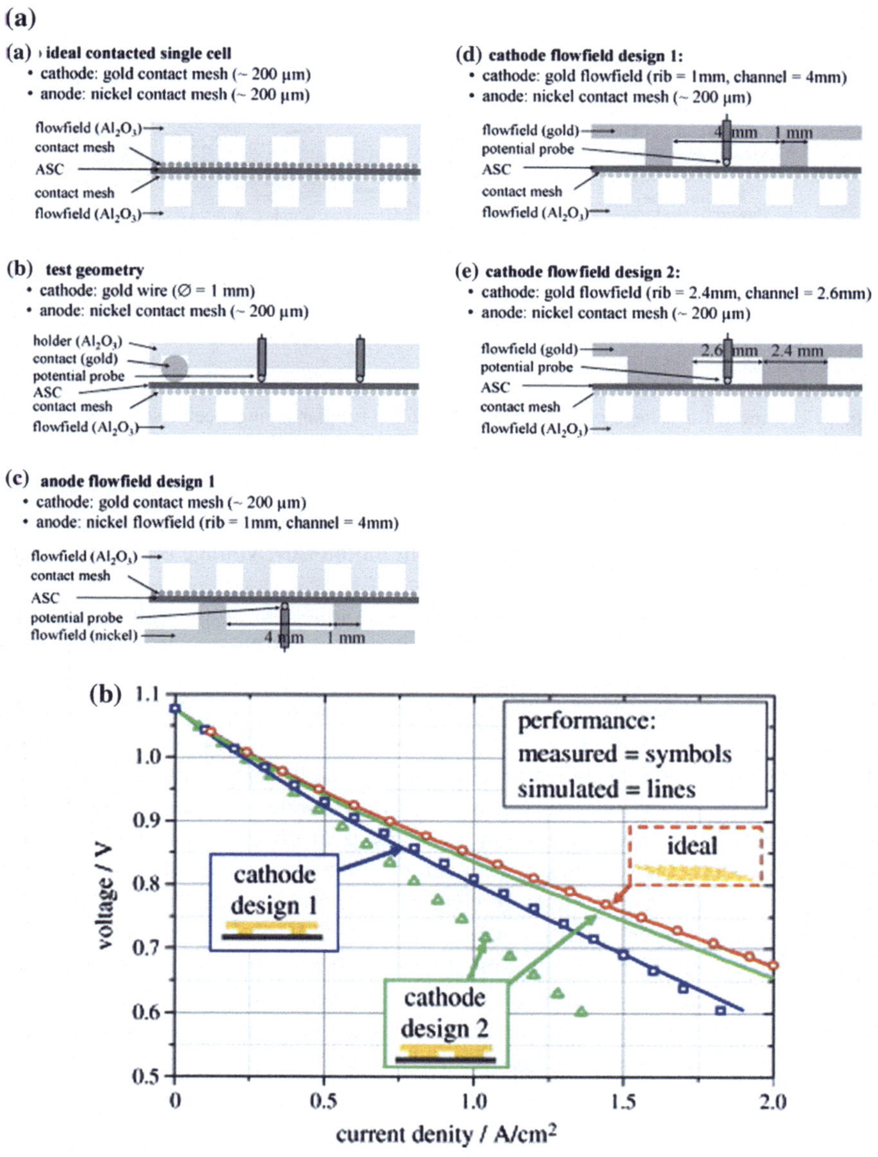

Fig. 9 Different flow field designs and their effect on cell performance (Kornely et al. 2011)

3.2 *Voltage Probes*

The drawing of alumina blocks in Fig. 10 shows the number and position of voltage probes, and probes for current pickup in the test house corresponding to the cell assembly shown in Fig. 8a. The voltage difference between voltage probe 1 and 3

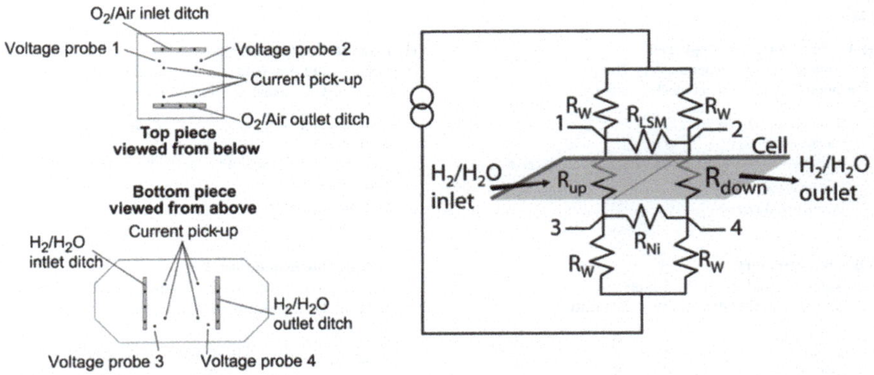

Fig. 10 Positions for voltage probes and current pickup for planar single cell test geometry given in Fig. 8a and sketch of corresponding electric circuit (Jensen et al. 2009a, b)

will monitor the voltage across the cell at the fuel inlet part of the cell while the voltage difference between probe 2 and 4 will monitor the voltage across the cell at the fuel outlet part of the cell.

Placing two or more voltage probes on the same side of the cell enables the measurements of the in-plane voltages, e.g., the voltage difference between probe 3 and 4 in Fig. 10 will measure the in-plane voltage difference between inlet and outlet for the fuel side of the cell. This voltage difference will of course be small compared to the cell voltage but in-plane voltage measurement will provide information on differences in current density from one region of the electrode compared to another region of the same electrode. Measuring, e.g., the fuel side in-plane voltage (probe 3 and 4 in Fig. 10) is advantageous in the analysis of differences in fuel electrode performance along the fuel flow, e.g., in cases where impurities in the gas phase are investigated as reported in literature (Jensen et al. 2009a, b; Ebbesen et al. 2010; Rasmussen and Hagen 2010; Hauch et al. 2014). Furthermore, such extended probing for single cell test as illustrated in Fig. 10 combined with the proper AC characterization of cell performance (see Sect. 5) enables detailed modeling of the cell performance and effects of the impurities when combined with models for the corresponding equivalent circuit as illustrated in Fig. 10 (Jensen et al. 2009a, b).

3.3 Temperature Measurements

Ideally, one would like to measure the temperature in the active electrodes upon cell testing; however, in practice, holes will typically be drilled in the alumina blocks for test housing and thermocouples will be placed similar to the voltage probes, i.e., in the cell assembly but not in the electrodes themselves. This enables measurements of the cell temperature during testing, e.g., during characterization via polarization curves as illustrated in bottom part of Fig. 11 (Ebbesen and Mogensen 2013).

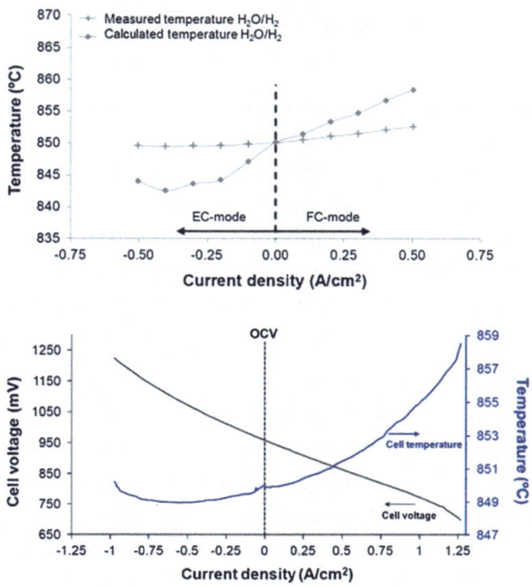

Fig. 11 Temperature measurements during IS at different currents (i.e. measurements of ohmic and polarization resistances at increasing current load) and temperatures calculated based on the changes in the ohmic resistance obtained from IS during step in current (*top*) and temperature measured during an IV-curve (Ebbesen and Mogensen 2013). Be aware of the different scaling for the two graphs

Notice how the cell temperature increases almost 10 °C during the fuel cell mode part of the polarization curve (exothermic reaction and joule heating) while slight decrease in cell temperature is observed during the majority of the electrolysis mode part of the polarization curve (cooling due to the endothermic reaction but partly counterbalanced by the increased joule heating upon increased current density). The left part of Fig. 11 clearly illustrates the issue of inaccurate temperature measurements due to the position of the thermocouples "away" from the spot of the electrochemical reactions. In Fig. 11, impedance spectra are recorded with intervals of 0.1 A/cm^2 between −0.5 A/cm^2 (electrolysis mode) and 0.5 A/cm^2 (fuel cell mode), and at each current density, the cell temperature was measured by thermocouples close to the cell via drilled holes in the alumina test house and plotted as "+" in Fig. 11 (top). Furthermore, the ohmic resistance was obtained from each of the recorded impedance spectra, and from the obtained ohmic resistance, the corresponding change in cell temperature was calculated (based on materials' conductivity). Applying this alternative way of calculating the cell temperature, Ebbesen and Mogensen (2013) clearly illustrate the differences between temperatures measured by thermocouples "outside" of the cell and the actual temperature that can be expected in the active electrodes and electrolytes during DC characterization.

3.4 Sealing, Contacting, and Current Collection

Sealing of the cell assembly can be obtained by glass seals as illustrated in Fig. 8a by glass bars or glass felt that softens upon heating. Appropriate glass composition will typically depend on start-up temperature profile (including reduction temperature). A variety of glass compositions have been tested and reviewed for SOC sealing applications as described in literature (Schilm 2010; Chou et al. 2007; Reis et al. 2010; Nielsen et al. 2007). From a test point of view, the aim of the glass seal is of course gas tightness; however, from an electrode characterization point of view, one should pay attention to the fact that glass seals can contribute with undesirable impurities being led in the gas phase to the electrodes during testing especially during electrolysis operation as described in literature (Hauch et al. 2007; Wiedenmann et al. 2010). Minimizing the glass area exposed to the gas stream is recommendable and alternative sealing materials (e.g., gold) can be considered.

Proper current collection and contacting of the cell for single cell testing is necessary in order to avoid misinterpretation of cell performance results. Resistance due to the contacting of the cell will contribute to the overall resistance of the cell as described in literature (Barfod et al. 2006). This will affect the quantification of the ohmic resistance and the corresponding activation energy, $E_a(R_{ohmic})$. The ohmic resistance contribution from contacting the cell will of course depend on the exact set-up but typically be in the range of 20 m Ω cm^2 in the temperature range 700–850 °C (Barfod et al. 2006).

Furthermore, non-optimal contacting where, for instance, part of a larger cell is not contacted at all will lead to higher current density in the contacted regions of the cell compared to the nominal current density set by the cell tester, i.e., only smaller regions of the cell will be tested but at different test conditions than those set by the operator/cell tester, inevitably giving misleading results for the cell performance.

3.5 Galvanostatic Test Versus Potentiostatic Test

Here, the important concepts of overpotential and polarization have to be defined before discussing the pros and cons of galvanostatic test versus potentiostatic test. The term "polarization" is used for the whole cell and for any of the cell component that gives rise to a loss in voltage, when the cell is electrically loaded. The concept "overpotential" (the same as "overvoltage") makes only sense for the electrodes that have an equilibrium potential at a given set of conditions. Thus, when the potential is moved away from this equilibrium, the electrode potential will change away from the equilibrium value, and the value of the change away from equilibrium is called the "overpotential."

The electrochemical driving force of an electrode process is the overpotential, and the current density is reflecting the resulting reaction rate. Thus, the natural test method should be potentiostatic (constant voltage), but due to historical reasons and the easiness of doing galvanostatic (constant current) tests, almost all reported cell tests have been galvanostatic. The galvanostatic test results may be significantly different from the potentiostatic, because a change in driving force (overpotential) may cause changes in electrode mechanisms during the test. Thus, at high current density, degradation over time may push the electrode overvoltages to a level where an electrode potential will exceed the stability limit of the electrode and/or the electrolyte material, and the electrode or the whole cell may be destroyed (Knibbe et al. 2010; Chen et al. 2013; Zhang et al. 2014; Hauch et al. 2016).

3.6 *Other Types of Testing*

Apart from simple performance tests and long-term durability tests, a number of other test variations may be performed using the same cell (and stack) test set-up. Some important types of tests are pressurized tests, redox tests, thermal cycling, electrical load cycling, and fuel-cell-electrolysis-cell-mode cycling, e.g., using a setup as shown in Fig. 8a.

Examples of pressurization tests have been published by Jensen et al. (2010) and Sun et al. (2015) and an illustration of polarization curves for pressurized tests is given in Fig. 12. The set-up is placed in a furnace inside an autoclave, which is approved to a pressure of 100 bar. The autoclave is filled with flowing pure nitrogen of the same pressure as the pressure in the electrode compartments, i.e., the pressure must be the same inside and outside the cell test house. Naturally, the testing with pressurized hydrogen, CO, and oxygen demands strong safety measures. A detailed description of this is comprehensive and too specialized to include in this text, but it should be noted that it is possible to satisfy even the very strict Danish safety rules for experiments inside a normal laboratory building. The strategy behind is that the flow rate and volume of the nitrogen are so that if all inflowing oxygen and hydrogen were mixed outside the cell house, then it would take an appreciable amount of time before dangerous concentration levels of CO, H_2, and O_2 would be reached. This is then combined with CO and H_2 sensors that stop the supply of H_2 and/or CO if a certain very low concentrations of these are detected. Furthermore, only so-called safety gas of N_2 with maximum 9 % H_2 is allowed in the setup until the temperature is above 600 °C. Over that temperature, H_2 plus O_2 will just burn with a simple flame, which will heat the test house. Then, a separate thermocouple with the purpose of surveillance only will trigger a valve that stops the H_2 and CO flow.

The same basic strategy of safety thinking is behind the testing with pure H_2 and CO in general in order to tolerate testing such as redox test and thermal cycling as these tests will, by nature, often end with a mechanical break down of the cell, which will allow oxygen or air to mix with the fuel gases. Examples of redox testing are given by e.g. Pihlatie et al. (2009) and of electrical load testing by e.g. Hagen et al. (2006), and examples of temperature cycling can be found in literature as well. Examples of fuel-cell-electrolysis-cell-mode cycling are more

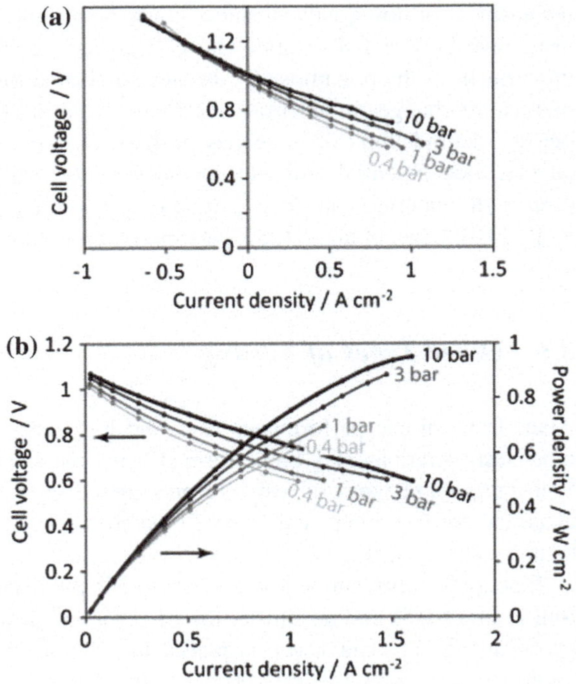

Fig. 12 Polarization curves (IV-curves) at various pressure at 850 °C, H_2/H_2O:50/50, air to the oxygen electrode. Ni/YSZ|YSZ|LSM/YSZ-based and fuel electrode-supported planar SOC (Jensen et al. 2010)

rare, but a recent study has been published by Graves et al. (2015). In case of switching between fuel cell and electrolysis modes, extra gas supplies of steam and CO_2 are necessary. Otherwise, it is mainly the careful test planning, which will be different for the different types of testing.

3.7 *Impurities*

Both fuel electrodes and oxygen electrodes of SOC may be poisoned by many compounds coming from many sources, and even though the set-up and gases used are believed to be very clean, realities have often revealed the contrary. Unfortunately, there is seldom a simple procedure, which can clarify the reason and origin for an observed degradation over time. It is even difficult to figure out whether it is a poisoning initiated or another reason that is the cause of an observed degradation. One source of impurities is the cell itself. No material is totally pure. At least some trace elements will be present, and practically, "pure" materials will most often contain some impurities on the level of 100 ppm or more. Some impurities will be able to segregate from the surface and/or interface of electrodes and electrolyte. A relative simple set-up to test materials for any impurities, which in particular tend to segregate to surfaces and triple phase boundaries (TPB), is

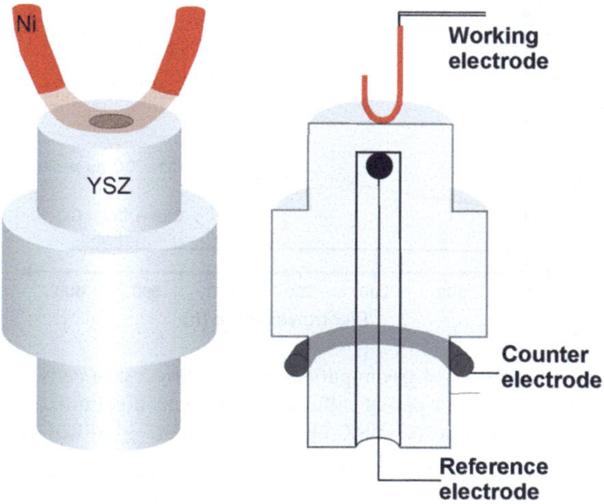

Fig. 13 An illustration of model electrode made by a bent Ni wire pressed onto an electrolyte pellet of YSZ as used at Technical University of Denmark. The YSZ have been a simple pellet shape instead of the slightly complicated pellet made for three-electrode setup. Modified from (Vels Jensen et al. 2001)

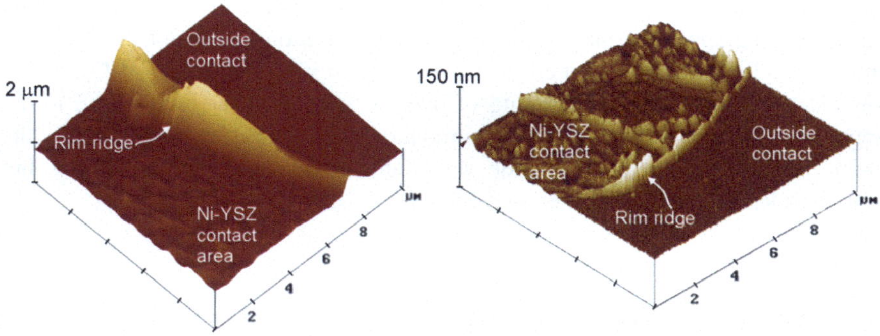

Fig. 14 AFM micrographs of the YSZ side of Ni–YSZ interfaces after few days of testing at 1000 ° C in wet hydrogen (97 % H_2/3 % H_2O); *left* 99.8 % pure Ni, *right* 99.995 % pure Ni. The electrolyte contained <100 ppm impurities. The rim ridges consist only of impurities like SiO_2, TiO_2, and Na_2O. Notice the different scale in the two figures. From (Hansen et al. 2004; Vels Jensen et al. 2001)

sketched in Fig. 13, and results for two different impurity levels of Ni-wire electrodes on a YSZ electrolyte are shown in Fig. 14. The main part of the rim ridge in the left part of Fig. 14 originates from the 99.8 % pure (i.e., impure) Ni wire, while the much smaller rim ridge in the right part of Fig. 14 probably originates from the YSZ and is believed to consist mainly of SiO_2 (Hansen et al. 2004).

In spite of the differences, both TPB look fully blocked, but actually some current can pass even in such model electrodes. Fortunately, the situation is much less critical in a Ni–YSZ cermet electrode, because the ratio between the TPB

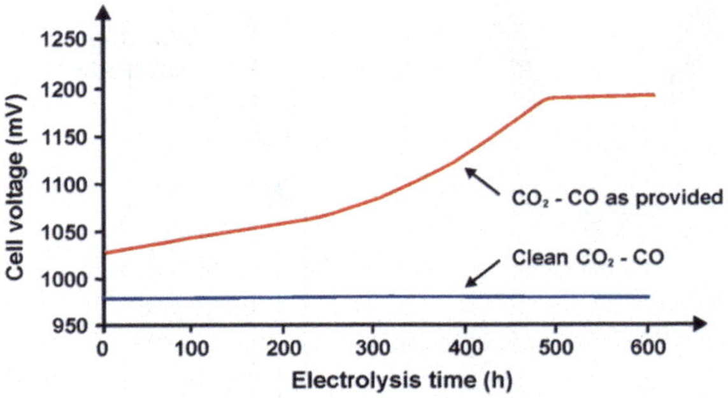

Fig. 15 A sketch of the influence of gas impurities in durability test of cells in electrolysis mode. The as provided gas contained a few ppb of sulfur (about the detection limit), and the other gas was cleaned using a scrubber of hot Ni–YSZ electrode powder. Modified from (Ebbesen and Mogensen 2010)

length and the volume of materials is much bigger for a cermet with submicron particles than in these model electrodes; yet, the TPB will probably not be really clean in any case, and the impurities seen using the model electrodes may also be found in Ni–YSZ electrodes using, e.g., TEM (Hauch et al. 2008).

An example of how to test for—and the effect of—impurities coming from a fuel electrode gas stream is presented in Fig. 15. The as-provided-gas was analyzed by gas chromatography, but only sulfur with a concentration just about the detection limit of ca. 20 ppb was measured. However, the test with clean CO + CO_2, which was cleaned using a filter consisting of fine electrode powder of Ni–YSZ in the "next experiment," proved that the degradation was caused by impurities in the as-provided-gases, and the electrode was stable under same test conditions but applying cleaned gases (Ebbesen and Mogensen 2010).

4 Area Specific Resistance (ASR)[1]

4.1 The Concept of Area Specific Resistance (ASR)

A fuel cell stack can be regarded as a "black box" into which hydrogen (gas) and oxygen (air) are inputs and electricity and exhaust gases are outputs. For such a stack, ASR is defined as:

$$\mathrm{ASR} = \frac{\mathrm{Emf} - U}{i} \tag{2}$$

[1] Based on previous work by Mogensen and Hendriksen (2003).

where Emf is the electromotive force with the inlet fuel and air, and U is the cell voltage at the current density, i, at the design point. A possible design point, might be 0.6 V at 1000 °C, a fuel utilization of 85 %, and an air flow of 4 times the stoichiometric amount ("4 stoichs").

The cell voltage, U, should be measured independently from the leads carrying the current, i.e., separate potential probes should be used. This ASR is in most cases not very sensitive to small variations in cell voltage and fuel utilization. By determining the ASR at a few different temperatures, apparent activation energy, E_A, may be derived. Thus, in a voltage interval (say from 0.5 to 0.7 V) and a temperature interval (say from 650 to 1050 °C), the cell may be characterized, with fair approximation, by only two characteristic numbers, namely ASR at one temperature and E_A. In case the I–V-curve is concave, it may be tempting to use a differential ASR (i.e., the tangent) at high current densities as this gives a nice low value. Such a number has the drawback that it does not reflect the cell performance over the full polarization range as does the quantity defined by Eq. (2).

Sometimes, cell tests are conducted with very low fuel and air utilizations, because these are easier to perform than the "realistic" tests with high fuel utilization. In case of insignificant fuel and oxygen utilizations, the relevant definition of ASR is again that of Eq. (2), but the insignificant utilization makes this value incomparable to ASR derived from experiments with high fuel utilization, because the concentration polarization resistance due to the fuel and air conversion can be a considerable fraction of the total cell resistance. Thus, it is, in general, necessary to specify also the fuel and oxygen utilization together with temperature and apparent activation energy. Furthermore, using almost dry hydrogen, as is the common practice for SOFC tests, it is not easy to conduct experiments with a real negligible fuel utilization, since even small current densities will create enough water to change significantly the Emf of the hydrogen/water fuel gas versus air, e.g., if the inlet gas contains 0.1 % H_2O and the fuel utilization is 0.1 %, this changes the H_2O/H_2 ratio by a factor of 2, which in turn changes the Emf by 34 mV at 850 °C. Therefore, in order to be able to compare results for different fuel utilizations, the ASR value should be corrected for the effect of fuel utilization. Before describing how this may be done, various contributions to the total ASR are examined below.

ASR may be divided into ohmic resistance, R_s, and electrode polarization resistance, R_p. The ohmic resistance originates from the electrolyte, the electrodes materials, and the current collection arrangement. This is very much dependent on geometric factors such as thickness of the cell components and the detailed geometry of the contact between the current collection and electrodes, and between electrodes and electrolyte as current constrictions may be important (Primdahl and Mogensen 1998). The electrode polarization resistance is further divided into contributions from the various rate-limiting steps. Thus, ASR can be broken down into five terms:

Table 1 Contributions to ASR for an anode-supported full cell having a multilayer tape cast anode half-cell sintered at 1295 °C and a screen printed LSCF/CGO cathode (Ni/YSZ|YSZ|$CGO_{barrier}$|LSCF/CGO)

Resistances (Ω cm^2)	850 °C OCV	750 °C OCV	650 °C OCV	750 °C 0.63 A/cm^2
R_s (R_{elyt} + $R_{connect}$)	0.051	0.113	0.328	0.109
$R_{p,electrochem;\ oxelectrode}$	0.017	0.035	0.099	0.025
$R_{p,electro.chem;\ fuelelectrode}$	0.029	0.076	0.287	0.054
$R_{p,diff}$	0.029	0.025	0.030	0.010
$R_{p,conv.}$	0.062	0.059	0.062	0.043
ASR_{Total}	0.188	0.308	0.806	0.241

The cell was characterized at 850, 750, and 650 °C in a single cell test setup as illustrated in Figs. 8a and 10. Values for ASR contributions are based on CNLS equivalent circuit model fitting to experimental impedance spectra (Birkl 2012) (Master thesis by Christoph Birkl, supervised by Dr. Anne Hauch. This thesis is available via DTU Library, Technical Information Center of Denmark, Anker Engelundsvej 1, building 101, DK-2800 Kgs. Lyngby, Denmark). Results are obtained at a $p(H_2)/p(H_2O)$ ratio of 80/20, air to the oxygen electrode, and a fuel utilization of ~23 % at 0.63 A/cm^2. Estimated error for the resistances ~ ±0.004 Ω cm^2

$$ASR = R_{elyt} + R_{connect} + R_{p,elchem} + R_{p,diff} + R_{p,conv} \quad (3)$$

where R_{elyt} is the electrolyte resistance calculated from the measured specific conductivity and the thickness; $R_{connect} = R_s - R_{elyt}$ is the resistance due to the non-optimized contact and current collection; $R_{p,elchem}$ is the electrode polarization originating from all the limiting chemical and electrochemical processes on the electrode surfaces, in the bulk electrode material and on the electrolyte–electrode interfaces, and $R_{p,elchem}$ contains contributions from the oxygen electrode ($R_{p,electrochem;\ ox.electrode}$) as well as the fuel electrode ($R_{p,electrochem;\ fuelelectrode}$); $R_{p,diff}$ is the contribution from the gas phase diffusion; and $R_{p,conv}$ is the contribution due to the gas conversion, i.e., fuel oxidation and oxygen reduction. This division of ASR is based on what is possible to measure and calculate reliably rather than on any physical or electrochemical basis. Some terms in Eq. (3) can therefore be thought of as "equivalent resistances," e.g., the Emf drop due to the changes in gas composition resulting from the fuel utilization is translated into an equivalent resistance. Depending on the exact type of electrode, different types of contributions are possible as derived from more basic electrochemical point of view. For example, current constriction may be important if the electrode has coarse porous structure but of less or no importance in case of a fine-structured electrode. In one type of oxygen electrode, the surface diffusion may be important, but in another the diffusion of oxide ions (and electrons) through the electrode particles may cause the main polarization loss. Values for "Eq. (3)"—ASR contributions for an fuel electrode-supported cell with a 300-μm-thick support fed with H_2 (with 20 % H_2O) are given in Table 1 for different temperatures at OCV and at a current density of

0.63 A/cm^2; all numbers based on impedance spectroscopy data recorded on the same cell (Birkl 2012).

It is observed that the contributions from the concentration polarization, i.e., $R_{\mathrm{p,diff}}$ and $R_{\mathrm{conv.}}$ is dominating at 850 °C but not at 750 and 650 °C. In an electrode-supported cell, the limitation of gas diffusion through the support is a cell relevant resistance, whereas $R_{\mathrm{conv.}}$ results from operation demands and choice of test set-up geometries, and it is thus of special interest to be able to correct ASR for the effect of fuel conversion.

If a significant amount of fuel is consumed in the cell (or stack) under test, a resistance derived on the basis of Emf of the inlet gas [c.f. Eq. (2)] will be an overestimation of the "true" cell resistance. The larger the fuel utilization, the larger will be the overestimation. A comparison between cell test results obtained under different and non-negligible fuel utilizations must thus, to be meaningful, be based on a resistance measure, $\mathrm{ASR_{cor}}$, where the effects of changes in gas composition over the cell area have been taken into account. How the correction is applied depends on how the gases are fed to the cell. Here, two idealized cases are considered, namely the case where the fuel compartment may be considered a continuously stirred tank reactor (CSTR) or a plug flow reactor.

If the fuel compartment can be considered CSTR-like due to effective mixing because of a turbulent gas stream and fast gas diffusion, $\mathrm{ASR_{cor}}$ can be calculated from the expression

$$\mathrm{ASR_{cor}} = \frac{\mathrm{Emf_{avg}} - U}{i} \tag{4}$$

where $\mathrm{Emf_{avg}}$ signifies the average Emf, which in this case is the same as the Emf of the outlet gas. An example of such conditions is reported elsewhere (Primdahl and Mogensen 1998).

The plug flow case is slightly more complex. Under the assumptions that the local ASR is independent of the position along the fuel and air flow channels, and the flow pattern is co-flow, $\mathrm{ASR_{cor}}$ may be calculated from the expression (Solheim et al. 1991):

$$\mathrm{ASR_{cor}} = \left\{ \frac{i}{X^i_{\mathrm{H_2}} - X^o_{\mathrm{H_2}}} \int\limits_{X^o_{H_2}}^{X^i_{H_2}} \frac{\mathrm{d}X_{\mathrm{H_2}}}{\mathrm{Emf}\,(X_{\mathrm{H_2}}) - U} \right\}^{-1} \tag{5}$$

where

$$\mathrm{Emf}\,(X_{\mathrm{H_2}}) = E^0 - \frac{\mathrm{RT}}{2F} \ln \left\{ \frac{X^i_{\mathrm{H_2O}} + X^i_{\mathrm{H_2}} - X_{\mathrm{H_2}}}{X_{\mathrm{H_2}} \sqrt{X^i_{\mathrm{O_2}} - \frac{N_{\mathrm{f}}}{2N_{\mathrm{a}}}(X^i_{\mathrm{H_2}} - X_{\mathrm{H_2}})}} \frac{1}{\sqrt{P_{\mathrm{a}}/\mathrm{atm}}} \right\} \tag{6}$$

here X is the molar fraction, P_{a} is the air pressure, and superscripts i and o signify inlet and outlet, respectively. N_{f} and N_{a} are the molar flows of fuel and air.

The assumption of a position-independent local ASR is an approximation, which is not always justifiable. Part of the fuel electrode polarization resistance is dependent on fuel composition. However, often this part is small. If the cell is not isothermal, the local resistance will vary with position due to its temperature dependence. Also, the actual flow pattern may be much more complex than just co-flow. Even so, if the fuel utilization is large, ASR_{cor} derived from Eqs. 4 to 5 will always be a better characteristic of a cell than a value derived neglecting the fuel utilization [Eq. (2)]. More precise evaluation of ASR_{cor} requires a rigorous 3-D modeling of the cell test.

For the purposes of evaluating Eq. (5), the integral may be approximated by a sum (Mogensen et al. 1999):

$$ASR_{cor} = \left\{ \frac{i}{N} \sum_{j=0}^{N-1} \frac{1}{\mathrm{Emf}\,(X_{H_2}(j)) - U} \right\}^{-1} \tag{7}$$

where $\mathrm{Emf}\,(X_{H_2}(j))$ is given by Eq. (5) with

$$X_{H_2}(j) = X^i_{H_2} + \frac{j + \frac{1}{2}}{N} \left(X^o_{H_2} - X^i_{H_2} \right) \tag{8}$$

The more terms are included in the sum, the better the approximation.

A "first-order" correction for the effects of finite fuel and air utilizations may be obtained by taking only one term in the sum in which case ASR_{cor} should be evaluated from Eq. (4) with

$$\mathrm{Emf}_{avg} = E^0 - \frac{RT}{2F} \ln \left\{ \frac{\bar{P}_{H_2O}}{\bar{P}_{H_2} \sqrt{\bar{P}_{O_2}/\mathrm{atm}}} \right\}, \tag{9}$$

where the bar indicates the "average," i.e.,:

$$\bar{P}_{H_2O} = \frac{P^i_{H_2O} + P^o_{H_2O}}{2}, \quad \bar{P}_{H_2} = \frac{P^i_{H_2} + P^o_{H_2}}{2} \quad \text{and} \quad \bar{P}_{O_2} = \frac{P^i_{O_2} + P^o_{O_2}}{2} \tag{10}$$

If there are no significant leaks in the cell or the test equipment, then both fuel and air utilization, and from this the compositions, may be calculated from the flow rates and the current using Faraday's law. Alternatively, the composition of the outlet fuel and air may be obtained by gas analysis. The conversion resistance, $R_{p,conver}$, may be calculated using the concept of Emf_{avg}:

$$R_{p,conver} = \frac{\mathrm{Emf}_{inlet} - \mathrm{Emf}_{avg}}{i} \tag{11}$$

However, the anisotropic nature (temperature, gas composition, and current flow) of a real cell under current flow often invalidates this simple approach. Gas conversion under electric load causes an uneven distribution of the current density with decreasing current density in the downstream direction. Fuel composition gradients over the cell originating from leaks cause different driving potentials at different points. In extreme cases, this may result in high Emf areas driving low Emf areas in electrolysis mode. Thus, internal currents may flow in the cell even at OCV. Gas leaks in the cell also affect the current density distribution under *load* and cause localized heating by combustion. For tubular cell designs with high in-plane resistance, the current density distribution may be affected; furthermore,

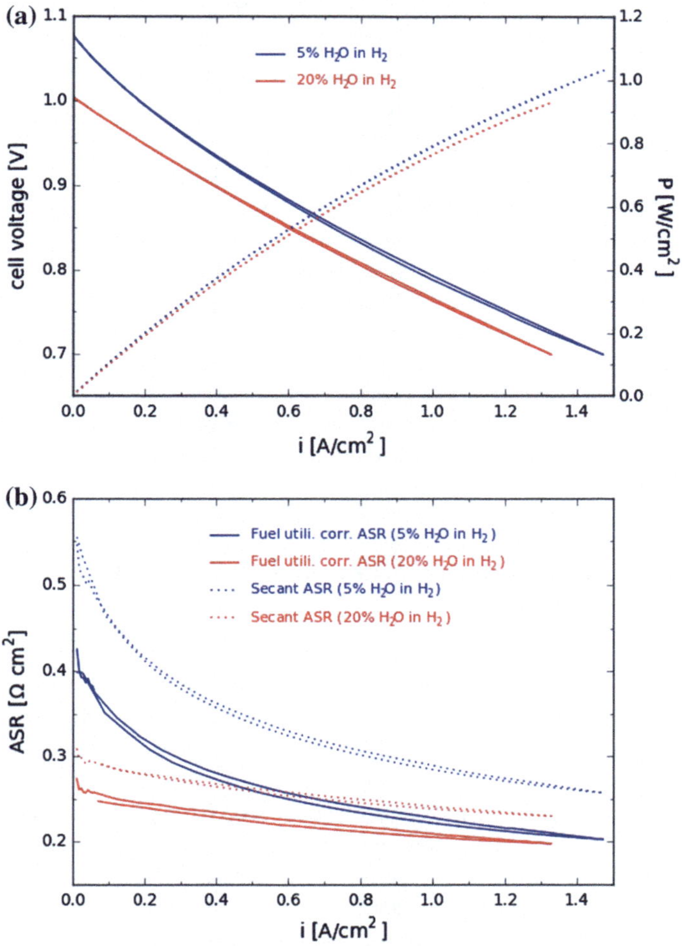

Fig. 16 IV-curve at 750 °C, air to the cathode and $p(H_2)/p(H_2O)$:80/20 and 95/5 to the fuel electrode for a planar Ni/YSZ|YSZ|CGO_{ba}|LSC/CGO-based cell. **a** IV-curve and the corresponding power density curves and **b** corresponding fuel utilization corrected ASR and ASR values based on the secant of the IV-curves (Data from DTU Energy, Technical University of Denmark)

the temperature may not be constant over the whole cell length with flowing gases adding to the inhomogeneity of the current density.

4.2 Fuel Utilization Corrected ASR

Modeling can simplify or reduce the extent of experimental tasks and predict likely behavior under a broad range of test conditions. However, subsequent validation by comparison with relevant cell and stack data is always important.

An example of the magnitude of $R_{conv.}$ is illustrated in Fig. 16, which shows I–V characteristics obtained in hydrogen with either 20 or 5 % water vapor for a fuel electrode-supported cell. Both the ASR deduced directly from the curves via secants of the IV-curves and ASR corrected for the fuel conversion [Eqs. (7) and (8)] are shown in Fig. 16. The correction for $R_{conv.}$ in the case of 20 % H_2O is in this specific case ca. 30–35 m Ω cm^2, reflecting the gas composition dependence of the Nernst voltage. After correcting for the effect of non-negligible fuel utilization, the cell resistance is still significantly smaller when measured with 20 % water in the feed than with 5 %. This reflects a gas composition dependence of some of the loss terms in Eq. (3). Further data and illustration of IV characteristic at different hydrogen steam ratios to illustrate this can be found in the work by Hendriksen et al. (2003), where it is argued that the observed composition dependence is primarily due to the composition dependence of the diffusive losses on the fuel electrode side (diffusion overvoltage), and it is shown how one may utilize characteristics obtained with different water vapor/hydrogen ratios to assess the magnitude of the diffusion loss (Hendriksen et al. 2003).

5 Methods for Characterization of Electrodes, Cells, and Short Stacks

This section will describe typical DC (polarization curves) and AC (impedance spectroscopy) characterization methods for electrodes and cells followed by a short section on complementary non-electrochemical characterization techniques. The focus will be on the practical aspects of obtaining proper and reliable impedance data and "intelligent" use of impedance spectroscopy for single cell characterization. For a comprehensive and more theoretical description of different polarization losses, which can be measured by impedance spectroscopy, the reader is referred to the literature (Nielsen and Hjelm 2014; Barfod et al. 2007; Kromp et al. 2011; Primdahl and Mogensen 1998; Primdahl and Mogensen 1999; Jacobsen et al. 2008; Jensen et al. 2009b; Huang et al. 2007).

5.1 DC Characterization/Polarization Curves

A typical result from a single cell test is a polarization curve also called current voltage curve (IV-curves). Figure 16 shows typical IV-curves for a planar Ni/YSZ|YSZ|CGO_{ba}|LSC/CGO-based SOC at 750 °C. The IV-curve will provide data for the performance of the SOC over a large polarization range; however, a complementary technique such as electrochemical impedance spectroscopy is necessary to analyze the different contributions to the total performance loss in detail. Figure 16 also shows the corresponding power density curve and the fuel utilization corrected ASR calculated from the IV-curve data. Often, the results of screening and performance tests are reported in terms of power density (W/cm^2) figures, and corresponding values for the maximum power densities are extensively quoted. However, this measure can be confusing, because power density varies greatly with fuel composition and with electrode polarization. The IV-curves for SOCs are often more or less linear, and therefore, allow an interpretation in terms of ASR. Even though ASR has no generally accepted definition, it is much less dependent on test conditions than power density, and it is preferable to use it to compare screening test results. The concept of ASR was discussed in the previous section. As stated in the previous section, the ASR value can be calculated in different ways, e.g., based on: (a) the chord from a given cell voltage, e.g., 700 mV to Emf using Eq. (2), (b) a differential ASR (i.e., the tangent) at, e.g., 700 mV, or (c) the fuel utilization corrected ASR at, e.g., 700 mV using Eq. (7).

From the IV-curves in Fig. 16, the fuel utilization can be calculated using Faradays law. The maximum current in this case of Fig. 16 only corresponds to approximately 56 % fuel conversion, and there is no notable sign of fuel starvation. The limit for maximum fuel utilization without notable loss due to the fuel starvation will of course depend on both the chosen design for the fuel flow and the cell microstructure. For cells as those applied for data for Figs. 16 and 22, the fuel starvation will typically only appear for fuel utilizations above 90 %. In the case illustrated in these two figures, the half-cell microstructure is comparable with previously reported, i.e., an approximately 300-μm-thick anode support layer with a porosity of 25–30 % and a 10–15-μm-active fuel electrode with a porosity down to 14 % (Hauch et al. 2012). Furthermore, IV-curves can be a useful tool in investigating leak problems in cell testing (Sect. 6).

5.2 AC Characterization/Electrochemical Impedance Spectroscopy

It is a non-trivial task to break down the total loss measured on a single cell into its components using the results from electrode testing and polarization curves of full cells. Impedance spectroscopy on full practical cells is, however, a technique by which a breakdown of losses can be made. General introduction to impedance spectroscopy can be found in literature (Macdonald and Barsoukov 2005), and the following text will focus on the application for SOC characterization purposes. Impedance spectra obtained on a full SOC (or even a half-cell or symmetric cell) are difficult to interpret due to many processes involved and partly overlap in frequency range for these processes. Therefore, at least the following three aspects should be considered when applying electrochemical impedance spectroscopy for analysis of SOC: (1) proper impedance spectroscopy measurements, (2) intelligent use of different test conditions to analyze differences in contribution of losses, and (3) modeling the experimental impedance data to obtain the quantification of the breakdown of losses.

Regarding aspect 1, this includes measurements in the relevant frequency range, i.e., typically from 100 kHz to 0.1 Hz for SOC with an appropriate number of points recorded per frequency decade (i.e., 10–15 points per decade) if quantitative analysis is desired. Frequencies should be set to minimize effects of electronic noise from other electronic devices in the cell test rig. Furthermore, the signal amplitude of the AC signal for the impedance measurements should be adjusted to be small enough to provide a small perturbation in the linear region of the system and still be large enough for a reasonable signal-to-noise ratio (Huang et al. 2007) e.g. for a cell with the performance as given in Fig. 16 an AC signal amplitude of $\sim$40 mV will be appropriate. The signal-to-noise ratio will of course also be optimized by running many cycles at the same frequency, e.g., 50–200 cycles depending on frequency range and time available for the measurement.

Regarding aspect 2, intelligent use of impedance spectroscopy at different test conditions provides valuable information on different processes constituting the total loss and gives necessary input to a subsequent quantitative modeling of the impedance response from an SOC. Impedance spectra at different temperatures but same gas compositions will enable the calculations of the activation energies and typically enable the separation of non-thermally activated processes (e.g., physical processes such gas diffusion) from thermally activated processes (e.g., electrochemical processes in the electrodes). Impedance spectra at different gas compositions can be used to analyze the different polarization losses' dependency of gas composition and "one electrode sided" gas changes can be used to separate which impedance contributions originate from each of the electrodes for a full cell test. Analysis of differences in impedance spectra [ADIS, (Jensen et al. 2009b, 2007)] can, e.g., by use of impedance spectra at different gas compositions, be used qualitatively to analyze which part of a full cell that degrades. Figure 17 provides examples of using analysis of differences in impedance spectra (ADIS) upon changing gas composition at one electrode at the time, before and after long-term fuel cell test. It is evident from Fig. 17 (top) that when changing from air to oxygen for the LSM/YSZ-based oxygen electrode, a difference in impedance occur at a frequency around 200 Hz, and this response is basically identical before and after long-term test, indicating that the impedance from the oxygen electrode before and after test is similar. The ADIS for the fuel electrode gas shift in the top graph of Fig. 17 has several small peaks before long-term test [see also (Barfod et al. 2007)], and the ADIS for the fuel electrode gas shift has changed dramatically after test,

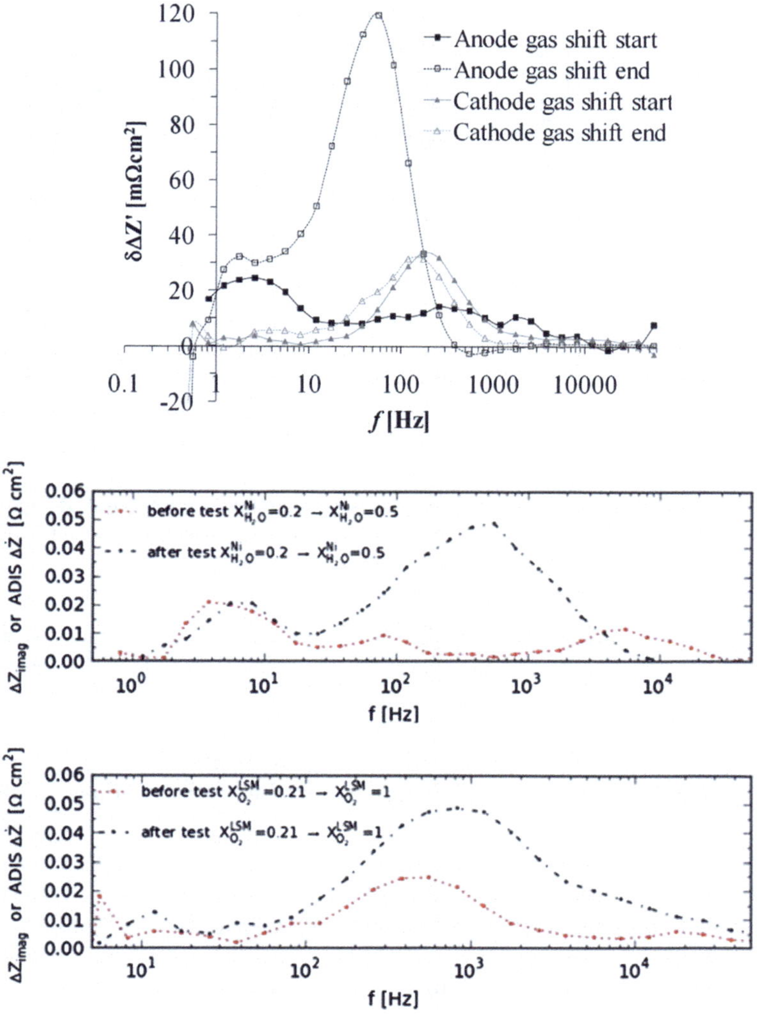

Fig. 17 *Top* Analysis of differences in impedance spectra (ADIS) before and after fuel cell test upon changing fuel gas composition between $p(H_2)/p(H_2O)$:96/4 and $p(H_2)/p(H_2O)$:80/20 and oxygen electrode gas between air and O_2. Gas change impedance spectra were recorded at OCV and 750 °C before and after fuel cell test (Hauch et al. Hauch et al. 2011a, b), and (*bottom*) ADIS before and after electrolysis testing (Hjalmarsson et al. 2013)

indicating that the fuel electrode has been severely affected by the long-term test, and the main peak has decreased from around 800 Hz before test to around 80 Hz after test. Figure 17 also provides an example of ADIS before and after long-term electrolysis testing using different gasses. In this example, significant changes for the impedance response for ADIS for both electrodes can be observed as a result of long-term electrolysis testing.

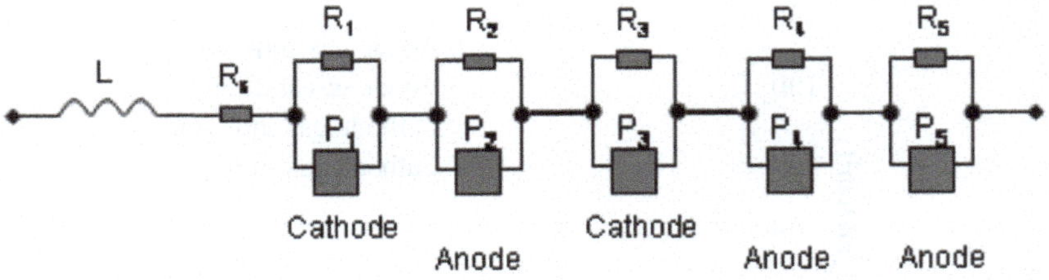

Table I. Arcs necessary to characterize losses of a full cell (approximate numbers).

	f_{sum} 700°C (Hz)	f_{sum} 850°C (Hz)	Characteristics
Arc 1	18.000	50.000	Small or no p_{O_2} dependence Small dependence on current
Arc 2	1000	8000	Dependent on fuel gas composition
Arc 3	100	1100	Strong p_{O_2} dependence Strong dependence on current
Arc 4	20	20	Changes with p_{H_2O}/p_{H_2}
Arc 5	3	3	Changes with p_{H_2O}/p_{H_2}

Fig. 18 Example of an applied equivalent circuit model for a full Ni/YSZ-YSZ-LSM/YSZ-based cell and the corresponding interpretation of processes and typical ranges for summit frequencies (Barfod et al. 2007)

The examples illustrated in Fig. 17 show how the use of impedance spectra at different gas compositions can provide qualitative but model-independent, electrode-specific information on different impedance contributions and assist in the analysis of which electrode—or both electrodes—that degraded during long-term testing, and provide information regarding the approximate summit frequencies for several of the different losses contributing to the total polarization loss for the full SOC.

Regarding aspect 3, to quantitatively characterize the different impedance contributions, the impedance spectra are typically analyzed in terms of equivalent circuit models applying a complex non-linear least squares (CNLS) fitting routine (Sonn et al. 2008; Nielsen and Hjelm 2014; Jacobsen et al. 2008; Orazem 2008). As an example, a resistor in parallel with a constant phase element (CPE) will typically be used as an approximation to model the charge transfer reaction process at the electrochemically active sites of the Ni/YSZ-based fuel electrode, and an example of an equivalent circuit model applied for a full SOC is given in Fig. 18 and many other examples can be found in literature. However, attention should be kept on the fact that such impedance data treatment can be highly affected by the "operator" and such impedance analysis should always be supported by complementary analyses such as ADIS using different gas compositions and temperatures. That being said, the results from the CNLS fitting of the impedance spectra applying equivalent circuit models will provide valuable quantitative information on, e.g., resistances and summit frequency for the different contributions to the total polarization loss.

Figure 19 illustrates a breakdown of losses for a fuel electrode-supported Ni/YSZ|YSZ|LSM/YSZ-based cell at 850 °C based on the equivalent model depicted in Fig. 18 in the current density interval from −0.5 to 0.5 A/cm^2. Notice

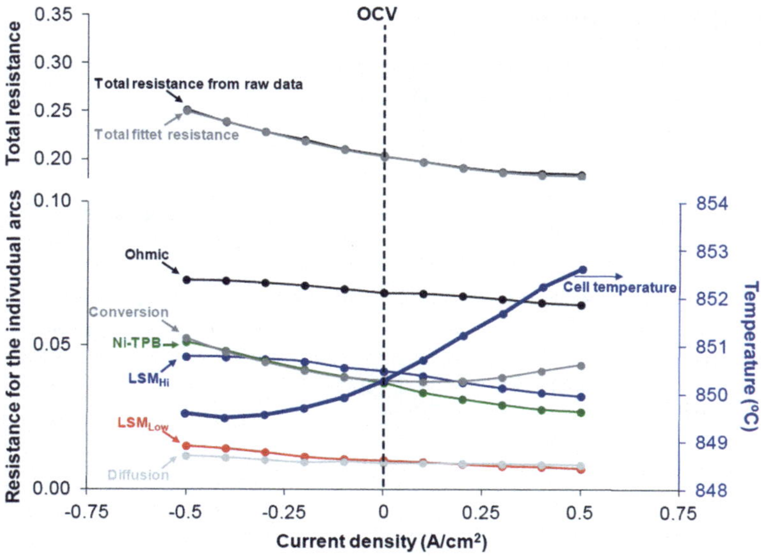

Fig. 19 Current density dependency of the different contributions to the total ASR of an fuel electrode-supported Ni/YSZ|YSZ|LSM/YSZ-based cell. The results are based on the analysis of impedance spectra at 850 °C, applying a H_2O/H_2:50/50 fuel (Ebbesen and Mogensen 2013)

the linearity of the IV-curve (DC characterization) also shown in Fig. 19 in the current density range, in which impedance spectra were recorded (Ebbesen and Mogensen 2013). The linearity of the cell response at the given conditions justify the comparison of results from DC and AC characterization of the cell, and the linearity of the IV-curve reveals that proper ASR can be calculated from the secant ASR from the IV-curve (DC characterization).

A complementary way of analyzing impedance spectra is via distribution of relaxation times (DRT) as described by Schichlein et al. (2002). This method can advantageously be combined with ADIS and equivalent circuit modeling; however, high quality impedance spectra are required for proper DRT analysis. Figure 20 shows an example of a Nyquist plot of impedance spectra and the corresponding DRT plot. As evident from Fig. 20, the DRT plot provides a high frequency resolution that enables clear distinction between the many processes contributing to the total polarization loss in the given full cell test. The area beneath the DRT curve represent the resistance of the cell.

There will be cases where it is not possible to perform a breakdown of the losses of a full SOC based on the measured impedance spectra, e.g., because of overlapping impedance contributions or quality of the spectra. However, there can still be useful information to obtain from the impedance spectra in terms of the ohmic resistance and the total polarization resistance of the SOC.

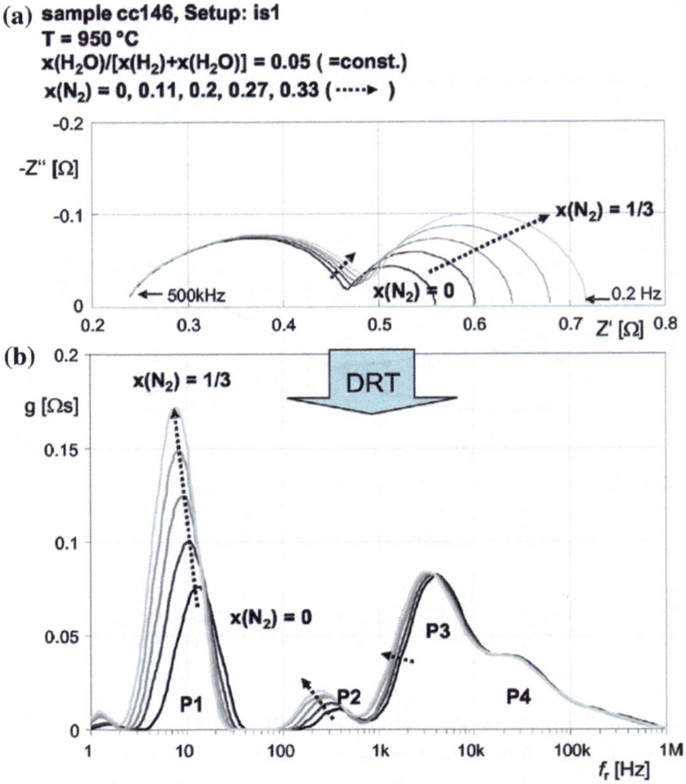

Fig. 20 Nyquist plot of impedance spectra and corresponding distribution of relaxation times (*DRT*) plot upon dilution of fuel electrode gas with nitrogen for a fuel electrode half-cell at 950 °C (Sonn et al. 2008)

5.3 Concentration Polarization Contributions to the Total Loss

Performance losses from processes like gas conversion and gas diffusion and resistance contributions with a metallic-type dependency (e.g., from metallic connections) are not thermally activated, i.e., they do not follow Arrhenius equation type of behavior. These contributions to the total cell resistance will be the most significant at higher temperature where contributions from the thermally activated processes are the smallest (see e.g., in Table 1). This fact can be used to separate and thereby also quantify, e.g., resistance contributions due to the gas diffusion (non-thermally activated) from resistance contributions due to the electrochemical processes in the oxygen electrode based on MIEC (thermally activated process) which frequency wise can be overlapping at certain test conditions (Leonide et al. 2008; Nielsen and Hjelm 2014).

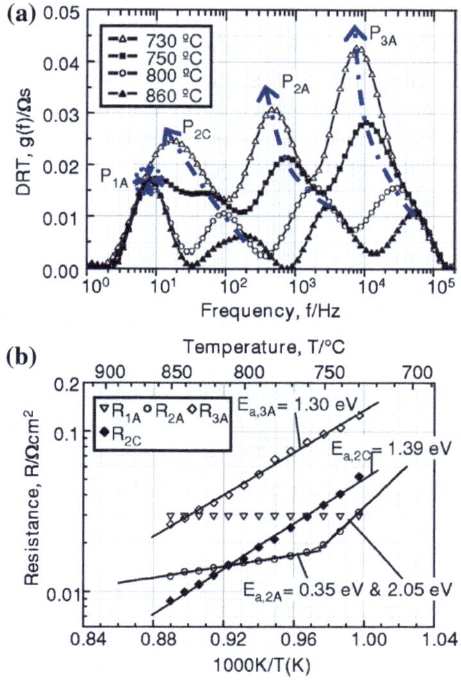

Fig. 21 Temperature dependency of the different contributions to the total ASR of an anode-supported Ni/YSZ|YSZ|LSCF/CGO-based cell (Leonide et al. 2008)

Losses due to gas diffusion (typically observed in the impedance spectrum in the frequency range 10–100 Hz) will of course dependent on several microstructural parameters such as porosities, tortuosity, and layer thicknesses as described and modeled by several authors (Jacobsen et al. 2008; Bessler 2005) as well as gas composition (Kromp et al. 2011). The gas diffusion impedance will typically be small, i.e., in the range of 10–40 m Ω cm^2 (Birkl 2012; Barfod et al. 2007) at relevant test conditions for cells with fuel electrode-supported layer thicknesses of ~300 μm and larger for thicker anode support layers.

5.4 *Thermally Activated Contributions to the Total Loss*

The thermally activated processes include resistance contributions originating from the electrolyte and the two electrodes. The resistance due to the electrolyte (conduction of ions) is often denoted ohmic resistance or series resistance, and from a cell testing and characterization point of view, it is worthwhile to notice that the activation energy for this electrolyte resistance can easily be affected by non-thermally activated resistance contributions from, e.g., contacting of the cell which will also contribute to the ohmic resistance. Figure 21 shows a breakdown of losses for a fuel electrode-supported Ni/YSZ|YSZ|LSCF/CGO-based cell at OCV (Leonide et al. 2008) in the temperature interval from 760 to 860 °C. The results shown in Fig. 21 are based on the analysis of impedance spectra and illustrate the thermal activation of the electrochemical processes in the fuel and oxygen electrode.

5.5 Complementary Characterization Techniques

The characterization techniques described above are in situ electrochemical characterization of the performance and characteristics of an SOC. It is desirable to supplement the in situ electrochemical characterization with non-electrochemical, and mainly ex situ, characterization of the cells which can provide complementary information, e.g., on cell degradation mechanisms, typically microscopic and spectroscopic methods (Leng 2008).

Electron microscopy is widely used for studying SOC on a nano- and micrometer scale both as feedback for processing optimization and start-up profile optimization (Hauch et al. 2012; Jiao and Shikazono 2015; Jørgensen et al. 2015) and for investigation of aging and degradation of cells (Chen et al. 2013; Liu and Jiao 2005; Hauch et al. 2008; Tietz et al. 2013; Knibbe et al. 2010; Shimura et al. 2014). Besides more traditional, SEM investigation of electrodes which can also include quantitative image analysis (Holzer et al. 2011; Iwanschitz et al. 2012; Faes et al. 2009), more specialized SEM imaging techniques can also be used to gain information on the percolation of, e.g., the Ni network as described by Thydén et al. (2008), and illustrated in Table 2.

In recent years, combined focused ion beam (FIB) milling and SEM imaging have also been used to make 3D reconstructions of electrode structures (Wilson et al. 2006; Shearing et al. 2010a, b; Jørgensen et al. 2015; Shimura et al. 2014; Jørgensen et al. 2010) leading to very detailed and quantitative information on the active TPB lengths, particle size distributions, tortuosity etc. in a given electrode structure. Such data may advantageously be used to investigate an electrode microstructure before and after long-term test and be correlated to the electrochemical characterization obtained during long-term testing. An example of the use of 3D reconstructions of electrodes is given in Table 2. Over the last decade, there have also been several examples of applying transmission electron microscopy (TEM) for the analysis of SOC which enables, e.g., investigation of nano-sized impurities in the electrode structures, oxygen "bubbles" formation in the electrolyte, and formation of undesirable secondary phases, (Knibbe et al. 2010; Hauch et al. 2008; Liu and Jiao 2005) and even examples of in situ studies of NiO reduction using TEM (Jeangros et al. 2012). Besides electron microscopy, techniques such as scanning tunneling microscopy (STM) (Hansen et al. 2013; Khabibulakh et al. 2009;

Table 2 Example of combination of electrochemical characterization techniques combined with complementary non-electrochemical characterization techniques such as SEM, 3D reconstructions of electrodes and ToF-SIMS analysis

Characterization\cell type	Technologically relevant fuel electrode-supported cell	Full cell	Model electrode/LSM microelectrode on YSZ
Illustration of electrochemical performance characterization	*p*[e]	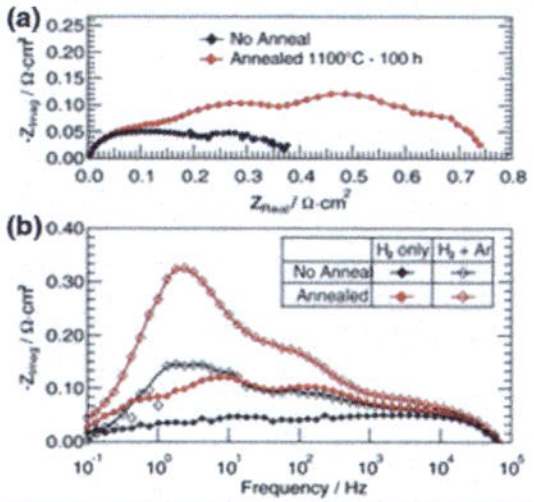	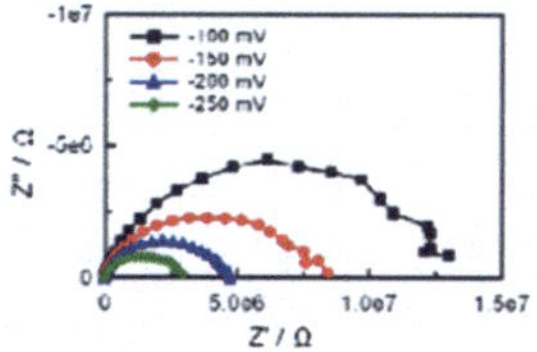
Description of electrochemical performance characterization	Simple cell voltage measurement over time under current load (technological relevance)	Impedance spectroscopy—for cells annealed differently—provides data for "overall"/average electrode	Impedance spectroscopy for a well-known electrode geometry (enables studies of electrode kinetics)
Illustration of complementary characterization	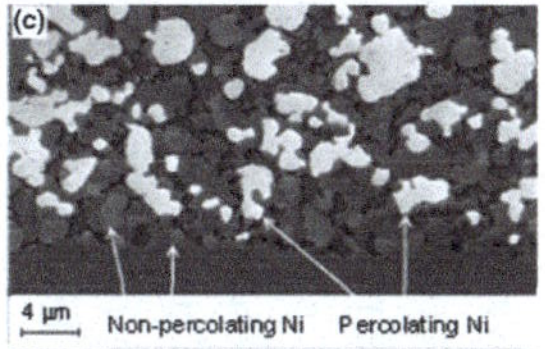	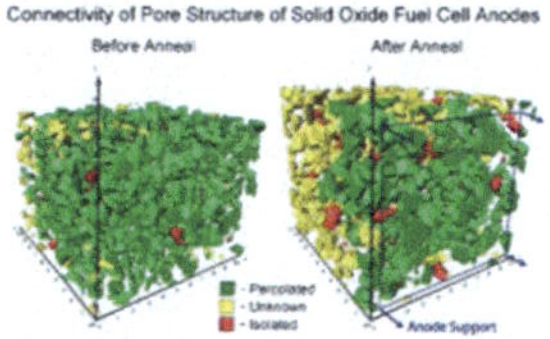	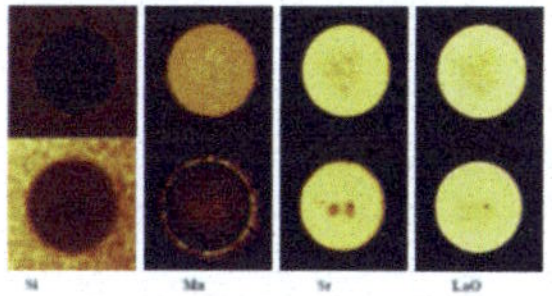
Description of complementary characterization	Fast/non-time-consuming microscopy for the analysis of Ni percolation (of test B)	Time-consuming 3D reconstruction of porous structure and its percolation via FIB milling and imaging	ToF–SIMS measurements for a reference (upper) and tested sample
References	Hauch et al. (2011a, b)	Cronin et al. (2011)	Wu et al. (2013)

Cai et al. 2011) and scanning probe microscopy (Wu et al. 2013) have also been used to investigate impurities and secondary phases at the electrode surface structures.

Furthermore, a variety of analysis techniques based on spectroscopy can be and have been applied for SOC characterization such as secondary ion mass spectroscopy, SIMS (Horita et al. 2012; Hansen et al. 2006; Schmidt et al. 2008; Kiebach et al. 2014) and Raman spectroscopy (Li et al. 2012; McIntyre et al. 2015; Kishimoto et al. 2012; Shimazu et al. 2011), and spectroscopy based on X-rays such as energy dispersive spectroscopy, EDS, wavelength dispersive spectroscopy, WDS (Wiedenmann et al. 2010), and X-ray photoelectron spectroscopy, XPS (Backhaus-Ricoult 2008; Utz et al. 2011). To illustrate and highlight how electrochemical measurements have recently been combined with complementary techniques for SOC characterization, a couple of examples are given in Table 2.

Table 2 illustrates the strength in applying both electrochemical characterization (DC and AC methods) and complementary characterization techniques, as well as illustrating the different levels of complexity both regarding the type of cells to test and characterize (model electrodes versus technological relevant full cells) but also with respect to choice of methods for characterization.

The data shown in the second column illustrate a rather extended and complex test matrix in the sense that the tests are performed on full cells—sister cells with identical microstructure but different test conditions. The corresponding analyses of impedance spectra (not shown here, reported elsewhere) during the fuel cell test, i.e., under current load strongly suggest that it is for all 3 tests the Ni/YSZ electrode that degrades. As the impedance data was obtained on full cells including oxygen electrodes, it is valuable to support the impedance analysis by postmortem microscopy investigations from which it was evident that the fuel electrodes degraded and also that the degradation was related to a decreased level of percolation for the Ni network, possibly due to impurities (Hauch et al. 2011a, b).

The data in the third column illustrate more complex electrochemical test in form of impedance, however, measured at OCV, and the impedance data is supplemented by 3D reconstruction of the electrode structure.

The data in the right most columns provide an example with a very simple test geometry (model electrode), and this enables a more direct comparison of the electrochemical characterization results with precise microstructural measurements of changes in the structure/composition upon polarization.

Comparing the results reported in Table 2 in horizontal direction, it also illustrates variety in electrochemical characterization techniques (in situ characterization) and microstructural characterization techniques (ex situ). Cell voltage curves (and polarization curves) represent simple but valuable overall information on the performance of a cell, whereas the AC characterization (impedance spectroscopy) provides more detailed information (3rd and 4th column); however, the analysis of the different contribution to the overall resistance of the cell can be complex when applied for full cells (3rd column). The examples on microstructure characterization in Table 2 illustrate different techniques having different pros and cons. The ToF–SIMS analysis (4th column) provides high accuracy measurement on a simple geometry, where as the 3D reconstruction provides very accurate information on a more complex structure; however, this technique can be somewhat time-consuming and only a tiny part of the electrode structure is investigated (which is hopefully representative for the entire electrode structures). The 2D SEM analysis as illustrated in the 2nd column provides a simple and fast investigation of the electrode structure and can be done for several hundred micrometer within less than a day, which provides the possibility of investigating, e.g., gradients in coarsening or impurity level along, e.g., a flow direction in a technologically relevant large scale cell.

6 The Problem of Gas Leakage in Cell Testing

A significant source of error in cell testing can be due to leakage. This section first describes possible origins and observed effects of leakage followed by a paragraph on assessment of the size of leakage.

6.1 Origin and Effects of Leakage

Gas leakage of air or oxygen from the oxygen electrode compartment to the fuel electrode compartment can happen either along the edges of the cells due to improper sealing or from the oxygen electrode through the cell via cracks or holes in the electrolyte (Rasmussen et al. 2008). Normally, the air flow is much higher than the fuel flow, and thus, the pressure is slightly higher in the oxygen electrode compartment than in the fuel electrode compartment leading to gas leakage from the oxygen electrode compartment to the fuel electrode compartment rather than vice versa. A gas leakage of air to the fuel electrode compartment will inevitably lead to lowering of the OCV, and attention should be paid to the measured OCV compared to the theoretically calculated Emf from the Nernst equation. Furthermore, one should notice that a large localized leak of air to the fuel electrode compartment, e.g., at the rim/edge of the cell causes a local increase in temperature due to the hydrogen combustion, and such temperature increase may not necessarily be registered if the thermocouple is positioned away from the leak (e.g., at the center of the cell). However, the increased temperature will decrease the cell resistance locally. Large leaks in the cell or in seals cause inhomogeneous gas composition over the cell, even at OCV. Another possible explanation for differences in OCV compared to the calculated Emf is in principle electronic leak through the electrolyte. If the electronic conductivity in such a leak decreases with cell voltage, a flat IV-curve may be obtained and care should be taken in the interpretation of such IV-curve measurements.

6.2 Assessment of the Size of Gas Leakage

The effect of gas leaks on cell tests may be separated into two components: (1) a loss of driving force (Emf) caused by oxygen entering to convert hydrogen into steam, and (2) a volumetric loss of fuel affecting the fuel utilization caused by a pressure gradient versus ambient. A gas leak may be before, in, or after the cell in a test set-up fuel gas line (see e.g., Fig. 10). Leaks before the cell affect the intended fuel composition and quantity; however, a true average composition of the gas may be obtained from the measured OCV for the cell. Leaks after the cell are seldom important to the cell test, unless of course analysis of the outlet gas composition is carried out. A cell test set-up including measurement of the pO_2 in the inlet gas stream, measurement of the cell's OCV, the pO_2 in the outlet gas stream, and appropriate temperature measurements provides reasonable monitoring of inlet gas composition (in relation to expected composition based on set gas flows via mass

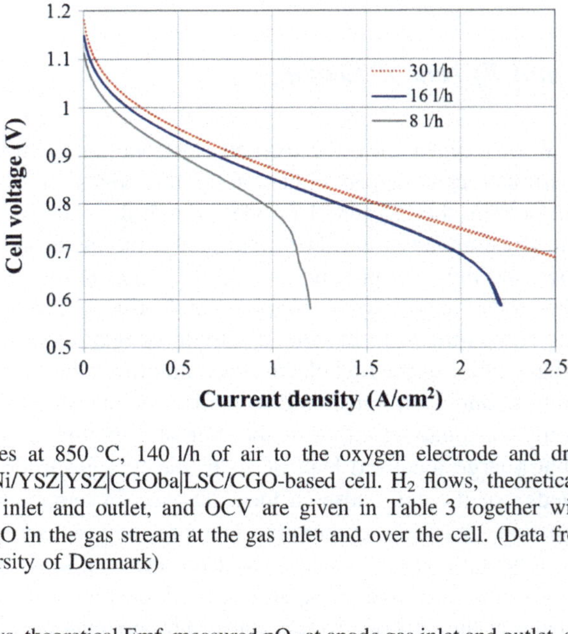

Fig. 22 IV-curves at 850 °C, 140 l/h of air to the oxygen electrode and dry H_2 to the fuel electrode for a Ni/YSZ|YSZ|CGOba|LSC/CGO-based cell. H_2 flows, theoretical Emf, measured pO_2 at fuel gas inlet and outlet, and OCV are given in Table 3 together with the calculated percentage of H_2O in the gas stream at the gas inlet and over the cell. (Data from DTU Energy; Technical University of Denmark)

Table 3 H_2 flows, theoretical Emf, measured pO_2 at anode gas inlet and outlet, and OCV together with the calculated percentage of H_2O in the gas stream at the gas inlet and over the cell for the two IV-curves depicted in Fig. 22

H_2 flow (l/h)	Emf (theo)	pO_2-inlet (mV)	OCV (mV)	pO_2-outlet (mV)	% H_2O in inlet gas	% H_2O in gas in electrode	FU at max. current
8	1218	1163	1131	1116	0.7	1.4	97
16	1218	1163	1163	1150	0.7	0.7	96

The pO_2 values are given as voltage against air, and FU is the fuel utilization

Fig. 23 Deviation of 5, 10, 20, 40, 60, and 100 mV between measured OCV and theoretical Emf translated into percent loss of feed hydrogen as a function of H_2O concentration in fuel gas, H_2/H_2O mixtures at 700 °C

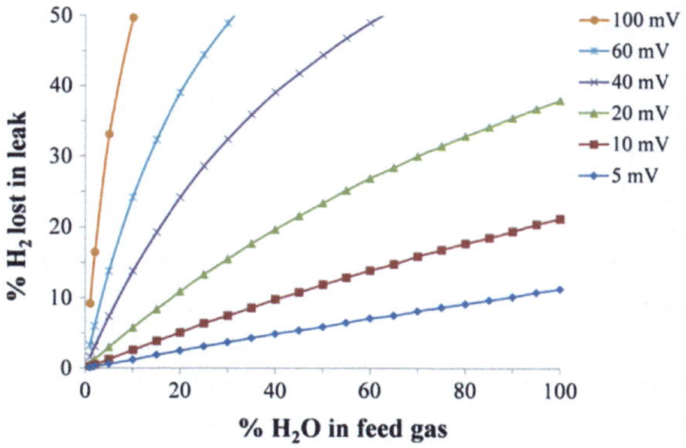

flow controllers), and leaks affecting the cell's OCV (deviation from Emf) and total gas leaks from inlet to outlet and enables the quantification of possible leaks. Figure 22 shows IV-curves measured on a planar fuel electrode-supported Ni/YSZ|YSZ|CGO_{ba}|LSC/CGO-based cell with different H_2 flows to the fuel electrode, and Table 3 lists the corresponding values for measured pO_2 in inlet and outlet gasses and corresponding percentage of H_2O in the fuel electrode gas stream at the inlet and over the cell. From such simple characterization, one can assess the size of the gas leak and a measure of the maximum fuel utilization obtainable for the tested SOC. A clear effect of the fuel starvation is observed at current densities of around 2.1 and 1.05 A/cm^2, respectively. Extended characterization should include similar flow variations for the gas to the oxygen electrode.

In more general terms, assuming oxygen leakage into the fuel compartment, one can estimate the loss of fuel based on observed OCV deviation from the theoretically expected Emf as calculated from the Nernst equation. Figure 23 shows a number of such curves at 700 °C for a H_2/H_2O gas feed. From a given feed percentage of H_2O, the loss of H_2 can be estimated from the Emf–OCV difference. On the other hand, reading the curve in another way, a given loss of H_2, e.g., 6 % of the feed causes a loss of driving force of 60 mV if the feed contains 2 % H_2O, whereas for a feed gas with 50 % H_2O the loss is only 5 mV. Such calculations illustrate the effects of gas leakage, and the effects are closely related to the chosen cell test conditions and highlight the importance of thorough measurements of gas compositions. There will rarely be leak from the "low-flow" fuel electrode side of the cell to the "high-flow" oxygen electrode side of the cell, and on top of that the potential of the oxygen electrode gas is far less sensitive to reaction of O_2 with H_2 and production of H_2O, e.g., an injection of 5 % H_2 in the air stream will only reduce the potential by less than 7 mV at 700 °C.

7 Summary

This chapter has considered the main types of electrochemical tests which have been applied to SOC and has outlined the main issues which require detailed attention for obtaining meaningful test results for electrodes and cells. One important aspect in the electrode testing is to assure a correct geometry in three-electrode set-ups. This is very difficult in practice in case of electrode-supported cells with thin electrolytes. Unfortunately, even testing the individual electrodes in sound geometry set-ups is not a perfect procedure either, because the sum of the contributions from individual cell components to the cell resistance does not add up to the actually measured total cell resistance. This is probably due to the differences in the fabrication of the special cells for electrode characterization and the practical full cells.

In the description of recommendable test geometries and set-up, such issues as proper probing of voltages and temperatures was also treated along with the complex matter of impurities, and their—often undesirable—effects on SOC cell test results.

As described in this chapter, it is recommended that cell test results be reported in a way that makes it easy to derive ASR from the IV-curves. Sufficient information should be provided so that the ASR values can be corrected for the effects of finite fuel utilization. Also, the choice of fuel composition should preferably reflect real cell operation conditions. The ASR should be derived using the Emf and a cell voltage in the range of 0.5–0.7 V and its corresponding current density. In case of a grossly non-linear IV-curve, a differential ASR value is of little practical use.

Furthermore, the description of electrochemical characterization methods for electrodes and cells was provided including characterization via DC methods (polarization curves) and AC methods (impedance spectroscopy), as well as an overview of complementary non-electrochemical characterization techniques relevant to SOC characterization. In the description of impedance spectroscopy for characterization of full SOC, introduction to and examples of break down of polarization losses were also given. Finally, the issue of leakage which can cause significant errors in performance data have been treated. Especially in the case of gas leakage, cells can easily be at a higher temperature than their surrounding environment, causing cells to give better apparent performance than individual electrodes tested under better controlled conditions. Also, the gas composition at the electrodes (usually at the fuel electrode) may be different from the intended composition in case of gas leakage. A method for estimating the size of the gas leakage has been presented here.

References

Adler, S. B. (2002). Reference electrode placement in thin solid electrolytes. *Journal of the Electrochemical Society, 149*(5), E166–E172.

Backhaus-Ricoult, M. (2008). SOFC-a playground for solid state chemistry. *Solid State Sciences, 10*(6), 670–688.

Barfod, R., Mogensen, M., Klemenso, T., Hagen, A., Liu, Y. L., & Hendriksen, P. V. (2007). Detailed characterization of anode-supported SOFCs by impedance spectroscopy. *Journal of the Electrochemical Society, 154*(4), B371–B378.

Barfod, R., Hagen, A., Ramousse, S., Hendriksen, P. V., & Mogensen, M. (2006). Break down of losses in thin electrolyte SOFCs. *Fuel Cells, 6*(2), 141–145.

Bessler, W. (2005). Gas dynamics impedance of solid oxide fuel cells. *Proceedings of the 26th Risoe international symposium on materials science*. Roskilde, Denmark: Risø National Laboratory.

Birkl, C. (2012). *Effect of anode sintering temperature on microstructure and performance of multilayer tape cast SOFC*. M.Sc. Engineering, Technical University of Denmark. Denmark: M.Sc. Thesis, Technical University of Denmark.

Bockris, J. O. M. (1970). *Modern electrochemistry. An introduction to an interdisciplinary area* (Vol. 1). New York: Plenum Press.

Boukamp, B. A. (2004). Electrochemical impedance spectroscopy in solid state ionics: recent advances. *Solid State Ionics, 169*, 65–73.

Cai, Z., Yener, K., Jeong, W. H., Yan, C., & Bilge, Y. (2011). Surface electronic structure transitions at high temperature on perovskite oxides: the case of strained La0. 8Sr0. 2CoO3 thin films. *Journal of the American Chemical Society, 133*(44), 17696–17704.

Carter, C. B. (2013). *Ceramic materials: Science and engineering*. Berlin: Springer.

Chen, M., Liu, Y. L., Bentzen, J. J., Zhang, W., Sun, X., Hauch, A., et al. (2013). Microstructural degradation of Ni/YSZ electrodes in solid oxide electrolysis cells under high current. *Journal of the Electrochemical Society, 160*(8), F883–F891.

Chou, Y.-S., Jeffry, W. S., & Robert, N. G. (2007). Novel alkaline earth silicate sealing glass for SOFC. *Journal of Power Sources, 168*(2), 426–433.

Cronin, J. S., James, R. W., & Scott, A. B. (2011). Impact of pore microstructure evolution on polarization resistance of Ni-Yttria-stabilized zirconia fuel cell anodes. *Journal of Power Sources, 196*(5), 2640–2643.

Ebbesen, S. D., Graves, C., Hauch, A., Jensen, S. H., & Mogensen, M. (2010). Poisoning of solid oxide electrolysis cells by impurities. *Journal of the Electrochemical Society, 157*(10), B1419.

Ebbesen, S. D., & Mogensen, M. (2010). Exceptional durability of solid oxide cells. *Electrochemical and Solid State Letters, 13*(9), D106–D108.

Ebbesen, S. D., & Mogensen, M. (2013). Kinetics of oxidation of H2 and reduction of H2O in Ni-YSZ based solid oxide cells. *ECS Transactions, 50*(49), 167–182.

Erning, J. W., Schaffrath, W., Stimming, U., Syskakis, E., & Wipperman, K. (1995). *Improvement in the performance of screen printed SOFC cathodes by noble metal catalysts.* Proceedings of SOFC IV. (p.492). Pennington, New Jersey, USA: The Electrochemical Society Inc. ISBN1-56677-095-5.

Faes, A., Hessler-Wyser, A., Presvytes, D., Vayenas, C. G., & Van Herle, J. (2009). Nickel–zirconia anode degradation and triple phase boundary quantification from microstructural analysis. *Fuel Cells, 9*(6), 841–851. ISI:000273394400008.

Gödickemeier, M., Sasaki, K., & Gauckler, L. J. (1995). Current-voltage characterics of fuel cells with ceria-based electrolytes. In M. Dokiya, O. Yamamoto, H. Tagawa, & S. C. Singhal (Eds.), *Solid oxide fuel cells IV, Electrochemical society proceedings series, PV 95-1:1072.* Pennington, NJ, USA: The Electrochemical Society Inc.

Graves, C., Ebbesen, S. D., Jensen, S. H., Simonsen, S. B., & Mogensen, M. B. (2015). Eliminating degradation in solid oxide electrochemical cells by reversible operation. *Nature Materials, 14*(2), 239–244.

Greef, R. (1985). *Instrumental methods in electrochemistry*. Chichester: Ellis Horwood.

Hagen, A., Barfod, R., Hendriksen, P. V., Liu, Y. L., & Ramousse, S. (2006). Degradation of anode supported SOFCS as a function of temperature and current load. *Journal of the Electrochemical Society, 153*(6), A1165–A1171.

Hamann, C. H. (1998). *Electrochemistry*. Weinheim: Wiley-VCH.

Hansen, K. V., Kion, N., & Mogens, M. (2006). TOF-SIMS studies of yttria-stabilised zirconia. *Surface and Interface Analysis, 38*(5), 911–916.

Hansen, K. V., Norrman, K., & Mogensen, M. (2004). H-2-H2O-Ni-YSZ electrode performance-effect of segregation to the interface. *Journal of the Electrochemical Society, 151*(9), A1436–A1444. ISI:000223622000021.

Hansen, K. V., Wu, Y., Jacobsen, T., Mogensen, M. B., & Theil Kuhn, L. (2013). Improved controlled atmosphere high temperature scanning probe microscope. *Review of Scientific Instruments, 84*(7), 073701.

Hauch, A., Birkl, C., Brodersen, K., & Jørgensen, P. S. (2012). Multilayer tape cast SOFC - effect of sintering temperature.*Proceedings—10th European Solid Oxide Fuel Cell Forum Chapter 8* (pp. 62–71). Switzerland: Lucerne.

Hauch, A., Bowen, J. R., Kuhn, L. T., & Mogensen, M. (2008). Nanoscale chemical analysis and imaging of solid oxide cells. *Electrochemical and Solid-State Letters, 11*(3), B38.

Hauch, A., Hagen, A., Hjelm, J., & Ramos, T. (2014). Sulfur poisoning of SOFC anodes: effect of overpotential on long-term degradation. *Journal of the Electrochemical Society, 161*(6), F734–F743.

Hauch, A., Jensen, S. H., Bilde-Soerensen, J. B., & Mogensen, M. (2007). Silica Segregation in the Ni/YSZ Electrode. *Journal of the Electrochemical Society, 154*(7), A619.

Hauch, A., Jørgensen, P. S., Brodersen, K., & Mogensen, M. (2011a). Ni/YSZ anode-effect of pre-treatments on Cell degradation and microstructures. *Journal of Power Sources, 196*(21), 8931–8941.

Hauch, A., Mogensen, M., & Hagen, A. (2011b). Ni/YSZ electrode degradation studied by impedance spectroscopy-effect of p(H2O). *Solid State Ionics, 192*, 547–551.

Hauch, A., K. Brodersen, M. Chen, and M.B. Mogensen. (2016). "Ni/YSZ Electrodes Structures Optimized for Increased Electrolysis Performance and Durability. *Solid State Ionics, 293*, 27–36.

Hendriksen, P. V., Koch, S., & Mogensen, M. (2003). Break-down of losses in thin electrolyte SOFCs. In S. C. Singhal & M. Dokiya (Eds.), *Solid Oxide Fuel Cells VI, PV2003-07* (p. 1147). Pennington, NJ, USA: The Electrochemical Society Proceedings Series.

Hjalmarsson, P., Sun, X., Liu, Y.-L., & Chen, M. (2013). Influence of the oxygen electrode and inter-diffusion barrier on the degradation of solid oxide electrolysis cells. *Journal of Power Sources, 223*, 349–357.

Holze, R. (2007). *Electrochemistry*. Berlin: Springer.

Holzer, L., Iwanschitz, B., Hocker, T., Münch, B., Prestat, M., Wiedenmann, D., et al. (2011). Microstructure degradation of cermet anodes for solid oxide fuel cells: Quantification of nickel grain growth in dry and in humid atmospheres. *Journal of Power Sources, 196*, 1279–1294.

Horita, T., Cho, D.-H., Wang, F., Shimonosono, T., Kishimoto, H., Yamaji, K., et al. (2012). Correlation between degradation of cathode performance and chromium concentration in (La, Sr)MnO3 cathode. *Solid state Ionics 18 proceedings of the 18th international conference on solid state Ionics Warsaw, Poland*, July 3–8, 2011 (Vol. 225(0), pp. 151–156).

Huang, Q. A., Hui, R., Wang, B. W., & Zhang, H. J. (2007). A review of AC impedance modeling and validation in SOFC diagnosis. *Electrochimica Acta, 52*(28), 8144–8164.

Ibach, H. (2009). *Solid-state physics: An introduction to principles of materials science*. Berlin: Springer.

Iwanschitz, B., Holzer, L., Mai, A., & Schuetze, M. (2012). Nickel agglomeration in solid oxide fuel cells: the influence of temperature. *Solid State Ionics, 211*, 69–73.

Jacobsen, T., Chatzichristodoulou, C., & Mogensen, M. B. (2014). Fermi potential across working solid oxide cells with zirconia or ceria electrolytes. *ECS Transactions, 61*(1), 203–214.

Jacobsen, T., Hendriksen, P. V., & Koch, S. (2008). Diffusion and conversion impedance in solid oxide fuel cells. *Electrochimica Acta, 53*(25), 7500–7508.

Jacobsen, T., & Mogensen, M. (2008). The course of oxygen partial pressure and electric potentials across an oxide electrolyte cell. *ECS Transactions, 13*(26), 259.

Jacobsen, T., & Skou, E. (1997). *Proceedings of IEA workshop on materials and processes*. Les Diablerets: Schweiz.

Jeangros, Q., Hansen, T. W., Wagner, J. B., Damsgaard, C. D., Dunin-Borkowski, R. E., Hébert, C., et al. (2012). Nickel oxide reduction studied by environmental TEM. *Proceedings of 10th European SOFC Forum 2012, Chapter 14* (p. 25). Switzerland: Lucerne.

Jensen, S. H., Hauch, A., Hendriksen, P. V., & Mogensen, M. (2009a). Advanced test method of solid oxide cells in a plug-flow setup. *Journal of the Electrochemical Society, 156*(6), B757.

Jensen, S. H., Hauch, A., Hendriksen, P. V., Mogensen, M., Bonanos, N., & Jacobsen, T. (2007). A method to separate process contributions in impedance spectra by variation of test conditions. *Journal of the Electrochemical Society, 154*(12), B1325.

Jensen, S. H., Hjelm, J., Hagen, A., & Mogensen, M. (2009b). Electrochemical impedance spectroscopy as diagnostic tool. In W. Vielstich, H. Yokokawa, & H. A. Gasteiger (Eds.), *Handbook of fuel cells—fundamental, technology and applications* (Vol. 6, pp. 1090–1102). Hoboken: John Wiley & Sons Ltd.

Jensen, S. H., Sun, X., Ebbesen, S. D., Knibbe, R., & Mogensen, M. (2010). Hydrogen and synthetic fuel production using pressurized solid oxide electrolysis cells. *International Journal of Hydrogen Energy, 35*(18), 9544–9549.

Jiao, Z., & Shikazono, N. (2015). Quantitative study on the correlation between solid oxide fuel cell Ni-YSZ composite anode performance and reduction temperature based on three-dimensional reconstruction. *Journal of the Electrochemical Society, 162*(6), F571–F578.

Jørgensen, M. J., Primdahl S., & Mogensen, M. (1999). Characterization of composite sofc cathodes using electrochemical impedance spectroscopy. *Electrochimica Acta, 44*(24), 4195–4201.

Jørgensen, P. S., Ebbehøj, S. L., & Hauch, A. (2015). Triple phase boundary specific pathway analysis for quantitative characterization of solid oxide cell electrode microstructure. *Journal of Power Sources, 279*(April), 686–693.

Jørgensen, P S, K V Hansen, R Larsen, and J R Bowen. (2010). A Framework for Automatic Segmentation in Three Dimensions of Microstructural Tomography Data. *Ultramicroscopy, 110*(3), 216–28.

Khabibulakh, K., Yildiz, B., Kavaipatti, B., & Salvador, P. (2009). Correlations of electronic and chemical state on la. *ECS Transactions, 25*, 2309–2318.

Kharton, V. V. (2009). *Solid state electrochemistry I: Fundamentals, materials and their applications*. Germany: Wiley-VCH.

Kharton, V. V. (2011). *Solid state electrochemistry/2, electrodes, interfaces and ceramic membranes*. Germany: Wiley-VCH.

Kiebach, R., Norrman, K., Chatzichristodoulou, C., Chen, M., Sun, X., Ebbesen, S. D., et al. (2014). TOF-SIMS characterization of impurity enrichment and redistribution in solid oxide electrolysis cells during operation. *Dalton Transactions, 43*(40), 14949–14958.

Kishimoto, H., Yashiro, K., Shimonosono, T., Brito, M. E., Yamaji, K., Horita, T., et al. (2012). In situ analysis on the electrical conductivity degradation of NiO doped yttria stabilized zirconia electrolyte by micro-Raman spectroscopy. *Electrochemical Frontiers in Global Environment and Energy, 82*, 263–267.

Kleinlogel, C., Gödickemeier, M., Honneger, K., & Gauckler, L. J. (1997). Solid oxide fuel cells operating with cathode supported thin film electrolyte. In T. A. Ramanarayanan, W. L. Worrell, H. L. Tuller, A. C. Khandkar, M. Mogensen, & W. Göpel (Eds.), *Third international symposium on ionic and mixed conducting ceramics, PV-97-24* (p. 1072). Pennington, NJ, USA: The Electrochemical Society Inc.

Knibbe, R., Traulsen, M. L., Hauch, A., Ebbesen, S. D., & Mogensen, M. (2010). Solid oxide electrolysis cells: degradation at high current densities. *Journal of the Electrochemical Society, 157*(8), B1209.

Kornely, M., Leonide, A., Weber, A., & Ivers-Tiffée, E. (2011). Performance limiting factors in anode-supported cells originating from metallic interconnector design. *Journal of Power Sources, 196*(17), 7209–7216.

Kromp, A., Dierickx, S., Leonide, A., Weber, A., & Ivers-Tiffee, E. (2012). Electrochemical analysis of sulfur-poisoning in anode supported SOFCs fuelled with a model reformate. *Journal of the Electrochemical Society, 159*(5), B597–B601.

Kromp, A., Leonide, A., Weber, A., & Ivers-Tiffee, E. (2011). Electrochemical analysis of reformate-fuelled anode supported SOFC. *Journal of the Electrochemical Society, 158*(8), B980–B986.

Lawlor, V. (2013). Experimental and numerical study of various MT-SOFC flow manifold techniques: single MT-SOFC analysis. *Journal of Fuel Cell Science and Technology, 10*(1), 011003.

Leng, Yang. (2008). *Materials characterization*. Chichester, UK: John Wiley & Sons Ltd.

Leonide, A., Sonn, V., Weber, A., & Ivers-Tiffee, E. (2008). Evaluation and modeling of the cell resistance in anode-supported solid oxide fuel cells. *Journal of the Electrochemical Society, 155*(1), B36–B41. ISI:000251241400020.

Li, X., Blinn, K., Fang, Y., Liu, M., Mahmoud, M. A., Cheng, S., et al. (2012). Application of surface enhanced raman spectroscopy to the study of SOFC electrode surfaces. *Physical Chemistry Chemical Physics: PCCP, 14*(17), 5919–5923.

Liu, Y.-L., & Jiao, C. (2005). Microstructure degradation of an anode/electrolyte interface in SOFC studied by transmission electron microscopy. *Solid State Ionics, 176*, 435–442.

Macdonald, J. R., & Barsoukov, E. (2005). *Impedance spectroscopy. Theory, experiment, and applications* (Vol. 2). Hoboken, US: John Wiley & Sons Inc.

McIntyre, M. D., Traulsen, M. L., Norrman, K., Sanna, S., & Walker, R. A. (2015). Polarization Induced Changes in LSM Thin Film Electrode Composition Observed by In Operando Raman Spectroscopy and TOF-SIMS. *ECS Transactions, 66*(2), 47–59.

Mogensen, M. B. (2002). Strategies for testing of solid oxide fuel cells and electrodes. In *Proceedings European fuel cell forum* (Vol. 2, pp. 893–902). Lucerne, Switzerland.

Mogensen, M., & Hendriksen, P. V. (2003). Testing of electrodes, cells and short stacks. In S. C. Singhal & K. Kendall (Eds.), *High temperature solid oxide fuel cells—fundamentals, design and applications* (Vol. 1, pp. 261–289). London: Elsevier.

Mogensen, M., & Jacobsen, T. (2009). Electromotive potential distribution and electronic leak currents in working YSZ based SOCs. *ECS Transactions, 25*, 1315–1320.

Mogensen, M., Larsen, P. H., Hendriksen, P. V., Kindl, B., Bagger, C., & Linderoth, S. (1999). Solid oxide fuel cell testing: results and interpretation. In S. C. Singhal & M. Dokiya (Eds.), Proceedings—the electrochemical society SOFC-VI, PV 99-19:904–15.

Mosbaek, R. R., Hjelm, J., Barfod, R., Høgh, J., & Hendriksen, P. V. (2013). Electrochemical characterization and degradation analysis of large SOFC stacks by impedance spectroscopy. *Fuel Cells, 13*(4), 605–611.

Nagata, M., Itoh, Y., & Iwaraha, H. (1994). Dependence of observed overvoltages on the positioning of the reference electrode on the solid electrolyte. *Solid State Ionics, 67*(3–4), 215–224.

Nielsen, J., Hagen, A., & Liu, Y. L. (2010). Effect of cathode gas humidification on performance and durability of solid oxide fuel cells. *Solid State Ionics, 181*(11–12), 517–524. ISI:000277926000009.

Nielsen, J., & Hjelm, J. (2014). A review and a comprehensive case study on the impedance of LSM: YSZ cathodes. *Electrochimica Acta, 115*, 31–45.

Nielsen, J., Jacobsen, T., & Wandel, M. (2011). Impedance of porous IT-SOFC LSCF:CGO composite cathodes. *Electrochimia Acta, 56*(23), 7963–7974.

Nielsen, K. A., Solvang, M., Nielsen, S. B. L., Dinesen, A. R., Beeaff, D., & Larsen, P. H. (2007). Glass composite seals for SOFC application. *Journal of the European Ceramic Society, 27*(2–3), 1817–1822.

Orazem, M. E. (2008). *Electrochemical impedance spectroscopy*. Hoboken: John Wiley & Sons.

Pihlatie, M., Ramos, T., & Kaiser, A. (2009). Testing and improving the redox stability of Ni-based solid oxide fuel cells. *Journal of Power Sources, 193*(1), 322–330.

Primdahl, S., & Mogensen, M. (1998). Gas conversion impedance: a test geometry effect in characterization of solid oxide fuel cell anodes. *Journal of the Electrochemical Society, 145*(7), 2431–2438.

Primdahl, S., & Mogensen, M. (1999). Gas diffusion impedance in characterization of solid oxide fuel cell anodes. *Journal of the Electrochemical Society, 146*(8), 2827–2833.

Ramos, T., Thydén, K., & Mogensen, M. (2010). Electrochemical characterization of Ni/(Sc) YSZ electrodes. *ECS Transactions, 28*(11), 123.

Rasmussen, J. F. B., & Hagen, A. (2010). The effect of H2S on the performance of SOFCs using methane containing fuel. *Fuel Cells, 10*(6), 1135–1142.

Rasmussen, J. F. B., Hendriksen, P. V., & Hagen, A. (2008). Study of internal and external leaks in tests of anode-supported SOFCs. *Fuel Cells, 8*(6), 385–393.

Reinhardt, G., & Göpel, W. (1998). Electrode reactions at solid electrolytes: finite difference calculations to describe geometric and electrtical properties of planar devices. In T. A. Ramanarayanan, W. L. Worrell, H. L. Tuller, A. C. Khandkar, M. Mogensen, & W. Göpel (Eds.), *Proceedings of third international symposium on ionic and mixed conducting ceramics* (p. 610). Pennington, NJ, USA: The Electrochemical Society Inc.

Reis, S. T., Pascual, M. J., Brow, R. K., Ray, C. S., & Zhang, T. (2010). Crystallization and processing of SOFC sealing glasses. *Journal of Non-Crystalline Solids, 356*(52–54), 3009–3012.

Schichlein, H., Muller, A. C., Voigts, M., Krugel, A., & Ivers-Tiffee, E. (2002). Deconvolution of electrochemical impedance spectra for the identification of electrode reaction mechanisms in solid oxide fuel cells. *Journal of Applied Electrochemistry, 32*(8), 875–882. ISI:0001784 72100007.

Schilm, J. (2010). Sealing glasses for SOFC-degradation behavior. *Advances in Solid Oxide Fuel Cells V, 30*(4), 185–193.

Schmidt, M. S., Hansen, K. V., Norrman, K., & Mogensen, M. (2008). Effects of trace elements at the Ni/ScYSZ interface in a model solid oxide fuel cell anode. *Solid State Ionics, 179*(27–32), 1436–1441.

Shearing, P. R., Brett, D., & Brandon, N. P. (2010a). Towards intelligent engineering of SOFC electrodes: a review of advanced microstructural characterisation techniques. *International Materials Reviews, 55*, 347.

Shearing, P. R., Cai, Q., Golbert, J. I., Yufit, V., Adjiman, C. S., & Brandon, N. P. (2010b). Microstructural analysis of a solid oxide fuel cell anode using focused ion beam techniques coupled with electrochemical simulation. *Journal of Power Sources, 195*(15), 4804–4810.

Shimazu, M., Yamaji, K., Isobe, T., Ueno, A., Kishimoto, H., Katsumata, K., et al. (2011). Stability of Sc2O3 and CeO2 co-doped ZrO2 electrolyte during the operation of solid oxide fuel cells: Part II the influences of Mn, Al and Si. *Solid State Ionics, 204–205*, 120–128.

Shimura, T., Jiao, Z., Hara, S., & Shikazono, N. (2014). Quantitative analysis of solid oxide fuel cell anode microstructure change during redox cycles. *Journal of Power Sources, 267*, 58–68.

Smart, L., & More, E. (1996). *Solid state chemistry—An introduction* (2nd ed.). Cheltenham, UK: T.J. International Ltd.

Solheim, A., Tunold, R., & Ødegård, R. (1991). The relationship between electrical energy output and energy efficiency in plug flow solid oxide fuel cells. In F. Gross, P. Zegers, S. C. Singhal, & O. Yamamoto (Eds.), *Solid oxide fuel cells II* (p. 297). Luxemburg: European Communities, EUR13564EN.

Sonn, V., Leonide, A., & Ivers-Tiffee, E. (2008). Combined deconvolution and CNLS fitting approach applied on the impedance response of technical Ni/8YSZ cermet electrodes. *Journal of the Electrochemical Society, 155*(7), B675–B679. ISI:000256198900016.

Sun, X., Bonaccorso, A. D., Graves, C., Ebbesen, S. D., Jensen, S. H., Hagen, A., et al. (2015). Performance characterization of solid oxide cells under high pressure. *Fuel Cells, 15*(5), 697–702.

Thydén, K., Liu, Y.-L., & Bilde-Sørensen, J. B. (2008). Microstructural characterization of SOFC Ni–YSZ anode composites by low-voltage scanning electron microscopy. *Solid State Ionics, 178*, 1984–1989.

Tietz, F., Sebold, D., Brisse, A., & Schefold, J. (2013). Degradation phenomena in a solid oxide electrolysis cell after 9000 H of operation. *Journal of Power Sources, 223*, 129–135.

Utz, A., Hansen, K. V., Norrman, K., Ivers-Tiffee, E., & Mogensen, M. (2011). Impurity features in Ni-YSZ-H-2-H2O electrodes. *Solid State Ionics, 183*(1), 60–70.

Van Heuveln, F. H., van Berkel, F. P. F., & Huijsmans, J. P. P. (1993). Electrochemical characterization of porous electrodes and application in SOFC. In F. W. Poulsen, J. J. Bentzen, T. Jacobsen, E. Skou, & M. J. L. Østergård (Eds.), *14th Risø international symposium on materials science* (p. 53). Roskilde, Denmark: Risø National Laboratory, Roskilde, Denmark.

Vels Jensen, K., Primdahl, S., Chorkendorff, I., & Mogensen, M. (2001). Microstructural and chemical changes at the Ni/YSZ interface. *Solid State Ionics, 144*, 197–209.

Wiedenmann, D., Hauch, A., Grobéty, B., Mogensen, M., & Vogt, U. F. (2010). Complementary techniques for solid oxide electrolysis cell characterisation at the micro-and nano-scale. *International Journal of Hydrogen Energy, 35*(10), 5053–5060.

Wilson, J. R., Kobsiriphat, W., Mendoza, R., Chen, H.-Y., Miller, J. M., Miller, D. J., et al. (2006). Three-dimensional reconstruction of a solid-oxide fuel cell anode. *Nature Materials, 5*, 541–544.

Winkler, J., Hendriksen, P. V., Bonanos, N., & Mogensen, M. (1998). Geometric requirements of solid electrolyte cells with a reference electrode. *Journal of the Electrochemical Society, 145*(4), 1184–1192.

Wu, Y., Hansen, K. V., Norrman, K., Jacobsen, T., & Mogensen, M. (2013). Oxygen electrode kinetics and surface composition of dense (La0. 75Sr0. 25) 0.95 MnO3 on YSZ. *ECS Transactions, 57*(1), 1673–1682.

Zhang, H. S. (2012). Performance comparison on cross-flow and counter-flow planar solid oxide fuel cells. *Proceedings of the ASME Turbo Expo 2012* (Vol. 3, pp. 91–97).

Zhang, W., Chen, M., Theil Kuhn, L., Bowen, J. R., & Bentzen, J. J. (2014). Electrochemistry unlocks wettability: epitaxial growth of oxide nanoparticles on rough metallic surfaces. *ChemElectroChem, 1*(3), 520–523.

Proton-Conducting Electrolytes for Solid Oxide Fuel Cell Applications

Dmitry Medvedev, Angeliki Brouzgou, Anatoly Demin and Panagiotis Tsiakaras

Abstract The necessity for reducing the operating temperatures of solid oxide fuel cells (SOFCs) below 800 °C is widely proposed in the last years. To this aim, proton-conducting oxides have gained widespread interest as the electrolyte materials, acting as an alternative to oxygen ion conductors. High-temperature proton conductors, owing to their lower activation energy for proton conduction, can achieve high conductivity at relatively low temperatures compared to their oxygen ion-conducting counterparts. In this chapter, the recent advances in the field of solid oxide proton-conducting materials that belong to the class of perovskite-based materials (such as doped $BaCeO_3$, $BaZrO_3$, $BaCeO_3$–$BaZrO_3$ $SrCeO_3$, and $LaScO_3$) and to other classes of materials (such as doped $Ba_2In_2O_5$, CeO_2, and $LaNbO_4$) are presented and analyzed.

1 Introduction

Solid oxide fuel cells (SOFCs) have been of interest the recent years due to the need for more efficient and cleaner electricity production (Mahato et al. 2015). It is well known that conventional SOFCs based on yttria-stabilized zirconia (YSZ) require high operating temperatures around 900 °C. However, such high operating

D. Medvedev · A. Demin · P. Tsiakaras (✉)
Laboratory of Electrochemical Devices based on Solid Oxide Proton Electrolytes, Institute of High Temperature Electrochemistry, Ural branch of the Russian Academy of Sciences, 20, Akademicheskaya St., 620990 Ekaterinburg, Russia
e-mail: tsiak@mie.uth.gr

A. Brouzgou · P. Tsiakaras
Laboratory of Alternative Energy Conversion Systems, Department of Mechanical Engineering, University of Thessaly, Pedion Areos, 38334 Volos, Greece

M. Boaro and A.S. Aricò (eds.), *Advances in Medium and High Temperature Solid Oxide Fuel Cell Technology*, CISM International Centre for Mechanical Sciences 574,
DOI 10.1007/978-3-319-46146-5_3

temperatures hinder the commercialization of SOFCs due to degradation and high-cost issues. One possible solution is the design and applications of new electrolytes with higher conductivity in comparison with that of YSZ. The high-temperature proton conductors represent such systems (Kreuer 1997).

High-temperature proton conductors (HTPCs) can achieve higher conductivity levels at relatively lower temperatures than those of oxygen-conducting counterparts due to lower activation energy of proton conductivity. It should be noted that the adjective "high-temperature" in HTPCs demonstrates the difference over other low-temperature proton-conducting materials in terms of stability of proton (protonic) groups rather than working conditions. HTPCs can operate in intermediate (300–600 °C) and high-temperature range (higher than 600 °C).

The advantages of SOFCs based on proton-conducting electrolytes over those based on oxygen-conducting electrolytes are discussed from the viewpoints of electric efficiency and fuel utilization in the works of Demin et al. (2004; Demin and Tsiakaras 2001).

The phenomenon of proton conductivity was presented and largely explained through the pioneering works of Iwahara and his collaborators (Iwahara et al. 1981; Takahashi et al. 1979). Working with of strontium cerate ($SrCeO_3$) doped with Sc or Mg or Yb, they revealed considerable proton conductivity. Subsequently, Iwahara et al. (Uchida et al. 1989) proposed that water plays an important role, interacting with oxygen vacancies existed in the solid oxide and then forming protons. Today, it is believed (Kreuer 1996) that the formed proton is associated with an oxygen ion. It has also been recognized that water interacts with oxygen vacancies to form proton defects. Then proton may jump to a neighboring oxygen ion through the rotational mechanism of a proton around an oxygen ion. In general, proton conduction in oxides is very much related to the presence of oxygen vacancies, although there are other possible reactions, contributing to the formation of protons.

Due to the unique transport properties, HTPCs have been studied considerably in the frame of fundamental and applied aspects. In the terms of applications, they can be used as electrolytes for a wide row of electrochemical devices, including solid oxide fuel and electrolysis cells, different kinds of sensors, hydrogen/water vapor pumps, and reactors based on hydrogen-permeable membranes for hydrogen and ammonia production. Although, information concerning the electrochemical devices working at very low temperatures has appeared (below 350 °C, Dual et al. 2015), the preparation of experimental cells is still carried out at very high temperatures of 1300–1400 °C. Thus, not only low-temperature proton transport features of HTPCs, but their ceramic, mechanical properties, stability, and compatibility, especially at high temperatures, are the matter of investigation.

2 Proton Conductivity in Oxide Materials

As well known, protons are elementary particles and due to their peculiarities strongly interact with electron clouds of other atoms. However, a proton does not exhibit many properties of the elementary particles. A free proton, such as other ions (cations and anions), is a charge carrier but contrary to them has extremely small radius, small mass, low coordination number, and missing electron cloud. This particular nature of proton determines some features of the corresponding transport (Chandra 1984).

The proton (H^+) is always covalently bonded to some electronegative atoms/ions (e.g., to oxygen ions in oxides) and can be shared between two electronegative atoms (e.g., O–H···O). Therefore, an activation transport mechanism may be realized.

In HTPCs based on complex oxides, the transport of protons roughly corresponds to the so-called Grotthuss mechanism. This mechanism has been postulated more than two hundred years ago for proton transfer in liquid water (Van de Grotthuss 2006). Bare H^+ ions do not exist in solids under equilibrium condition; they strongly interact with the electron density of electronegative oxygen ions. However, the formed O–H bond is weaker than that in the hydroxyl ion OH^- (Hibino et al. 1992). The proton migration is a result of the thermally activated hopping process requiring the breaking of the O–H bond. The realization of the proton hopping Grotthuss-type mechanism rather than hydroxyl ion migration (so-called vehicle mechanism) was directly proved by the H^+/D^+ isotope effect measurements in perovskite oxides (Kreuer et al. 1995; Kreuer 1999). The hopping mechanism of protons, schematically shown in Fig. 1, consists of the following two steps: (i) fast rotation of the proton which is localized on the oxygen ion and (ii) proton transfer by hopping toward a neighboring oxygen ion.

Proton transfer (migration) is slower and more difficult than proton rotation, which is associated with very low activation barriers (below 0.1 eV) (Uchida et al. 1989). Proton jump constitutes the rate-determining step and the activation energy of the proton migration, according the Grotthuss-type mechanism in complex oxides, is usually about 0.5 eV.

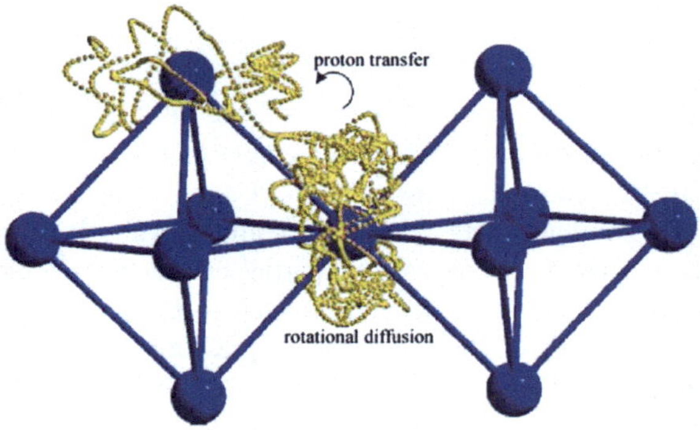

Fig. 1 Proton migration mechanism in the perovskite structure (Münch et al. 1996)

The presence of oxygen vacancies in the oxide materials is one of the main conditions for the accomplishment of proton transport. The oxygen vacancies in proton-conducting materials can be created by acceptor doping (impurity defects) or can be a structural element of oxides (structural defects). Upon hydration, the oxygen vacancies are filled by oxygen from water according to the dissociative mechanism of water dissolution. This process can be described in terms of quasi-chemical formalism (Kröger–Vink nomenclature) (Kröger 1964).

The interaction of the most proton-conducting materials with water vapor produces new structure elements:

$$H_2O + V_O^{\cdot\cdot} + O_O^x = 2OH_O^{\cdot} \tag{1}$$

where O_O^x is the oxygen ion in the normal lattice site, $OH_O^{\cdot}$ is the proton localized on the oxygen ion (proton defect), and $V_O^{\cdot\cdot}$ is the oxygen vacancy created in the crystal structure during acceptor substitution. For example, for Y-doped $BaZrO_3$ the formation of oxygen vacancies can be presented in the following form:

$$Y_2O_3 \xrightarrow{BaZrO_3} 2Y'_{Zr} + V_O^{\cdot\cdot} + 3O_O^x. \tag{2}$$

3 Basic Requirements for Proton-Conducting Electrolytes

The solid oxide electrolytes are considered as a heart of SOFCs, which conduct oxide ions or protons and thus provide the electrochemical reactions on the opposite sides (cathode and anode) of SOFCs (Mahato et al. 2015). The electrolytes are used in ensemble with electrode, interconnect, and glass sealant functional materials (so-called sandwich-structure). Therefore, the electrolytes, including HTPCs, should satisfy the different requirements for effective applications, starting from individual features of protonic electrolytes and ending their compatibility with other components of SOFCs.

3.1 Transport Properties

The high ionic conductivity of the protonic electrolytes is one of the main parameters required for achievement of excellent output characteristics of electrochemical devices. However, HTPCs are complex objects due to the fact that they can simultaneously possess protonic, oxygen ionic, and electronic (n- or p-type) conductivity. Moreover, the partial conductivities and, correspondingly, their sum (total conductivity) depend on such factors as chemical composition, temperature, oxygen and water vapor partial pressures, and grain (bulk) and grain-boundary transport. These factors determine some complexity in describing the transport nature of proton-conducting materials. However, the following trend can be noted for the most investigated materials based on $BaCeO_3$ and $BaZrO_3$ (Medvedev et al. 2014a,b, 2016):

1. Proton conductivity appears under the conditions of relatively low temperatures (<700 °C) and reducing atmospheres containing a high water vapor pressure.
2. Oxygen ionic conductivity dominates in a wide range of temperatures in dry reducing atmospheres and at relatively low temperatures in oxidizing conditions.
3. Hole transport is realized under oxidizing atmospheres at high temperatures.
   The validity of points (1)–(3) is clearly shown in Table 1 for Ln-doped barium cerates. This table lists the transport numbers (or in other words—the

Table 1 Transport numbers of ions, oxygen ions, protons, and holes for $BaCeO_3$-based materials in different atmospheres and temperatures

Materials	*T* (°C)	Atmosphere	*t*(O)	*t*(H)	*t*(i)	*t*(p)	References
$BaCeO_3$	715	Wet O_2	0.12	0.36	0.48	0.52	Sharova and Gorelov (2005)
	800	Wet O_2	–	–	0.08	0.92	Virkar and Maiti (1985)
	800	$pO_2 = 10^{-8}$ atm	–	–	1	0	
$BaCe_{0.85}Y_{0.15}O_{3-\delta}$	600	Wet air	0.10	0.80	0.90	0.10	Lim et al. (2011)
	700		0.17	0.61	0.79	0.21	
	800		0.22	0.32	0.54	0.46	
	700	Air	0.29	0.41	0.70	0.30	Sharova and Gorelov (2003)
		Wet H_2	0.35	0.65	1	0	
$BaCe_{0.8}Pr_{0.2}O_{3-\delta}$	500–900	Wet air	0.06–0.20	0.0–0.15	0.06–0.35	0.65–0.94	Wang and Qiu (2008)
	500–900	Dry O_2	0.1–0.25	–	–	0.75–0.90	
	500–600	Wet H_2	–	–	0.93–0.96	0.04–0.07	
$BaCe_{0.97}Nd_{0.03}O_{3-\delta}$	715	Wet O_2	0.17	0.32	0.49	0.51	Sharova and Gorelov (2004)
	900		0.23	0.07	0.30	0.70	
$BaCe_{0.9}Nd_{0.1}O_{3-\delta}$	600–850	Wet H_2 + Ar	0.03–0.68	0.32–0.97	1	0	Bannykh and Kuzin (2003)
	600	Wet H_2	0.04	0.96	1	0	Kuzin et al. (2000)
	850	Wet O_2	0.66	0.34	1	0	
$BaCe_{0.8}Ho_{0.2}O_{3-\delta}$	600–1000	Wet H_2	–	0.99–1	0.99–1	0.00–0.01	Qiu et al. (2005)
		Wet air	0.27–0.32	0.01–0.09	0.28–0.41	0.59–0.72	
$BaCe_{0.8}Tm_{0.2}O_{3-\delta}$	600	Wet H_2	–	0.95–0.93	–	–	Qiu and Wang (2010)
	500–900	Wet air	0.38–0.47	0.01–0.02	0.40–0.47	0.53–0.60	

Table 2 Bulk (σ_{bulk}) and grain-boundary ($\sigma_{g.b.}$) conductivities of materials based on $BaCeO_3$

Materials	°C/atmosphere	σ_{bulk} ($\times 10^4$ S cm^{-1})	$\sigma_{g.b.}$ ($\times 10^7$ S cm^{-1})	References
$BaCeO_3$	200/7 %H_2/Ar	0.05	0.3	Pasierb et al. (2009)
	300/7 %H_2/Ar	0.33	17.4	
	350/7 %H_2/Ar	0.57	75.7	
	570/7 %H_2/Ar	2.78	7071.8	
$BaCe_{0.99}Sm_{0.01}O_{3-\delta}$	300/wet 5 %H_2/Ar	1.22	0.2	Zhang et al. (2011)
	400/wet 5 %H_2/Ar	2.92	35.6	
	500/wet 5 %H_2/Ar	5.15	3562.3	
	600/wet 5 %H_2/Ar	6.85	37,923.7	
$BaCe_{0.97}Sm_{0.03}O_{3-\delta}$	300/wet 5 %H_2/Ar	3.17	187.0	
$BaCe_{0.9}Sm_{0.1}O_{3-\delta}$	300/wet 5 %H_2/Ar	6.79	2247.7	
$BaCe_{0.8}Sm_{0.2}O_{3-\delta}$	300/wet 5 %H_2/Ar	8.72	19,940.3	
$BaCe_{0.9}La_{0.1}O_{3-\delta}$	200/air	0.33	0.4	Amsif et al. (2011)
$BaCe_{0.9}Nd_{0.1}O_{3-\delta}$	200/air	0.36	2.2	
$BaCe_{0.9}Sm_{0.1}O_{3-\delta}$	200/air	1.51	4.1	
$BaCe_{0.9}Gd_{0.1}O_{3-\delta}$	200/air	2.89	6.1	
$BaCe_{0.9}Y_{0.1}O_{3-\delta}$	200/air	2.12	4.6	
$BaCe_{0.9}Yb_{0.1}O_{3-\delta}$	200/air	1.48	2.1	

contributions of partial conductivities) under different conditions. The presented data are extremely important, since they determine the area of applications of such materials as ionic electrolytes for SOFCs or mixed conductors for membrane processes.

4. Grain-boundary conductivity is a determining parameter at low temperatures, whereas the transport is controlled by bulk conductivity at higher temperatures (Table 2).

For other representatives of proton-conducting materials, not only p-type electronic conductivity but also n-type one may dominate under specified conditions (low oxygen partial pressure values). The examples of such systems are doped $SrCeO_3$ (Sammes et al. 2004) and $LaNbO_4$ (Huse et al. 2012) materials.

3.2 *Thermomechanical Properties*

In order to guarantee the mechanical strength of the sandwiched-structure of half-cell or SOFC elements during co-sintering, thermocycles, and working processes, the adjoining components should possess thermal affinity, i.e., similar behavior of their expansion and close values of thermal expansion coefficient (TEC). Considering expansion and TEC characteristics, the preference should be given to material possessing: (i) a stable crystal structure from room temperature to at least SOFCs' working temperatures and (ii) relatively high level of TEC, because of practically all the electrode materials are characterized by TECs higher than 10×10^{-6} K^{-1} (Tsipis and Kharton 2008; Pelosato et al. 2015). Unfortunately, many perspective proton-conducting systems demonstrate either the presence of one or more phase transitions ($BaCeO_3$, $Ba_2In_2O_5$, and $LaNbO_4$), affecting significantly the thermal behavior, or low TEC values ($BaZrO_3$). These two negative factors seriously obstruct the selection of thermally compatible electrodes and the development of multifunctional materials and devices.

3.3 Chemomechanical Properties

Because electrolyte membrane simultaneously contacts with oxidizing and reducing atmospheres, the partial pressure differences of potential-determined components provoke a material chemical expansion, which can lead to mechanical instability or failure of SOFC devices during operation (Bishop et al. 2014). This fact, for example, prevents the direct use of doped CeO_2 electrolytes in SOFCs without protective layers at the anode side. Reducing atmospheres results in Ce^{4+} ion reduction associated with the appearance of noticeable electronic conductivity in the CeO_2-based electrolytes and significant their chemical expansion causing mechanical stress (Bishop et al. 2009). The chemical expansion induced by oxygen partial pressure (pO_2) variation is barely researched for proton-conducting electrolytes. There are only some works on the pO_2 effect on the thermochemical properties of $BaCeO_3$-based materials. For example, Wang et al. (2012), using dilatometry technique (Fig. 2), showed that atmospheres with different pO_2 insignificantly change the TEC values of $BaCe_{0.7}Zr_{0.1}Y_{0.1}Yb_{0.1}O_{3-\delta}$ ceramic material. More precisely, this parameter is equal to 9.1×10^{-6} and 9.2×10^{-6} K^{-1} in air and 5 % H_2/N_2, respectively. The similar results were obtained by Medvedev et al. (Medvedev et al. 2014a,b) for Nd-doped $BaCeO_3$ electrolytes, which demonstrated no significant response of material linear change at pO_2 variation (Fig. 3).

However, being proton conductors, these materials are able to absorb water vapor from the ambient air and to form proton defects (reaction (1)). The formation of proton defects in the structure expands the crystal lattice (and ceramics in general) since this process can be considered as filling the vacancies of the anion sublattice ($V_O^{\bullet\bullet}$) with oxygen of water and subsequent formation of protons ($OH_O^{\bullet}$)

Fig. 2 Thermal expansion behavior of $BaCe_{0.7}Zr_{0.1}Y_{0.1}Yb_{0.1}O_{3-\delta}$ (BCZYY) sintered pellets in dynamic air, wet air, and 5 % H_2 balanced with 95 % N_2 (Wang et al., 2012)

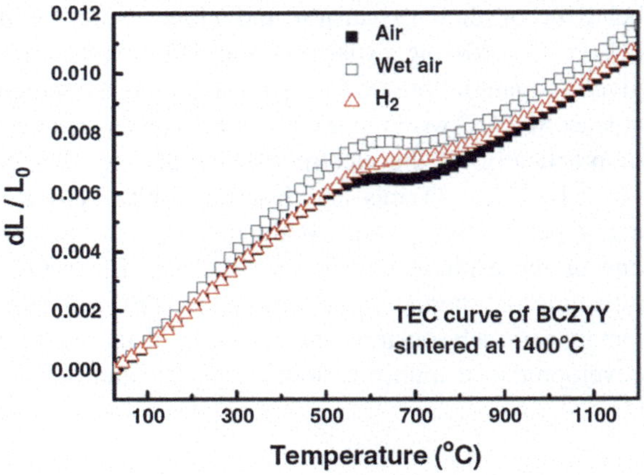

Fig. 3 Oxygen partial pressure dependences of the isothermal expansion of (1−x) $Ce_{0.8}Nd_{0.2}O_{2-\delta}$−x$BaCe_{0.8}Nd_{0.2}O_{3-\delta}$ samples at 900 °C (Medvedev et al. 2014a,b)

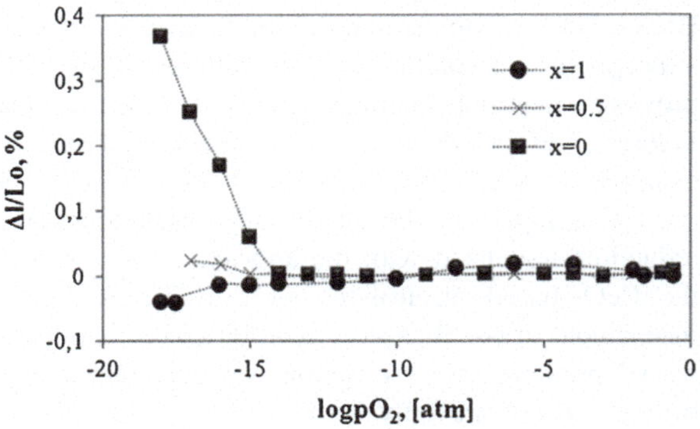

and O_O^x particles. Therefore, the chemical expansion induced by water vapor partial pressure (pH_2O) variation is a unique feature of considered materials. Recently, many research groups have turned the careful attention to the nature of pH_2O-induced chemical expansion (Hiraiwa et al. 2013; Andersson et al. 2014; Han et al. 2014; Lyagaeva et al. 2015). Figure 4 demonstrates humidification effect on size (unit cell parameter) characteristics of $BaZrO_3$-based ceramics evaluated by the aid of high-temperature XRD analysis in controlled atmospheres.

It can be seen that unit cell parameters of investigated zirconates (except, Dy-containing one) in dry air and argon atmospheres are actually close, which confirms the above-mentioned data on weak pO_2 influence on the expansion. On the other hand, an increase in pH_2O results in a significant growth of unit cell parameters due to the realization of hydration process.

As can be seen above, negligible pO_2-induced chemical expansion of the materials based on $BaCeO_3$ and $BaZrO_3$ is caused by a relative stability of Ce^{4+}-

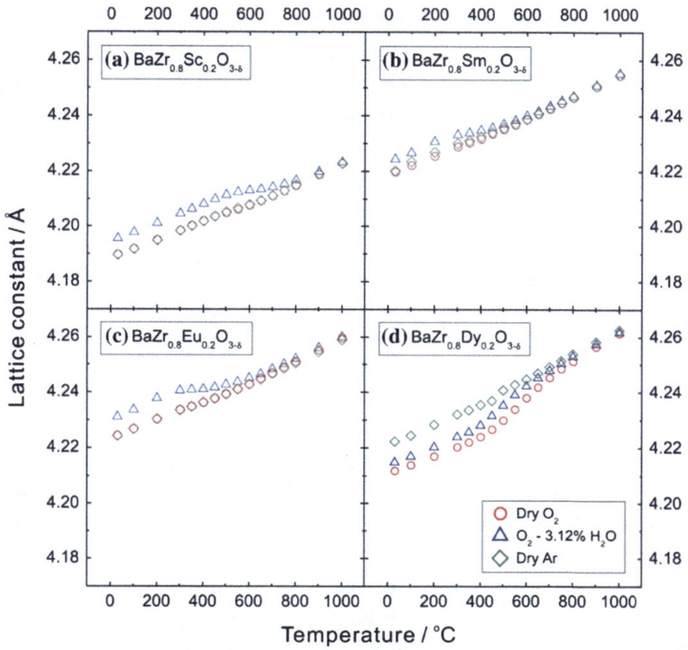

Fig. 4 Variation in unit cell parameters of **a** $BaZr_{0.8}Sc_{0.2}O_{3-\delta}$, **b** $BaZr_{0.8}Sm_{0.2}O_{3-\delta}$, **c** $BaZr_{0.8}Eu_{0.2}O_{3-\delta}$, and **d** $BaZr_{0.8}Dy_{0.2}O_{3-\delta}$ in dry O_2, wet O_2 (~3 % H_2O), and dry Ar with temperature. All samples were quenched to room temperature after the heat treatment at 1600 °C (Han et al. 2014)

and Zr^{4+} ions against to partial reduction. The HTPCs containing elements with variable oxidation state are seem to be more sensitive to pO_2 change from the viewpoint of possible chemical expansion/contraction.

3.4 Stability

Significant disadvantage of SOFCs in front of other types of fuel cells is high working temperatures, which is the cause of expedited diffusion processes. This can result in different negative factors such as chemical interaction of functional components with each other or with gas atmospheres, impurity phase formation, or gradual decomposition of single-phase materials. The high sintering temperatures required for densification of electrolyte materials are also considered to be decreased in order to avoid some technological challenges. Summarizing, the proton-conducting materials must possess good stability during the procedures of ceramics formation and long-time SOFCs operation.

Material stability under ceramic preparation. The high-temperature synthesis and sintering are considered among the main problems associated with the formation of highly dense proton-conducting ceramics. This problem is particularly relevant for complex oxide systems containing alkaline earth elements (e.g., barium). The barium volatility (or evaporation) is well-known fact for $BaCeO_3$- and $BaZrO_3$-based electrolytes, which leads to a colossal decrease in conductivity due to the change of the compositions in the grain (volume) and grain-boundary regions associated with the nonstoichiometry of barium (Medvedev et al. 2016). The BaO volatility under high-temperature treatment can be presented as:

$$Ba(Ce,Zr)_{1-y}M_yO_3 \rightarrow Ba_{1-x}(Ce,Zr)_{1-y}M_yO_3 + xBaO\uparrow \tag{3}$$

or in terms of Kröger–Vink nomenclature as:

$$xBa_{Ba}^{x} + xO_{O}^{x} + xM'_{Ce,Zr} + xV_{O}^{\cdot\cdot} \rightarrow xM_{Ba}^{\cdot} + xBaO\uparrow \tag{4}$$

where M is the acceptor dopant (e.g., Y, Sc, Yb, Gd), and x is the Ba-deficiency.

Firstly, as shown in Eq. (4), BaO loss results in a redistribution of the acceptor dopant, occupied B-position on both A- and B-sublattices of ABO_3 perovskite, which violates the given (nominal) stoichiometry of the material (Shima and Haile 1997). Secondly, the partial Ba^{2+} by M^{3+} substitution is accompanied with a decrease in oxygen vacancy concentration responsible for oxygen ionic and protonic transport (Bonanos and Poulsen 1999). Thirdly, nonstoichiometric compounds can decompose due to a limited range of solid solution existence. If the range of perovskite's stability (x) for Ba_xCeO_3-based oxides varies from 0.8 to 1.2 (Ma et al. 1998), then that for Ba_xZrO_3-based oxides narrows from 0.96 to 1.06 (Yamazaki et al. 2010). The exceeded nonstoichiometry of M-doped $Ba(Ce, Zr)O_3$ is the reason of the appearance of poor-conducting impurity phases M_2O_3 (Fisher et al. 2013), CeO_2 (Wu et al. 2004), ZrO_2 (Osman et al. 2010), and BaO (Fisher et al. 2013) segregated on the grain-boundary region. Finally, the loss of barium from the grain volume of ceramic leads not only Ba-volatility, but also localization of Ba-rich impurities (BaO, $Ba(OH)_2$) on the grain boundaries (Yamazaki et al. 2009). These impurities, reacting with the trace of CO_2 containing in air atmosphere, form $BaCO_3$ phase which causes the degradation of mechanical and electrical properties.

The literature results demonstrate that even small loss of Ba results in significant decrease in grain and grain-boundary conductivity. Therefore, a careful cationic control of stoichiometry is required for the achievement of ceramic materials with desirable compositions. The most common strategy, preventing cation misbalance during ceramics formation, is the sintering of a material in a powder of the same composition (covering method). However, Choi et al. recently proposed another strategy (Choi et al. 2014). They controlled a surrender atmosphere during sintering and found that the high pO_2 values stipulate barium oxide evaporation from zirconate ceramics followed by the formation of multi-phase materials, whereas the change of air atmosphere with Ar suppresses the Ba-volatility and leads to the formation of single-phase ceramics (Fig. 5).

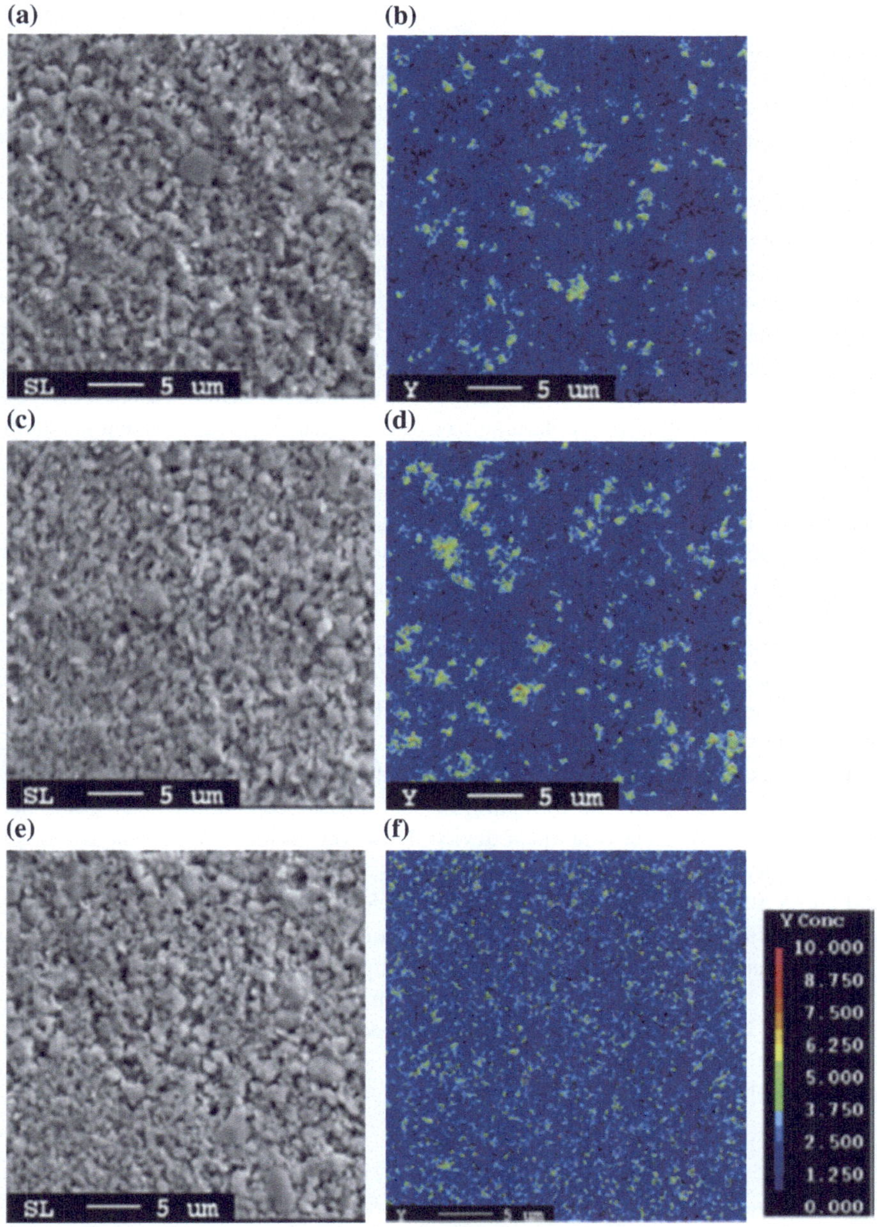

Fig. 5 Morphologies (**a**, **c**, **e**) and maps of yttrium distribution (**b**, **d**, **f**) for $BaZr_{0.84}Y_{0.15}Cu_{0.01}O_{3-\delta}$ ceramics sintered in oxygen (**a**, **b**), air (**c**, **d**), or argon (**e**, **f**) flow (Choi et al. 2014)

It should be noted that the impurity phase formation can be realized under sintering procedure not only for Ba-containing systems, but also for alkaline rare element-free compounds. However, the reasons of materials instability can be different: Ba-volatility for cerates and zirconates or weak interval of solid solution existence (in the case of $LaNbO_4$-doped compounds), which will considered in detail below.

Material stability against gas components. The chemical stability of proton-conducting oxides is an important aspect for their effective usage in high-temperature electrochemical devices. Among HTPCs, doped $BaCeO_3$ materials demonstrate highest value of protonic conductivity. However, they exhibit low chemical stability against CO_2, H_2S, and even H_2O due to the formation of different Ba-containing compounds. Consequently, many studies have focused on searching more stable systems. As a result of such activity, modified cerates ($BaCeO_3$ doped by Zr, In, Sn, Ti), other Ba-containing oxides ($BaZrO_3$, $Ba_2In_2O_5$) or alkaline earth element-free materials ($LaScO_3$, $LaNbO_4$, $La_2Ce_2O_7$) are extensively studied during the last decades. Such materials possess, as a rule, lower ionic (protonic in particular) conductivity than simple doped cerates. Moreover, some of them demonstrated unacceptable stability under reducing atmospheres (Ti- and Sn-based compounds). Therefore, a compromise between transport and stabilizing properties is not reached yet.

Compatibility with other functional materials. During the high-temperature co-sintering process of the electrodes on the solid-state electrolyte surface, or the long-term operation of a SOFC, a possible diffusion of cations of transition elements from the electrodes into the electrolyte can be the reason for the formation of contact phases. These phases are chemically different from the functional layers of an electrochemical cell. The formation of such phases can significantly affect the performance of the electrochemical device. However, in our opinion, the chemical compatibility of proton-conducting electrolytes with cathode, anode, interconnect, and glass sealant systems is the most unexplored field. There are only a few systematic investigations showing reliable data on compatibility of doped $SrCeO_3$ (Tolchard and Grande 2007) and $BaZrO_3$ (Goupil et al. 2012) with some cathode systems. In other investigations, the conclusion about acceptable compatibility of electrolytes with cathode or anode materials is postulated only on the base of XRD analysis for two-component powders, which were calcined at a specified temperature for rather short period of time.

3.5 Densification

The drawback of some materials (cerates, zirconates, and niobates) is their poor densification under relatively low sintering temperatures (less than 1500 °C), if a standard solid-state synthesis is used. This is due to a refractory nature of constituent cations. For example, the sintering temperature and soaking time can respectively reach 1700 °C and 120 h for materials based on $BaCeO_3$, whereas higher temperatures and longer soaking times are required for the formation of gastight $BaZrO_3$-based materials (Medvedev et al. 2015a, b). Therefore, the cost-effective and simple methods are highly needed in order to decrease energy consumption and to avoid materials instability and stoichiometric misbalance associated with Ba-volatility. Two main strategies are currently used for successful ceramics densification at lower sintering temperatures: wet chemical technologies and solid-state reactive sintering.

Chemical (or solution) methods, an integral part of which is the use of solutions at one of the process stages, are an alternative to the solid-state methods of powders and materials preparation. The main advantages of these technologies are the considerably smaller size of particles of the prepared powders and their high homogeneity, which favors the decrease in synthesis temperature and duration of phase formation of multicomponent compounds, as well as the preparation of gas tight ceramic products at lower sintering temperatures.

Solution technologies include precipitation (co-precipitation), combustion methods (glycine–nitrate, citrate–nitrate, sol–gel, and Pechini methods), cryochemical and hydrothermal syntheses, and others. Combustion and co-precipitation methods have become the most widely used methods for materials based on barium cerate and zirconate and they include the formation of submicron-sized and nano-sized powders. The activity of such powders to sintering should be higher than that of powders obtained by the solid-state method. However, on the basis of results of different studies (Medvedev et al. 2015a, b), the formation of the dense ceramics at lower sintering temperature is possible only for the system based on barium cerate. For materials based on barium zirconate, the utilization of solution methods does not favor the preparation of gastight materials at rational processing temperatures (below 1500 °C). Therefore, it is necessary to search for new less power-consuming methods and technologies for the formation of a ceramic with the desired set of properties.

Along with solution methods, the traditional solid-state method with addition of small amounts of sintering additives (e.g., certain oxides of 3d transition elements or low-melting salts) can also be used for preparing gastight ceramics at lowered temperatures. This method is called as a solid-state reactive sintering and its features will discuss below at consideration of $BaZrO_3$ system.

4 Brief Overview of Proton-Conducting Oxides

After the discovery of proton conduction in doped $LaYO_3$ and $SrZrO_3$ by Iwahara et al. in `80 s, there were some other early reports on proton-conducting materials based on doped $SrCeO_3$ (Uchida et al. 1982; Iwahara et al. 1982; Uchida et al. 1983; Iwahara et al. 1983). In 1985, Virkar and Maiti (Virkar and Maiti 1985) reported results for the ionic conduction of pure and Y-doped $BaCeO_3$ and they believed that the conductivity of these oxides becomes predominantly ionic at $pO_2 < 10^{-6}$ atm at

600–1000 °C. In 1987, for the first time Mitsui et al. (Mitsui et al. 1987) revealed proton transport in $BaCe_{0.95}Yb_{0.05}O_{3-\delta}$ and $BaZr_{0.95}Yb_{0.05}O_{3-\delta}$ materials.

These materials belong to the class of perovskite-type oxides (ABO_3) and among them, $BaCeO_3$-based materials exhibit the highest conductivity values ($\sim 10^{-2}$ Scm^{-1} at 600 °C). However, as it will be below mentioned, in the presence of CO_2 they are relatively susceptible to carbonation. On the other hand, the proton conductors based on $BaZrO_3$ exhibit better chemical stability against the carbonation, but high sintering temperatures (1700 °C) are required to obtain them in the form of dense ceramic samples. It should be noticed that barium zirconates exhibit lower total conductivity compared to $BaCeO_3$, mainly due to their high grain-boundary resistance.

In the last years, alternative proton-conducting materials have been also proposed, such as oxides based on $LaScO_3$ and $LaNbO_4$ that offer high CO_2 tolerance. However, the values of conductivity are lower by one (and more) order of magnitude than those reported for the above-mentioned HTPCs.

More recently, other classes of materials (e.g., rare-earth tungstates) have attracted great attention due to their relatively high mixed proton–electron conductivity. In general, these materials show predominant proton conductivity at low temperature, while they are mixed ionic–electronic conductors at high temperatures. Most of them exhibit good chemical stability in CO_2 and sulfur-containing gases, which represents one of the main advantages of these materials.

In the following paragraphs, the recent advances in the field of solid oxide proton-conducting materials (regarding composition, structure–properties correlations, their advantages and disadvantages, and stability and their applications in SOFCs) are presented and analyzed.

4.1 Perovskite Materials

$BaCeO_3$-based materials. Among the protonic conductors, materials based on barium cerate ($BaCeO_3$) exhibit the highest proton conductivity (Medvedev et al. 2014a,b; Kreuer 1997, 1999). This may be explained mainly taking into account the following: (i) the structural features caused by large ionic radii of cations, which occupies A and B positions of ABO_3 perovskite, (ii) its relatively low electronegativity, and (iii) the microstructural features consist in a rather low grain-boundary resistance effect on the total resistance of ceramic samples. The most investigated systems based on barium cerate are oxides doped with acceptors of trivalent elements.

The influence of dopants on the conductivity of $BaCeO_3$-based materials is quite complicated. The conductivity of $BaCe_{1-x}Ln_xO_{3-\delta}$ increases with an increase of x reaching the maximal values at $x = 0.10–0.25$ depending on the nature of lanthanide (Fig. 6), which is a typical behavior for other doped oxides.

This can be explained according the following steps (Medvedev et al. 2014a,b):

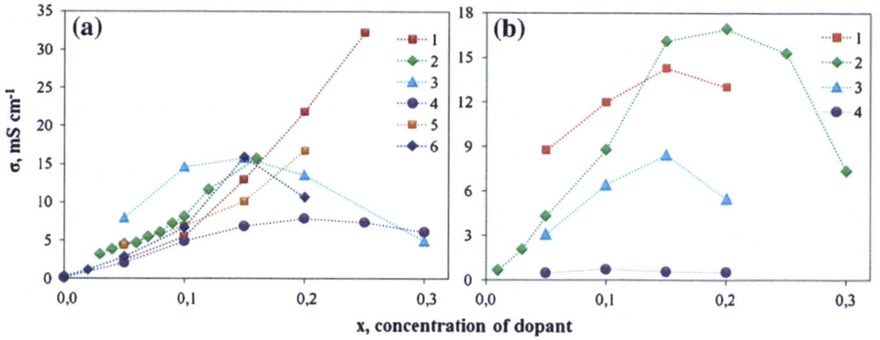

Fig. 6 Electrical conductivity of $BaCe_{1-x}Ln_xO_{3-\delta}$ materials at 600 °C (Medvedev et al. 2014a,b): **a** Wet air: *1* Ln = Gd, *2* Ln = Nd, *3* Ln = Gd, *4* Ln = Gd, *5* Ln = Gd, *6* Ln = Sm; **b** wet hydrogen: *1* Ln = Gd, *2* Ln = Sm, *3* Ln = Dy, *4* Ln = Ca

1. dissolution of the relatively low Ln_2O_3 concentrations in perovskite structure leads to the formation of free oxygen vacancies;
2. near the point corresponding to maximum conductivity the increase of amount of free vacancies practically ends and further dissolution takes place with the formation of $(R'_{Ce} - V_O^{\cdot\cdot})^{\cdot}$ and $(R'_{Ce} - V_O^{\cdot\cdot} - R'_{Ce})^x$ associates;
3. subsequent decrease in conductivity with a further increase in Ln_2O_3 was associated with a decrease in the amount of free vacancies due to the growth of the associates' concentration.

The solubility limit of most of the rare earth elements is between 20 and 30 mol % in the Ce-sublattice. At higher concentration of lanthanides, alongside with main perovskite structure, X-ray diffraction (XRD) analysis can register an impurity phase on the base of $BaLn_2O_4$ (Gorbova et al. 2008; Bonanos et al. 1989).

It deserves to be highlighted that ionic transport properties in $BaCe_{1-x}Ln_xO_{3-\delta}$ are controlled, as a rule, by the volume (grain) properties of electrolytes at temperatures higher than 600–650 °C. The influence of grain boundaries becomes more significant for low-temperature interval ($T < 600$ °C), as it is shown in Table 2.

The level of protonic (or ionic) conductivity is not the main parameter of HTPCs because the most highly conductive $BaCeO_3$ of them show low thermodynamic stability in the presence of H_2O and other acidic compounds (Medvedev et al. 2014a,b):

$$BaCeO_3 + H_2O \rightarrow Ba(OH)_2 + CeO_2 \tag{5}$$

$$BaCeO_3 + CO_2 \rightarrow BaCO_3 + CeO_2 \tag{6}$$

$$BaCeO_3 + H_2S \xrightarrow{[H]} BaS + CeO_2 + H_2O \tag{7}$$

As a result, research interest arose toward developing materials with high ionic conductivity and satisfactory stability against atmospheres of real SOFC operation conditions. The double doping (co-doping) of $BaCeO_3$ has been widely studied, because this strategy permits simultaneously influencing a number of key properties of the materials, including stability (Table 3).

Unfortunately, the co-doping with stabilizing dopants inevitably results in a decrease of ionic conductivity. The researchers highlighted some main reasons of the observed behavior (Radojković et al. 2013; Chi et al. 2013a, b; Liu et al. 2012; Xie et al. 2009; Bi et al. 2009):

1. The decrease of the free volume of unit cell and the connected with it decrease of migration channels size. This can be the reason of concentration and migration drop of charge carriers (oxygen ions and even protons).
2. The poor densification and/or low average grain size, which results in the significant increase of grain-boundary resistance.
3. The drop concentration of free oxygen vacancies in the case of doping by donor elements.

From the viewpoint of thermomechanical properties, $BaCeO_3$ oxide exhibits three phase transitions under heating from room temperature to 1000 °C. One of which is first-order phase transition (realized at ~400 °C), while the others are second-order transitions (~260 and 890 °C) (Medvedev et al. 2014a,b). Consequently, phase transitions significantly affect the thermal expansion and TEC characteristics of $BaCeO_3$ and its doped analogs. This is clearly shown in Fig. 3, where a sharp bend of the dilatometric curve is realized. Such nonmonotonic behavior of thermal expansion can be the reason of undesirable mechanical stress of SOFCs. Fortunately, the strategy of Ce-substitution by cations with lower ionic radii (e.g., Zr) helps to stabilize the cubic perovskite structure of cerates even at room temperature. Therefore, further heating does not result in sharp changes of expansion curves.

$BaZrO_3$-based materials. The crystal features of $BaZrO_3$ consist of shorter and stronger Zr–O bonds (in comparison with Ce–O ones) and practically ideal combination between Zr–O and Ba–O distances causes its excellent chemical stability of cubic perovskite-structured materials (Bi and Traversa 2014). However, their total proton conductivity is much lower than that of cerates due to a grain-boundary effect. The possible reasons of the high grain-boundary resistance of doped $BaZrO_3$ are not clarified yet, but can be associated with the change of chemical composition in grain boundary regions (Ba volatilization, the presence of traces of nano-sized impurity phases, increasing concentration of point defects and their associates, and other local inhomogeneity).

One of the main problems of the present materials is the use of very high sintering temperatures (1800–2000 °C) to obtain high dense and single-phase ceramic samples, when solid-state reaction method is used (Table 4).

Table 3 Comparison of stability of In-, Nb-, Sn-, Ta-, and Bi-containing materials based on $BaCeO_3$

System	Treatment conditions	Decomposition	Stability	References
$BaCe_{0.8-x}In_xLn_{0.2}O_{3-\delta}$ (Ln = Gd, Y)	Boiling water (2 h) → XRD	For $x = 0$, the impurity phases are observed with decomposition of perovskite	For $x = 0.1$, no impurity phases are observed	Liu et al. (2012)
$BaCe_{0.95-x}In_xTi_{0.05}O_{3-\delta}$	94 vol.% N_2 + 3 vol.% CO_2 + 3 vol.% H_2O at 700 °C for different soaking time (τ) → XRD	The materials are unstable at $x = 0$, $\tau > 20$ h; $x = 0.1$ and 0.2, $\tau > 20$ h; $x = 0.3$ and 0.4, $\tau > 80$ h	The materials are stable at $x = 0$, $\tau < 20$ h; $x = 0.1$ and 0.2, $\tau < 20$ h; $x = 0.3$ and 0.4, $\tau < 80$ h	Chi et al. (2013b)
$BaCe_{0.9-x}Nb_xY_{0.1}O_{3-\delta}$	100 % CO_2 at 700 °C for 5 h → XRD	At $0 \leq x \leq 0.01$, the significant interaction is observed	For $0.03 \leq x \leq 0.05$, the conservation of perovskite structure with the presence of impurity phases is observed	Radojković et al. (2013)
$BaCe_{0.8-x}Sn_xY_{0.2}O_{3-\delta}$	Boling water (3 h) → XRD	For $x = 0$, a significant interaction is observed with the formation of $BaCO_3$, CeO_2 and Y_2O_3 impurity phases For $x = 0.05$, the conservation of perovskite structure with the presence of impurity phases is observed	At $0.1 \leq x \leq 0.25$, no impurity phases are observed	Xie et al. (2009)
$BaCe_{0.8-x}Ta_xY_{0.2}O_{3-\delta}$	100 % CO_2 at 700 °C(3 h); 100 % CO_2 at 900 °C (3 h); boiling water (6 h) → XRD	–	For $x = 0.1$, the high stability is noted	Bi et al. (2009)
$BaCe_{0.8-x}Bi_xY_{0.2}O_{3-\delta}$	Boiling water → XRD	For $x = 0$ and $x = 0.1$, the perovskite decomposition, the formation of $Ba(OH)_2$ and CeO_2 interaction phases and amorphous phase are detected	For $x = 0.3$ and $x = 0.5$, no impurity phases are observed	(Chi et al. 2013a)

Table 4 Relative density of $BaZr_{1-x}Y_xO_{3-\delta}$ ceramics

x	Synthesis method	T_{synth} (°C/ τ, h)	T_{sint}(°C/ τ, h)	ρ_{rel} (%)	References
0.1	Hydroxide co-precipitation + solid-state reaction method	–	1600/3	80	Gorelov and Balakireva (2009)
0.2	Sol–gel method	1000/5	1500/5	65.7	Guo et al. (2009)
0.2	Sol–gel method	1100/5	1600/8	80	Fabbri et al. (2008)
0.2	Sol–gel method	1000/12	1400/10	78.3	Sawant et al. (2012)
0.1	Sol–gel method	1000/48	1600/6	74	Dahl et al. (2011)
0.15	Solid-state reaction method	1300/2	1670/24	84.1	Park et al. (2015)
0.15	Sol–gel method	900/2	1670/24	95.5	Park et al. (2015)
0.15	Glycine–nitrate synthesis	1300/2	1670/24	80.1	Park et al. (2015)
0.15	Solid-state reaction method	1300/2	1670/24	86.5	Park et al. (2015)
0.2	Solid-state reaction method	–	1700/–	92	Kreuer et al. (2001)
0.1	Solid-state reaction method	1250/10	1715/30	96	Schober and Bohn (2000)
0.1	Hydroxide co-precipitation + solid-state reaction method	1200/–	1850/1	99.7	Gorelov and Balakireva (2009)
0.1	Solid-state reaction method	1400/10	2200/–	98	Duval et al. (2007)

T_{synth} is the synthesis temperature, τ is the soaking time, T_{sint} is the sintering temperature, and ρ_{rel} is the relative density

High-temperature sintering, as was noted above, is closely related to the following problems:

1. The barium volatility, which leads to a decrease in conductivity due to the change of the composition in grain (volume) and grain-boundary regions associated with the stoichiometric deficiency of barium.
2. The difficulty of forming fuel cells with a thin layer $BaZrO_3$-based electrolyte. For obtaining the half-cell structure «electrolyte/electrode» or SOFC structure «anode/electrolyte/cathode», the green multilayer structures are co-sintered for good adhesion and to achieve a gastight electrolyte. As shown in Table 4, in the case of $BaZrO_3$-based materials the sintering temperatures exceed 1450 °C. This in turn should lead to a strong chemical interaction of the electrolyte with the electrodes followed by the formation of low conductive phases in interfacial region.

In order to decrease the sintering temperatures of zirconates, the wet chemical methods (sol–gel method, citrate–nitrate combustion, co-precipitations, etc.) are usually used. These methods aim at obtaining the nanopowders which possesses enhanced sinterability and results in, to some degree, decrease in technological temperatures. Another possible way is the use of small amounts of MO_n sintering additives (MO_n = CuO, NiO, ZnO, LiF, etc., Table 5). This alternative way involves: (i) the formation of liquid phases (1124 °C for Cu_2O, 800 °C for

Table 5 Relative density of $BaZrO_3$-based ceramics obtained using sintering additives

System	x	Synthesis method	$T_{sint.}$ (°C)	$\rho_{rel.}$ (%)	References
$BaZr_{0.9}Y_{0.1}O_{3-\delta}$ + x mol% CuO	0	Solid-state reaction method	1400	55	Gao et al. (2010)
	1		1400	70	
	2		1400	93	
	3		1400	96	
	4		1400	96	
	4		1300	88	
$BaZr_{0.8}Y_{0.2}O_{3-\delta}$ + x wt% NiO	0	Solid-state reaction method	1500	<35	Tong et al. (2010)
	0.5		1500	70	
	1		1500	95	
	2		1500	97	
$BaZr_{0.8}Y_{0.2}O_{3-\delta}$ + x wt % ZnO	0	Citrate–nitrate synthesis	1700	>90	Peng et al. (2010)
	1		1450	>90	
$BaZr_{0.9}Y_{0.1}O_{3-\delta}$ + x mol% ZnO	0	Freeze drying	1500	84	Amsif et al. (2014)
	4		1300	88	
$BaZr_{0.8}Y_{0.2}O_{3-\delta}$ +x wt % LiF	0	Glycine–nitrate synthesis	1400	61	Tsai et al. (2010)
	7		1400	91	

T_{sint} is the sintering temperature and ρ_{rel} is the relative density

Ba_2CuO_3, 1099 °C for BaO·ZnO, 857 °C for $BaLiF_3$) alongside with the solid one and (ii) the formation of highly defective M-containing phases. Both of these ways assist ceramics densification at lower sintering temperatures due to the realization of high-diffusion processes (Medvedev et al. 2015a, b).

In the oxide systems based on zirconates, a decrease in total conductivity with increasing ionic radius of lanthanides is observed. Due to this fact, the dopants with ionic radii (Sc^{3+}, Y^{3+}) close to that of zirconium one are more preferred. However, $BaZr_{1-x}Y_xO_{3-\delta}$ (BZY) materials are the most investigated systems among other zirconates, including Sc-doped ones. The conductivity of BZY continuously increases when concentration of yttrium grows (Kuzmin et al. 2009); this behavior can be associated with the formation of oxygen vacancies in the perovskite structure (Fig. 7).

Grain boundaries have a significant impact on the transport processes. Therefore, the total conductivity of ceramics is sensitive even to a weak change of the technological (or technical) factors, such as the properties of precursors, synthesis methods, and temperature regimes. Consequently, total conductivity can be changed by one order of magnitude and more. For example, for similarly prepared $BaZr_{0.8}Y_{0.2}O_{3-\delta}$ ceramic materials, as reported in literature, conductivity ranges from 0.7 to 8.1 S cm^{-1} at 600 °C in a humidified atmosphere of inert gas (Table 6).

$BaZrO_3$-based systems from room temperature to melting temperature exhibit a cubic structure. Therefore, linear expansion dependences of these materials are

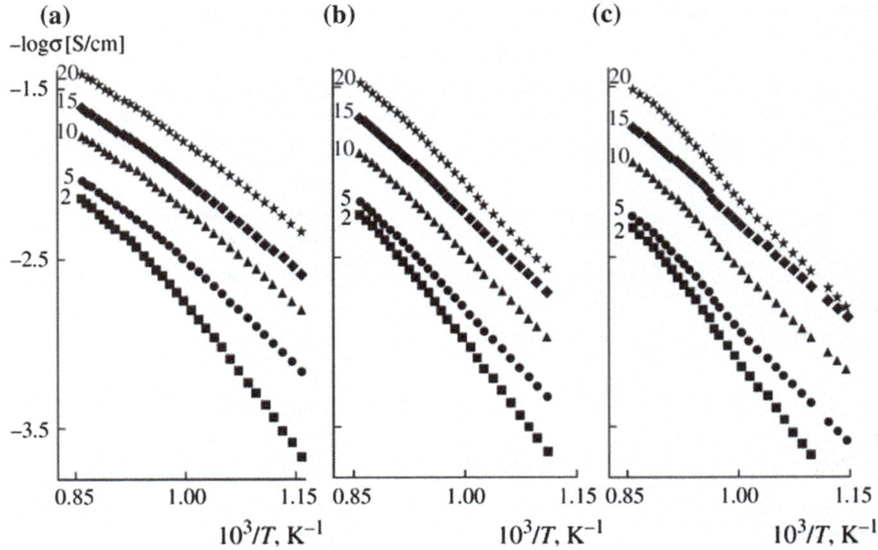

Fig. 7 Total conductivity of $BaZr_{1-x}Y_xO_{3-\delta}$ in air with different humidification levels: 0.04 kPa (**a**), 0.61 kPa (**b**), and 3.17 kPa (**c**) (Kuzmin et al. 2009)

Table 6 Ionic conductivity of $BaZr_{1-x}Y_xO_{3-\delta}$ ceramic materials in wet N_2 or wet Ar as a function of the quality of ceramics (mean grain size, D_{av}, and relative density, $\rho_{rel.}$) and technological regimes of ceramics preparation (method preparation, temperatures of synthesis, $T_{synth.}$, and sintering, $T_{sint.}$)

x	Preparation method	T_{synth}/T_{sint} (°C)	ρ_{rel} (%)	D_{av} (μm)	T (°C)	σ_{total} (mS cm^{-1})	E_a (eV)	References
0.01	Pechini method	1100/1700	95	3.5	700	0.6	0.61	Park (2011)
0.2	Sol–gel method	1000/1400	78.3	–	600	10.4	0.66	Sawant et al. (2012)
0.2	Solid-state reaction method + NiO addition	1200/1450	95	0.6	700	5.4	0.54	Yoo et al. (2015)
0.2	Glycine–nitrate synthesis	1250/1600	99	0.46	600	8.0	–	Babilo et al. (2007)
0.2	Combustion synthesis	1200/1600	>90	1.00	600	8.1	–	Yamazaki et al. (2009)
0.2	Combustion synthesis	1200/1600	>90	0.44	600	5.6	–	Yamazaki et al. (2009)
0.2	Alkoxides hydrolysis	–/1250	98.1	0.05	600	1.0	0.59	Cervera et al. (2008)
0.2	Alkoxides hydrolysis	–/1500	99.4	0.2	600	0.7	0.54	Cervera et al. (2008)

Table 7 Thermal expansion coefficients of zirconates in ambient or dry (*) and wet (**) air atmospheres

Composition	Temperature range (°C)	αav ($\times 10^6 K^{-1}$)	References
$BaZrO_3$	50–1100	7.1*	Yamanaka et al. (2003)
$BaZr_{0.95}Y_{0.05}O_{3-\delta}$	100–800	7.8*, (8.8**)	Andersson et al. (2014)
$BaZr_{0.9}Y_{0.1}O_{3-\delta}$	100–800	7.4*, (8.8**)	Andersson et al. (2014)
$BaZr_{0.8}Y_{0.2}O_{3-\delta}$	100–900	8.2*	Lyagaeva et al. (2015)
$BaZr_{0.8}Y_{0.2}O_{3-\delta}$	100–800	8.0*, (9.7**)	Andersson et al. (2014)

smoother than those for $BaCeO_3$-based systems due to the absence of phase transitions associated with the change in the structure symmetry. However, the TEC levels of zirconates are rather low (Table 7) in comparison with traditional ZrO_2, CeO_2, and $LaGaO_3$ electrolytes (Tsipis and Kharton 2008), which can complicate the selection of compatible electrode systems.

$BaCeO_3$–$BaZrO_3$ mixed systems. The solid solutions based on $BaCeO_3$ and $BaZrO_3$ attracted much attention mainly because of their high chemical stability in the presence of H_2O- and CO_2-containing atmospheres as well as in the presence of S-containing components (e.g., H_2S). According to thermodynamic calculations, barium cerate is the most unstable material among other cerates or zirconates of alkaline earth elements in the presence of CO_2 and H_2O, whereas barium zirconates are the most stable (Kreuer 2003; Medvedev et al. 2014a,b). For example, the water vapor exposition of $BaCeO_3$ at $T < 400$ °C results in decomposition of this phase being followed by the formation of barium hydroxide and cerium oxides (Haile et al. 2001; Wu and Liu 1997; Matsumoto et al. 2007). This is testified by the thermodynamic calculations (Medvedev et al. 2015a, b). It should be pointed out that such materials become stable to H_2O at $T > 400$ °C and, hence, may be used for high-temperature applications.

The stability of $BaCeO_3$ in pure CO_2 is lower: According to results reported in literature (Matsumoto et al. 2007), the chemical interaction of perovskite phase with CO_2 takes place up to 1041 °C. According to the existing experimental data, the interaction between H_2S- and $BaZrO_3$-based materials was found to be almost negligible, which may be explained by kinetic peculiarities, but for the $BaCeO_3$-based ones the presence of H_2S has a destructive impact (Li et al. 2008; Fang et al. 2008).

Among other doped systems, Y-doped $BaCeO_3$–$BaZrO_3$ materials are the most studied in literature. Single-phased materials can be formed in the system of $BaCe_{1-x-y}Zr_xY_yO_{3-\delta}$ ($0 \leq x \leq 1-y$; $0 < y \leq 0.2$). By increasing zirconium concentration (x), synthesis and sintering temperatures become higher and they can amount 1400 and 2200 °C, respectively (Table 4). Therefore, the use of wet chemical methods (such as co-precipitation, combustion method, sol–gel, and Pechini) and the introduction of sintering aids are the most actual strategies for the formation of dense ceramics with a non-porous microstructure.

It should be noted that the crystal structure of $BaCe_{1-x-y}Zr_xY_yO_{3-\delta}$ materials strongly depends on many factors, such as method of powder preparation, synthesis/sintering technological regimes, and concentration of uncontrolled impurities in precursors. Due to this fact, the same Y-doped $BaCeO_3$–$BaZrO_3$ materials can possess different crystal structures, including monoclinic, orthorhombic, rhombohedral, and cubic (see Table 7 in Medvedev et al. 2016). At the same time, with the increase in zirconium concentration an increase in the perovskite symmetry is observed.

The transport properties of $BaCeO_3$–$BaZrO_3$ materials were thoroughly investigated by different research groups (Sawant et al. 2012; Fabbri et al. 2008; Ricote et al. 2012; Katahira et al. 2000; Lagaeva et al. 2015). Despite that the conductivity for the same oxides greatly differs (by half-order of magnitude or more), a general trend can be noted:

1. The conductivity decreases with increasing Zr content in cerate-zirconates—most probably due to the decrease in grain and grain-boundary conductivities.
2. The grain-boundary conductivity deterioration (Figs. 8 and 9) can be associated with the fact that grains size becomes smaller and relative density of ceramics drops.
3. The grain conductivity can be also decreased (Fig. 9) due to several reasons, such as: narrowing of migration channels, which is defined by the free volume of unit cell and the higher electrostatic repulsion in the Ce–OH pair, rather than in Zr–OH ones, that follows from the electronegativity concept.

On the other hand, among the reasons that can negatively affect grain conductivity are (Fig. 10): (i) the narrowing of migration channels, which is defined by the free volume of unit cell and (ii) the higher electrostatic repulsion in the Ce–OH pair, rather than in Zr–OH ones, according to electronegativity concept.

Ca- and Sr-based zirconates. From the viewpoint of electrical properties, the oxide systems based on $SrZrO_3$ are less conductive in comparison with $BaZrO_3$ analogs. However, the scarcity of literature information does not allow one to mark out the most appropriate dopants for this perovskite system. For example, Omata

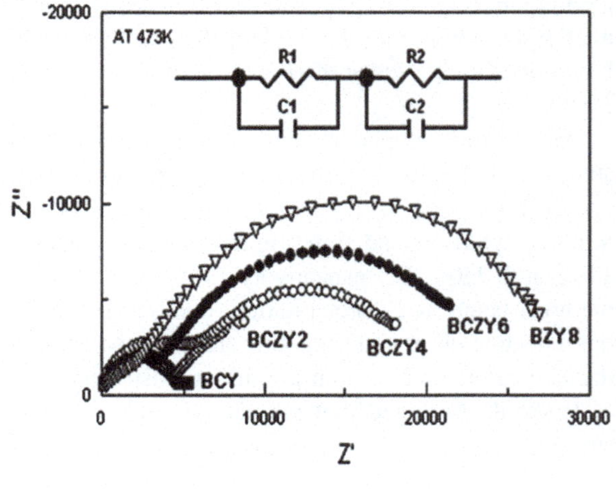

Fig. 8 Nyquist plots of BCZY samples and equivalent circuit employed for evaluating the grain and grain-boundary resistance (Sawant et al. 2012)

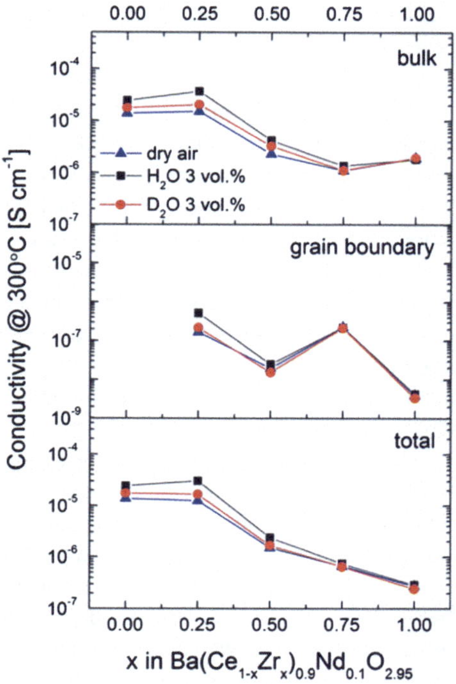

Fig. 9 Bulk, grain-boundary and total electrical conductivity of $Ba(Ce_{1-x}Zr_x)_{0.9}Nd_{0.1}O_{2.95}$ solid solutions as a function of Zr content (Zając et al. 2012)

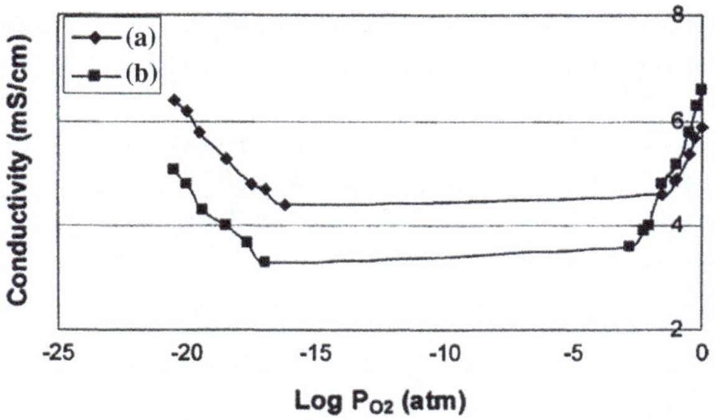

Fig. 10 Conductivity of $SrCe_{1-x}Y_xO_{3-\delta}$ materials as a function of pO_2 at 800 °C: $x = 0.1$ (**a**) and $x = 0.15$ (**b**) (Sammes et al. 2004)

et al. (Omata et al. 2008) found that the conductivity of $SrZr_{1-x}Sc_xO_{3-\delta}$ materials is higher than that of $SrZr_{1-x}Y_xO_{3-\delta}$, whereas the work of Huang and Petric (1995) showed opposite results.

In- and Sc-doped $CaZrO_3$ demonstrate the better transport properties in contrast to the materials doped by other elements. For example, the conductivity values of $CaZr_{1-x}Sc_xO_{3-\delta}$ are comparable with those of zirconates (Gorelov et al. 2014). It should be noted that a decrease in the ionic radius of twelve-coordinated elements A in $AZrO_3$ is the cause of the decrease in conductivity. However, at the same time, the character of conductivity changes from mixed (ionic-hole) to predominately protonic.

For example, according to results of Gorelov et al. (2014), the $CaZr_{1-x}Sc_xO_{3-\delta}$ materials are purely protonic electrolytes at the conditions of wet air atmosphere and low temperatures (less than 600 °C).

Materials based on $SrCeO_3$. $SrCeO_3$-based oxides are isostructural to $BaCeO_3$ and as already mentioned exhibit also protonic conductivity. Their conductivity compete the values reached for barium cerates. As shown in section $BaCeO_3$-based materials, doped $BaCeO_3$ demonstrates a pure ionic conductivity in reducing atmospheres and mixed ionic–electronic conductivity in oxidizing ones. A decrease in ionic radius of $ACeO_3$ alkaline earth element results not only the deterioration of conductivity (due to the decrease in size of migration channels), but also in the appearance of undesirable electronic conductivity in reducing atmospheres (due to easier Ce ions reduction, Fig. 10).

This feature does not allow one to use such materials as electrolytes for solid oxide fuel cells; however, it opens the possibility for their applications as non-electrode membranes for hydrogen production or carrying out of conversion processes (Li et al. 2011; Shrivastava et al. 2012; Xing et al. 2015).

Materials with $A^{3+}B^{3+}O_3$ structure. Some oxides with perovskite structure belonging to $A^{3+}B^{3+}O_3$-structured perovskite class (A, B = Ln or trivalent p- and d-elements with constant oxidation state) also exhibit proton transport and demonstrate high chemical stability. The proton transport was found to be the characteristic feature of oxides based on $LaScO_3$, $LaInO_3$, $LaLuO_3$, $LaErO_3$, $LaYbO_3$, and $LaYO_3$ (Okuyama et al. 2014). Among this series, the Sc- and Yb-containing perovskites are highlighted as most promising mainly due to their good electrical properties (Fig. 11).

Considering the transport properties of these systems with the variation of pO_2, it can be stated that all of such $A^{3+}B^{3+}O_3$ perovskites are characterized by a mixed ionic–electronic conductivity in oxidizing atmospheres and a pure ionic in reducing one (Fig. 12).

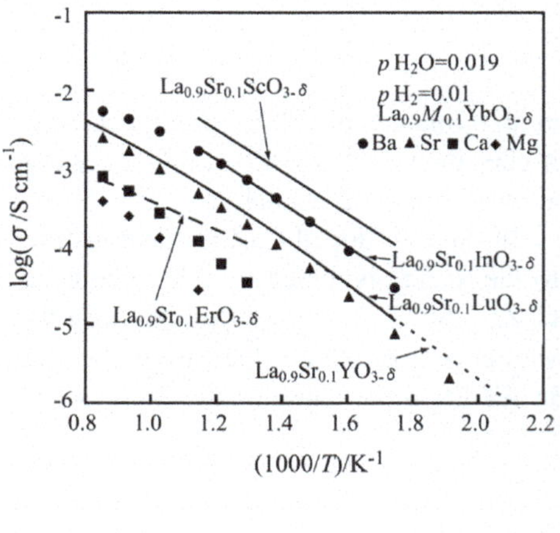

Fig. 11 Total conductivity of $A^{3+}B^{3+}O_3$ materials in wet H_2 atmosphere (Okuyama et al. 2013)

Fig. 12 Oxygen partial pressure dependences of conductivity for $La_{0.9}Ba_{0.1}YbO_{3-\delta}$ (Okuyama et al. 2013)

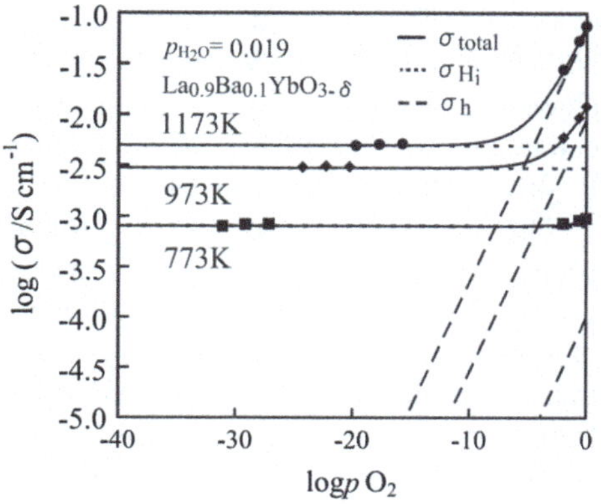

The same tendency is maintained for other representatives of $A^{3+}B^{3+}O_3$ protonic materials [$La_{1-x}Sr_xSc_{1-x}Mg_xO_{3-\delta}$ (Stroeva et al. 2010) and $La_{0.9}Sr_{0.1}M_{0.9}Mg_{0.1}O_{3-\delta}$ (Lybye et al. 2000)]. It is found that proton concentration and, correspondingly, level of proton conductivity grow with increasing ionic radius of B element, which is in agreement with size factors. Due to this fact, $LaYO_3$- and $LaYbO_3$-based materials should be considered as the most conductive systems. For example, Okuyama et al. (2014) revealed that yttrates and ytterbiates possessed unipolar protonic conductivity under wet reducing atmospheres, whereas scandates, indates, and aluminates demonstrated also noticeable oxygen ionic transport.

Because $A^{3+}B^{3+}O_3$ perovskites exhibit lower level of conductivity in comparison with $A^{2+}B^{4+}O_3$ systems, their applications in SOFCs should be verified. However, Y- and Yb-based oxides possess a wide electrolyte domain boundary and can be successfully used for sensors (Kalyakin et al. 2016) and pumps (Sakai et al. 2013) devices.

4.2 Other Materials

$Ba_2In_2O_5$-based materials. Two decades ago it was shown for the first time that proton transport can also be realized in perovskite-related compounds with structural oxygen vacancies (Nowick and Du 1995; Krug and Schober 1996). Contrary to simple perovskites, for which vacancies are created by acceptor doping (*impurity oxygen vacancies*), these compounds contain oxygen vacancies, which are inherently structural components (*structural oxygen vacancies*).

The brownmillerite structure of $A_2B_2O_5$ (or $A_2BB'O_5$) incorporates a maximal amount of oxygen vacancies: ~16.7 % (or 1/6) for each formula unit $ABO_{2.5}[V_O]_{0.5} \equiv A_2B_2O_5[V_O]_1$, whereas for simple perovskites the content of oxygen vacancies $([V_O^{\bullet\bullet}]/([V_O^{\bullet\bullet}]+[O_O^x]))$ is typically less than 4 % (Kochetova et al. 2016). As well-known, $Ba_2In_2O_5$ brownmillerite compounds also ehibit proton transport (Jankovic et al. 2012; Islam et al. 2001; Animitsa et al. 2012; Tarasova et al. 2015; Tarasova and Animitsa 2015). Upon exposure to a wet atmosphere near 300 °C, $Ba_2In_2O_5$ undergoes a reversible phase transformation. This hydrated phase was determined, by Zhang and Smyth in 1995a, b, to be a reasonable proton conductor though it has been found that the proton conductivity of the $Ba_2In_2O_4(OH)_2$ is less than 10^{-5} S cm^{-1} at 400 °C.

The presence of the phase transition around 930 °C is the main problem associated with $Ba_2In_2O_5$ applications. More precisely, this phase transition significantly affects the target properties of ceramic materials, including value and nature of conductivity, mechanical properties, and stability. The order/disorder phase transition is accompanied by disordering oxygen vacancies and changing symmetry from an orthorhombic to tetragonal at 930 °C and then to cubic symmetry around at 1040 °C (Hashimoto et al. 2002). The ionic conductivity of $Ba_2In_2O_5$ at temperature values higher than 930 °C is comparable with YSZ but below transition temperature the conductivity dramatically decreases (by 1–1.5 orders of magnitude). For this reason, the main doping strategy deals with the stabilization of the cubic phase at such low temperatures as possible.

Materials science aspects are aimed at the realization of *isovalent- or donor*-doping strategies, which contribute to the stabilization of cubic phase of barium indate at lower temperatures. Numerous cationic substitutions were performed in Ba^{2+}, In^{3+}, or both sublattices in order to improve the oxygen ion and the proton conductivities in $Ba_2In_2O_5$ (Rolle et al. 2008; Ta et al. 2006; Kochetova et al. 2013; Ito et al. 2012; Shin et al. 2011). For example, the partial isovalent substitution of In^{3+} by an element with smaller ionic radius (Sc^{3+}) results in a decrease of the phase transition (T_{pt}) temperature, whereas elements with higher ionic radius (Y^{3+}, Ln^{3+}) lead to an increase of T_{pt} (Fig. 13) (Ta et al. 2006).

The stabilization of cubic structure of indates can be realized even at 300 °C when Ba^{2+} or In^{3+} is partially substituted by donor elements (Kakinuma et al. 2001, Niwa et al. 2003). Because of this, the conductivity of donor-doped $Ba_2In_2O_5$ can be enhanced by 1–2 orders of magnitude than that of basic barium indate at the intermediate-temperature range (Fig. 13). The main reason of the phase stabilization and the conductivity enhancement of donor-doped indates is the slight decrease

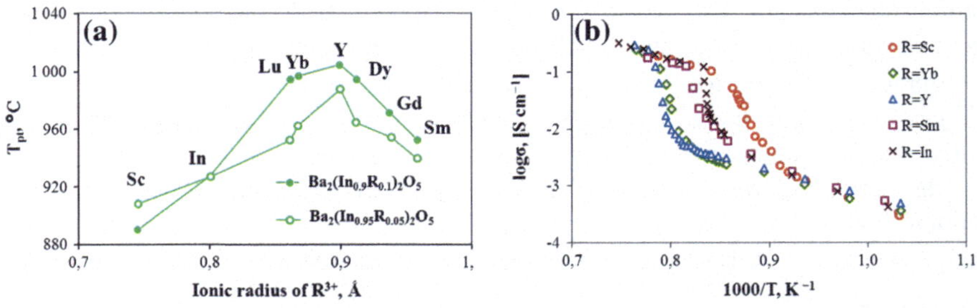

Fig. 13 Temperature of phase transition for $Ba_2(In_{1-x}R_x)_2O_5$ (**a**) and temperature dependences of conductivity for $Ba_2(In_{0.9}R_{0.1})_2O_5$ in air (**b**) (Ta et al. 2006)

in the excess concentration of oxygen vacancies, which results in the conservation of the disorder structure.

Another problem of these materials is the sharp (non-monotonic) change of linear and volume parameters of ceramic samples under the heating or cooling procedure (Yoshinaga et al. 2004; Hui et al. 2006), preventing their utilization as an ensemble with other functional materials (cathode, anode, interconnect, and seal). For example, in Fig. 14 the dilatometric results of $Ba_{2-x}Sr_xIn_2O_5$ and $Ba_2In_{2-y}Ga_yO_5$ ceramic materials are shown. As one can distinguish the investigated temperature ranges, the obtained dependences are nonlinear and exhibit an obvious inflection point that corresponds to the phase transition point.

$LaNbO_4$-based materials. Materials based on $LaNbO_4$ are the representatives of a relatively new class of HTPCs, which were firstly investigated by (Norby et al. 2013). The analysis of $LaNbO_4$ reveals that this material exhibits a phase transition that pass from the low-temperature monoclinic fergusonite-type structure to the high-temperature tetragonal sheelite-type one (Norby and Magrasó 2015). Consequently, this phase transition is coupled with a significant variation of both unit cell volume and linear parameters. The average TEC of low-temperature $LaNbO_4$ phase is two times higher than that for the high-temperature phase (Table 8).

Practically, the stabilization of $LaNbO_4$ tetragonal phase at low temperatures is difficult, since such system possesses a narrow interval of solid solution existence. The amount of acceptor dopants, resulting in the formation of oxygen defects and the appearance of oxygen ion and proton conductivities, is usually less than 1–5 mol% (Haugsrud and Norby 2006). Higher concentrations of dopants result in the formation of multiphase systems (Haugsrud and Norby 2006; Syvertsen et al. 2012).

Although the conductivity of $LaNbO_4$-based materials under humidification increases by 1–1.5 orders of magnitude, their absolute level of total conductivity is rather low (Fig. 15a, c), mostly because of the high contribution of grain-boundary

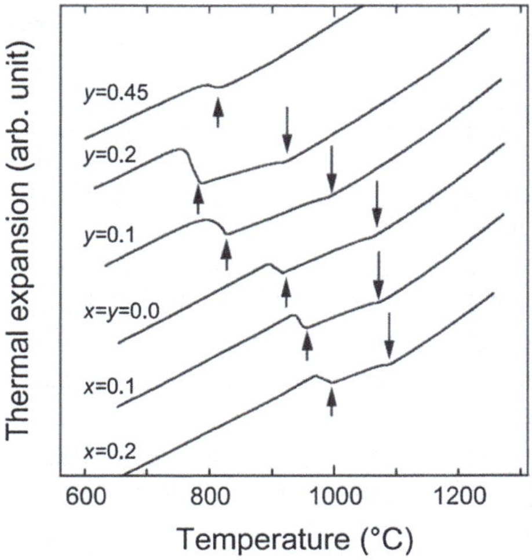

Fig. 14 Thermal expansion of $Ba_{2-x}Sr_xIn_2O_5$ and $Ba_2In_{2-y}Ga_yO_5$ in the cooling regime in air (Yoshinaga et al. 2004)

Table 8 Thermal expansion coefficients (α) and transition temperatures (T_{trans}) for $LaNbO_4$-based oxides

Composition	$\alpha_{monoclinic}$ ($\times 10^6$, K^{-1})	$\alpha_{tetragonal}$ ($\times 10^6$, K^{-1})	T_{trans} (°C)	References
$LaNbO_4$	15	8.6	450	Norby and Magrasó (2015)
$LaNbO_4$	14	8.4	504	Mokkelbost et al. (2009)
$LaNbO_4$	17.3	7.1	510	Vullum et al. (2008)
$La_{0.98}Ca_{0.02}NbO_{4-\delta}$	12	8.3	504	Haugsrud and Norby (2006)
$La_{0.98}Sr_{0.02}NbO_{4-\delta}$	14	9.1	492	Haugsrud and Norby (2006)
$La_{0.98}Sr_{0.02}NbO_{4-\delta}$	20	8	500	Brandao et al. (2011)
$La_{0.98}Ba_{0.02}NbO_{4-\delta}$	14	8.8	514	Haugsrud and Norby (2006)
$LaNb_{0.8}Ta_{0.2}O_4$	15.7	9.1	670	Syvertsen et al. (2012)
$La_{0.99}Mg_{0.01}Nb_{0.99}Al_{0.01}O_{4-\delta}$	12.9	7.8	500	Ivanova et al. (2015)
$La_{0.99}Ca_{0.01}Nb_{0.99}Al_{0.01}O_{4-\delta}$	15.1	8.9	490	Vullum et al. (2008)
$La_{0.99}Sr_{0.01}Nb_{0.99}Al_{0.01}O_{4-\delta}$	12.8	6.4	470	Vullum et al. (2008)
$La_{0.99}Ba_{0.01}Nb_{0.99}Al_{0.01}O_{4-\delta}$	10.3	7.0	450	Vullum et al. (2008)

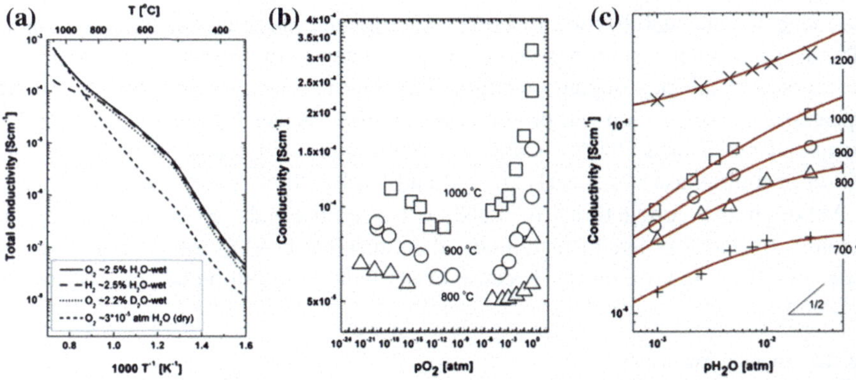

Fig. 15 Conductivity of $LaNb_{0.99}Ti_{0.01}O_4$ as a function of temperature (**a**), water vapor partial pressure (**b**), and oxygen partial pressure (**c**) (Huse et al. 2012)

conductivity (Norby and Magrasó 2015) and low substitution level (Ivanova et al. 2015; Bi et al. 2012; Solís and Serra 2011).

The above fact, alongside with the undesirable high values of n- and p-type conductivities (Fig. 15b) and the considerable change of mechanical properties (associated with monoclinic → tetragonal phase transition), prevents the application of $LaNbO_4$-based materials as electrolytes for SOFCs. Either using 15 μm Ca-doped $LaNbO_4$ (Magrasó et al. 2014) or 30 μm Sr-doped $LaNbO_4$ (Magrasó et al. 2011) electrolytes, power density results of SOFCs are close to 5 $mWcm^{-2}$ at 800 °C, that is, about 1–2 orders of magnitude lower than that of a SOFCs based on $BaCeO_3$ and $BaZrO_3$ proton-conducting electrolytes.

Fluorite-based materials. As well known, materials with fluorite structure, such as doped CeO_2, are considered mainly as oxygen ion-conducting electrolytes for different types of electrochemical devices (Patakangas et al. 2014; Figueiredo and Marques 2013).

However, even in such oxides proton transport can also be realized under certain conditions (Yokokawa et al. 2004; Nigara et al. 2003; Ruiz-Trejo and Kilner 2009; Sun et al. 2012). For example, the estimation of proton conductivity (1.2×10^{-6} S cm^{-1} at 800 °C) in Gd-doped CeO_2 system has been carried out on the base of hydrogen permeability experiments. Ruiz-Trejo and Kilner (2009), using electrochemical impedance spectroscopy, measured the response of conductivity on the pH_2O change at low temperatures. More precisely, the total conductivity of $Ce_{0.9}Gd_{0.1}O_{2-\delta}$ does not change with the change of pH_2O in air at temperatures higher than 200 °C and slightly differs at the lower ones. For example, the value of proton conductivity reaches 1.0×10^{-8} and 3.4×10^{-8} S cm^{-1} at 150 °C and 2.2×10^{-8} and 4.7×10^{-8} S cm^{-1} at 200 °C in dry (pH_2O = 0.003 atm) and wet (pH_2O = 0.029 atm) air, respectively. The separation of total conductivity in grain and grain-boundary components allowed the authors to conclude that bulk conductivity of $Ce_{0.9}Gd_{0.1}O_{2-\delta}$ is independent on the steam concentration, demonstrating the oxygen ionic transport. Therefore, proton transport in such and in similar oxides under the mentioned conditions should take place in grain-boundary regions (Ruiz-Trejo and Kilner 2009).

In the case of equimolar substitution of Ce with La in $Ce_{1-x}La_xO_{2-\delta}$, the formation of $La_2Ce_2O_7$ phase with cubic fluorite structure occurs (Sun et al. 2012). Absolute value of proton conductivity and proton transport numbers was calculated by the aid of AC conductivity and EMF measurements. The total conductivity in wet hydrogen is equal to 5.6×10^{-7} and 5.1×10^{-4} S cm^{-1} for 250 and 550 °C, respectively (Fig. 16), and at the same time, the proton transport numbers decrease from 0.89 to 0.13 (Fig. 17). Thus, the simple calculations show that proton conductivity in the $La_2Ce_2O_7$ ceramic amounts 5.0×10^{-7} and 6.7×10^{-5} S cm^{-1} for 250 and 550 °C, respectively.

Pyrochlore-based materials. Oxides with a general $A_2B_2O_7$ (A = Ln, B = Ti, Sn, Zr) formula belong to pyrochlore-structured materials. As for the most proton-conducting materials, both oxygen ion and proton conductivity arise after doping by acceptor impurities (Mg^{2+}, Ca^{2+}, Sr^{2+}, Ba^{2+}), resulting in anion disorder of pyrochlore structures. Principally, the partial substitution can be carried out for both A- and B-sublattices of $A_2B_2O_7$.

However, A-substitution is more favorable due to the possibility of achieving a higher conductivity, including the proton component (Eurenius et al. 2010b, c; Omata et al. 2004). This phenomenon is attributed to different oxygen sites with different basicities in the pyrochlore crystal. The analysis of the crystallochemical factors also shows that proton transport increases with increasing the ionic radius of Ln^{3+} (Eurenius et al. 2010b) and acceptor dopant (Omata et al. 2004).

In these systems, a relatively high hydration degree can be realized (Eurenius et al. 2010a), leading to predominantly proton transport in wet reducing

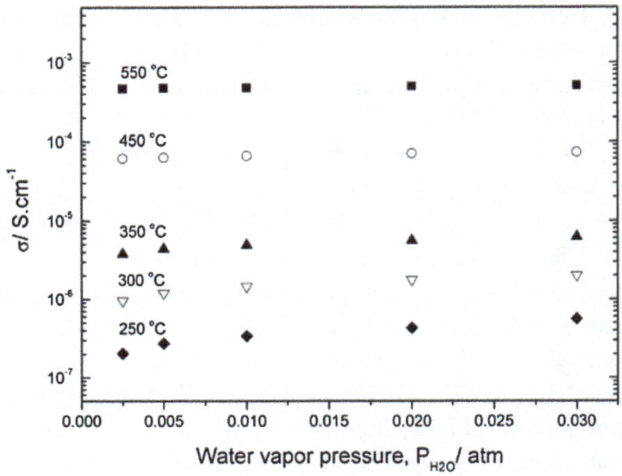

Fig. 16 Total conductivity of $La_2Ce_2O_7$ in air depending on pH_2O (Sun et al. 2012)

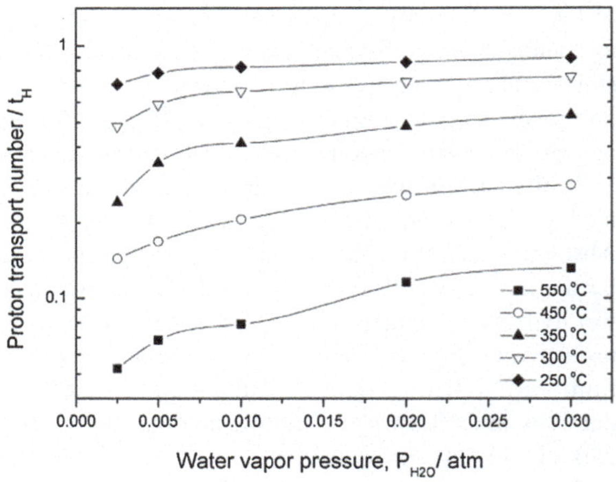

Fig. 17 Proton transport numbers of $La_2Ce_2O_7$ in air depending on pH_2O (Sun et al. 2012)

atmospheres, even at relatively high temperatures (for $La_{1.98}Ca_{0.02}Zr_{1.98}Ca_{0.02}O_{7-\delta}$, $t_H = 1$ at 200–600 °C and $t_H = 0.82$ at 800 °C) (Omata et al. 1997). In spite of this, pyrochlore systems have not been widely used as components for intermediate-temperature electrochemical devices, mostly because of their (Eurenius et al. 2010a, b, c; Omata et al. 2004; Omata et al. 1997): (i) very low absolute value of proton conductivity (10^{-7}–10^{-4} S cm^{-1} between 300 and 800°C); (ii) low grain-boundary conductivity; and (iii) unstable oxides in reducing atmospheres due to the material decomposition in the case of Sn-containing pyrochlores and the increase of unwanted electron conductivity as a result of Ti-ions reduction in the case of Ti-based ones.

5 Useful Concluding Remarks

Proton transport is found to be a characteristic feature of many materials with different structures and chemical compositions. Many research groups devoted the last decade particular attention in designing and developing this kind of oxide materials, examining the proton conductivity phenomenon and their perspective applications in electrochemical devices.

Among the different high-temperature proton-conducting systems, $BaCe_{1-x}Ln_xO_{3-\delta}$ materials (lanthanide-doped $BaCeO_3$) possess the highest ionic (oxygen ionic and protonic) conductivity. At first sight, this is favorable for their applications as electrolytes in solid oxide fuel cells. However, these materials exhibit a low chemical stability, which—conversely—limits potential applications. The double doping (co-doping) of $BaCeO_3$ has been widely studied because this strategy permits simultaneously affecting a number of key properties of the materials, including stability. The acceptor (In^{3+}), isovalent (Sn^{4+}, Ti^{4+}), and donor (Nb^{5+}, Ta^{5+}) dopants can be used in order to enhance chemical stability of $BaCeO_3$-based oxides. But the optimal concentration of such dopants should be verified due to negative impact on transport properties of the $BaCeO_3$ system.

The electrolytes based on $BaZrO_3$ are characterized by excellent thermodynamic stability and relatively high proton (bulk) conductivity in the temperature range 300–600 °C. However, the large grain-boundary resistance has a significant impact on the transport processes. For this reason, the total conductivity of ceramics is sensitive even to weak changes of the technological factors (such as the properties of precursors, synthesis methods, and temperature regimes), varying sometimes by ~ 1 order of magnitude for nominally the same compositions.

Partial substitution of cerium with zirconium in $BaCeO_3$ is one of the best ways of optimizing transport and mechanical properties enhancing also stability toward CO_2 H_2O and H_2S. The $BaCeO_3$–$BaZrO_3$ proton-conducting materials are currently state-of-the-art electrolytes possessing the most optimal combination of functional properties required for applied aspects. The SOFCs based on this system demonstrate promising electrochemical characteristics (Table 9) coupled with splendid long-term stability.

$LaScO_3$, $Ba_2In_2O_5$, $LaNbO_4$ also can be considered as electrolyte materials for solid oxide electrochemical devices, although their conductivity is much lower than those for the above-mentioned systems. On the other hand, these oxides demonstrate some advantages which consist of a high chemical stability and, in some cases, a purely ionic (co-ionic) transport under wide ranges of temperatures and oxygen partial pressures. The material optimization strategies should be used for extending the scope of their possible application. The proton transport is also possible for systems (CeO_2 with fluorite structure and $La_2Zr_2O_5$ with pyrochlore structure) that have traditionally been considered as oxygen ionic electrolytes. Generally, the proton conductivity for their doped analogs does not exceed 10^{-5} S cm^{-1} at 600–1000 °C.

Table 9 Parameters and performance characteristics of anode-supported SOFCs based on proton-conducting electrolytes at 600 °C

Cathode	Electrolyte	h (μm)	Anode substrate	OCV (V)	P_{max} (mW cm^{-2})	R_o (Ω cm^2)	R_p (Ω cm^2)	R_t (Ω cm^2)	References
$BaCeO_3$-based electrolytes									
$LaSr_3Co_{1.5}Fe_{1.5}O_{10}$–$BaCe_{0.7}Zr_{0.1}Y_{0.2}O_{3-\delta}$	$BaCe_{0.8}Ga_{0.2}O_{3-\delta}$ (BCG)	50	NiO–BCG	1.04	99	1.22	2.09	3.29	Tao et al. (2010)
$La_{0.6}Sr_{0.4}Co_{0.2}Fe_{0.8}O_{3-\delta}$	$BaCe_{0.8}Sm_{0.2}O_{3-\delta}$ (BCS)	39	NiO–BCZY	1.05	185	0.62	0.60	1.22	Su et al. (2015)
$Ba_{0.5}Sr_{0.5}Co_{0.8}Fe_{0.2}O_{3-\delta}$	BCS	50	NiO–BCS	1.05	132	1.22	2.28	3.43	Ranran et al. (2006)
$LaSr_3Co_{1.5}Fe_{1.5}O_{10}$	$BaCe_{0.7}In_{0.3}O_{3-\delta}$ (BCI)	20	NiO–BCI	1.02	213	–	0.81	–	Bi et al. (2009)
$L_{a0.6}Sr_{0.4}Co_{0.2}Fe_{0.8}O_{3-\delta}$	$BaCe_{0.95}Tb_{0.05}O_{3-\delta}$ (BCT)	12	Ni-BCT	1.09	218	0.18	0.84	1.02	Meng et al. (2011)
$BaZrO_3$-based electrolytes									
$La_{0.6}Sr_{0.4}Co_{0.2}Fe_{0.8}O_{3-\delta}$–BZY	$BaZr_{0.8}Y_{0.2}O_{3-\delta}$ (BZY) + Li	25	NiO–BZY	1.03	15	3.9	6.8	10.7	Sun et al. (2011)
$Sm_{0.5}Sr_{0.5}CoO_{3-\delta}$–$Ce_{0.8}Sm_{0.2}O_{2-\delta}$	$BaZr_{0.8}Y_{0.2}O_{3-\delta}$ (BZY)	20	NiO–BCZY	1.01	70	1.4	1.3	2.7	Sun et al. (2010)
$La_{0.6}Sr_{0.4}Co_{0.2}Fe_{0.8}O_{3-\delta}$–BZPY	$BaZr_{0.7}Pr_{0.1}Y_{0.2}O_{3-\delta}$ (BZPY)	20	NiO–BCZY	0.93	81	1.3	1.3	2.6	Fabbri et al. (2011)
$PrBaCo_2O_{5+\delta}$–$BaZr_{0.7}Pr_{0.1}Y_{0.2}O_{3-\delta}$	$BaZr_{0.7}In_{0.3}O_{3-\delta}$ (BZI)	15	NiO–BZI	0.95	84	2.0	0.3	2.3	Bi et al. (2011)
$La_{0.6}Sr_{0.4}Co_{0.2}Fe_{0.8}O_{3-\delta}$–$BaCe_{0.9}Yb_{0.1}O_{3-\delta}$	BZY	4	NiO–BZY	0.99	110	1.85	0.56	2.41	Pergolesi et al. (2010)
$Sm_{0.5}Sr_{0.5}CoO_{3-\delta}$–$Ce_{0.8}Sm_{0.2}O_{2-\delta}$	$BaZr_{0.7}Sn_{0.1}Y_{0.2}O_{3-\delta}$ (BZSY)	12	NiO–BZSY	0.98	214	0.46	0.41	0.87	Sun et al. (2013)
$La_{0.6}Sr_{0.4}Co_{0.2}Fe_{0.8}O_{3-\delta}$–$Ce_{0.8}Sm_{0.2}O_{2-\delta}$	BZY	6.6	NiO–BZY	0.95	246	0.7	0.2	0.9	Bae and Choi (2015)

(continued)

Table 9 (continued)

Cathode	Electrolyte	h (μm)	Anode substrate	OCV (V)	P_{max} (mW cm^{-2})	R_o (Ω cm^2)	R_p (Ω cm^2)	R_t (Ω cm^2)	References
Ba(Ce, Zr)O_3-based electrolytes									
$GdBaCo_2O_{5+\delta}$	$BaCe_{0.7}Zr_{0.1}Y_{0.2}O_{3-\delta}$ (BCZY)	10	NiO–BCZY	1.02	146	1.00	0.60	1.60	Lin et al. (2008)
$NdBaFe_{1.9}Mn_{0.1}O_{5+\delta}$	BCZY	30	NiO–BCZY	1.04	211	0.62	0.17	0.79	Mao et al. (2015)
$Y_{0.8}Ca_{0.2}BaCo_4O_{7+\delta}$	BCZY	20	NiO–BCZY	1.04	256	0.42	0.71	1.13	Shao et al. (2015)
$GdBa_{0.5}Sr_{0.5}Co_2O_{5+\delta}$–BCZY	BCZY	20	NiO–BCZY	1.02	335	0.45	–	–	Zhang et al. (2013)
$Sm_{0.5}Sr_{0.5}CoO_{3-\delta}$	BCZY	17	NiO–BCZY	1.04	747	0.37	0.13	0.50	Nien et al. (2011)

The basic type of cells, as a rule, is: wet H_2, anode|electrolyte|cathode, air. h represents the electrolyte thickness, OCV is the open circuit voltage, P_{max} is the peak power density, and R_o, R_p and R_t are the cell's ohmic, polarization, and total resistance. Anodes are produced from H_2-reduced anode substrates, containing 60–65 wt% NiO (continue)

Table 10 Comparison of target properties of the proton-conducting electrolytes

System	Advantages	Disadvantages
$BaCeO_3$	High oxygen ionic and protonic conductivity, enough densification	Low chemical stability, and the presence of phase transitions and non-monotonic thermal expansion
$BaZrO_3$	High chemical stability, high bulk conductivity	Poor sinterability, high sintering temperatures, low grain-boundary conductivity, low TEC value, high contribution of p-type electronic conductivity
$BaCeO_3$–$BaZrO_3$	Acceptable stability, thermal, and transport properties	There is a few information concerning chemical compatibility with other components (cathode, anode, interconnect, and seal materials)
$A(Ce,Zr)O_3$, A = Ca, Sr	Acceptable protonic conductivity	High contribution of p-type electronic conductivity, easy Ce^{4+}-reduction, narrow electrolytic boundary
$LaMeO_3$ (Me = Sc, In, Y…)	Wide electrolytic domain boundary, low contribution of n- or p-type conductivity, high chemical stability	Low absolute value of protonic conductivity, no information concerning chemical compatibility with other functional materials

It should be noticed that the materials considered here still possess certain drawbacks (Table 10), which at the moment limit their real applications. Future investigations should encompass the following key areas: material optimization, novel synthesis/processing routes and dense thin film preparation, chemical and thermal compatibility with other functional elements of electrochemical cells, and long-term and RedOx stability.

Acknowledgments This chapter is prepared in the framework of the following projects: the *Ministry of Education and Science of the Russian Federation* (Mega-grant contract no. 14. Z50.31.0001), and the *Russian Foundation for Basic Research* (projects no. 16-33-00006, 17-53-560008, and 17-03-00047).

References

Amsif, M., Marrero-Lopez, D., Ruiz-Morales, J., Savvin, S., Gabás, M., & Nunez, P. (2011). Influence of rare-earth doping on the microstructure and conductivity of $BaCe_{0.9}Ln_{0.1}O_{3-\delta}$ proton conductors. *Journal of Power Sources, 196*, 3461–3469.

Amsif, M., Marrero-López, D., Ruiz-Morales, J. C., Savvin, S. N., & Núñez, P. (2014). The effect of Zn addition on the structure and transport properties of $BaCe_{0.9-x}Zr_xY_{0.1}O_{3-\delta}$. *Journal of the European Ceramic Society, 34*, 1553–1562.

Andersson, A. K., Selbach, S. M., Knee, C. S., & Grande, T. (2014). Chemical Expansion Due to Hydration of Proton-Conducting Perovskite Oxide Ceramics. *Journal of the American Ceramic Society, 97*, 2654–2661.

Animitsa, I., Tarasova, N., & Filinkova, Y. (2012). Electrical properties of the fluorine-doped $Ba_2In_2O_5$. *Solid State Ionics, 207*, 29–37.
Babilo, P., Uda, T., & Haile, S. M. (2007). Processing of yttrium-doped barium zirconate for high proton conductivity. *Journal of Materials Research, 22*, 1322–1330.
Bae, H., & Choi, G. M. (2015). Novel modification of anode microstructure for proton-conducting solid oxide fuel cells with $BaZr_{0.8}Y_{0.2}O_{3-\delta}$ electrolytes. *Journal of Power Sources, 285*, 431–438.
Bannykh, A., & Kuzin, B. (2003). Electrical conductivity of $BaCe_{0.9}Nd_{0.1}O_{3-\alpha}$ in $H_2 + H_2O + Ar$ gas mixture. *Ionics, 9*, 134–139.
Bi, L., Fabbri, E., Sun, Z., & Traversa, E. (2011). Sinteractivity, proton conductivity and chemical stability of $BaZr_{0.7}In_{0.3}O_{3-\delta}$ for solid oxide fuel cells (SOFCs). *Solid State Ionics, 196*, 59–64.
Bi, L., Fang, S., Tao, Z., Zhang, S., Peng, R., & Liu, W. (2009a). Influence of anode pore forming additives on the densification of supported $BaCe_{0.7}Ta_{0.1}Y_{0.2}O_{3-\delta}$ electrolyte membranes based on a solid state reaction. *Journal of the European Ceramic Society, 29*, 2567–2573.
Bi, L., & Traversa, E. (2014). Synthesis strategies for improving the performance of doped-$BaZrO_3$ materials in solid oxide fuel cell applications. *Journal of Materials Research, 29*, 1–15.
Bi, L., Zhang, S., Zhang, L., Tao, Z., Wang, H., & Liu, W. (2009). Indium as an ideal functional dopant for a proton-conducting solid oxide fuel cell. *International Journal of Hydrogen Energy, 34*, 2421–2425.
Bi, Z., Pena-Martinez, J., Kim, J.-H., Bridges, C. A., Huq, A., Hodges, J. P., et al. (2012). Effect of Ca doping on the electrical conductivity of the high temperature proton conductor $LaNbO_4$. *International Journal of Hydrogen Energy, 37*, 12751–12759.
Bishop, S., Duncan, K., & Wachsman, E. (2009). Surface and bulk oxygen non-stoichiometry and bulk chemical expansion in gadolinium-doped cerium oxide. *Acta Materialia, 57*, 3596–3605.
Bishop, S., Marrocchelli, D., Chatzichristodoulou, C., Perry, N., Mogensen, M. B., Tuller, H., et al. (2014). Chemical expansion: Implications for electrochemical energy storage and conversion devices. *Annual Review of Materials Research, 44*, 205–239.
Bonanos, N., Ellis, B., Knight, K., & Mahmood, M. (1989). Ionic conductivity of gadolinium-doped barium cerate perovskites. *Solid State Ionics, 35*, 179–188.
Bonanos, N., & Poulsen, F. W. (1999). Considerations of defect equilibria in high temperature proton-conducting cerates. *Journal of Materials Chemistry, 9*, 431–434.
Brandao, A., Gracio, J., Mather, G., Kharton, V., & Fagg, D. (2011). B-site substitutions in $LaNb_{1-x}M_xO_{4-\delta}$ materials in the search for potential proton conductors (M = Ga, Ge, Si, B, Ti, Zr, P, Al). *Journal of Solid State Chemistry, 184*, 863–870.
Cervera, R. B., Oyama, Y., Miyoshi, S., Kobayashi, K., Yagi, T., & Yamaguchi, S. (2008). Structural study and proton transport of bulk nanograined Y-doped $BaZrO_3$ oxide protonics materials. *Solid State Ionics, 179*, 236–242.
Chandra, S. (1984). Fast proton transport in solids. Materials Science Forum. *Trans Tech Publ*, 153–169.
Chi, X., Zhang, J., Wen, Z., Liu, Y., Wu, M., & Wu, X. (2013a). Influence of In doping on the structure, stability and electrical conduction behavior of Ba $(Ce, Ti)O_3$ solid solution. *Journal of Alloys and Compounds, 554*, 378–384.
Chi, X., Zhang, J., Wu, M., Liu, Y., & Wen, Z. (2013b). Study on stability and electrical performance of yttrium and bismuth co-doped $BaCeO_3$. *Ceramics International, 39*, 4899–4906.
Choi, S. M., Lee, J.-H., Hong, J., Kim, H., Yoon, K. J., Kim, B.-K., et al. (2014). Effect of sintering atmosphere on phase stability, and electrical conductivity of proton-conducting Ba $(Zr_{0.84}Y_{0.15}Cu_{0.01})O_{3-\delta}$. *International Journal of Hydrogen Energy, 39*, 7100–7108.
Dahl, P. I., Lein, H. L., Yu, Y., Tolchard, J., Grande, T., Einarsrud, M.-A., et al. (2011). Microstructural characterization and electrical properties of spray pyrolyzed conventionally sintered or hot-pressed $BaZrO_3$ and $BaZr_{0.9}Y_{0.1}O_{3-\delta}$. *Solid State Ionics, 182*, 32–40.
Demin, A., & Tsiakaras, P. (2001). Thermodynamic analysis of a hydrogen fed solid oxide fuel cell based on a proton conductor. *International Journal of Hydrogen Energy, 26*, 1103–1108.

Demin, A., Tsiakaras, P., Gorbova, E., & Hramova, S. (2004). A SOFC based on a co-ionic electrolyte. *Journal of Power Sources, 131*, 231–236.

Duval, S., Holtappels, P., Vogt, U., Pomjakushina, E., Conder, K., Stimming, U., et al. (2007). Electrical conductivity of the proton conductor $BaZr_{0.9}Y_{0.1}O_{3-\delta}$ obtained by high temperature annealing. *Solid State Ionics, 178*, 1437–1441.

Eurenius, K., Ahlberg, E., Ahmed, I., Eriksson, S. G., & Knee, C. S. (2010a). Investigation of proton conductivity in $Sm_{1.92}Ca_{0.08}Ti_2O_{7-\delta}$ and $Sm_2Ti_{1.92}Y_{0.08}O_{7-\delta}$ pyrochlores. *Solid State Ionics, 181*, 148–153.

Eurenius, K., Ahlberg, E., & Knee, C. (2010b). Proton conductivity in $Ln_{1.96}Ca_{0.04}Sn_2O_{7-\delta}$ (Ln = La, Sm, Yb) pyrochlores as a function of the lanthanide size. *Solid State Ionics, 181*, 1258–1263.

Eurenius, K., Ahlberg, E., & Knee, C. (2010c). Proton conductivity in $Sm_2Sn_2O_7$ pyrochlores. *Solid State Ionics, 181*, 1577–1585.

Fabbri, E., Bi, L., Tanaka, H., Pergolesi, D., & Traversa, E. (2011). Chemically stable Pr and Y Co-doped barium zirconate electrolytes with high proton conductivity for intermediate-temperature solid oxide fuel cells. *Advanced Functional Materials, 21*, 158–166.

Fabbri, E., D'Epifanio, A., Di Bartolomeo, E., Licoccia, S., & Traversa, E. (2008). Tailoring the chemical stability of $Ba(Ce_{0.8-x}Zr_x)Y_{0.2}O_{3-\delta}$ protonic conductors for intermediate temperature solid oxide fuel cells (IT-SOFCs). *Solid State Ionics, 179*, 558–564.

Fang, S., Bi, L., Wu, X., Gao, H., Chen, C., & Liu, W. (2008). Chemical stability and hydrogen permeation performance of Ni–$BaZr_{0.1}Ce_{0.7}Y_{0.2}O_{3-\delta}$ in an H_2S-containing atmosphere. *Journal of Power Sources, 183*, 126–132.

Figueiredo, F., & Marques, F. (2013). Electrolytes for solid oxide fuel cells. *Wiley Interdisciplinary Reviews: Energy and Environment, 2*, 52–72.

Fisher, J. G., Kim, D.-H., Lee, S., Nguyen, D., & Lee, J.-S. (2013). Reactive sintering of $BaY_{0.1}Zr_{0.9}O_{3-\delta}$ proton conducting ceramics with CuO liquid phase sintering aid. *Journal of Ceramic Processing Research, 14*, 703–706.

Gao, D., & Guo, R. (2010). Structural and electrochemical properties of yttrium-doped barium zirconate by addition of CuO. *Journal of Alloys and Compounds, 493*, 288–293.

Gorbova, E., Maragou, V., Medvedev, D., Demin, A., & Tsiakaras, P. (2008). Investigation of the protonic conduction in Sm doped $BaCeO_3$. *Journal of Power Sources, 181*, 207–213.

Gorelov, V., & Balakireva, V. (2009). Synthesis and properties of high-density protonic solid electrolyte $BaZr_{0.9}Y_{0.1}O_{3-\alpha}$. *Russian Journal of Electrochemistry, 45*, 476–482.

Gorelov, V., Balakireva, V., Kuz'min, A., & Plaksin, S. (2014). Electrical conductivity of $CaZr_{1-x}$ (x = 0.01–0.20) in dry and humid air. *Inorganic Materials, 50*, 495–502.

Goupil, G., Delahaye, T., Gauthier, G., Sala, B., & Joud, F. L. (2012). Stability study of possible air electrode materials for proton conducting electrochemical cells. *Solid State Ionics, 209*, 36–42.

Guo, Y., Lin, Y., Ran, R., & Shao, Z. (2009). Zirconium doping effect on the performance of proton-conducting $BaZr_yCe_{0.8-y}Y_{0.2}O_{3-\delta}$ ($0.0 \leq y \leq 0.8$) for fuel cell applications. *Journal of Power Sources, 193*, 400–407.

Haile, S., Staneff, G., & Ryu, K. (2001). Non-stoichiometry, grain boundary transport and chemical stability of proton conducting perovskites. *Journal of Materials Science, 36*, 1149–1160.

Han, D., Shinoda, K., & Uda, T. (2014). Dopant site occupancy and chemical expansion in rare earth-doped barium zirconate. *Journal of the American Ceramic Society, 97*, 643–650.

Hashimoto, T., Ueda, Y., Yoshinaga, M., Komazaki, K., Asaoka, K., & Wang, S. (2002). Observation of two kinds of structural phase transitions in the $Ba_2In_2O_5$ system. *Journal of the Electrochemical Society, 149*, A1381–A1384.

Haugsrud, R., & Norby, T. (2006). High-temperature proton conductivity in acceptor-doped $LaNbO_4$. *Solid State Ionics, 177*, 1129–1135.

Hibino, T., Mizutani, K., Yajima, T., & Iwahara, H. (1992). Characterization of proton in Y-doped SrZrO 3 polycrystal by IR spectroscopy. *Solid State Ionics, 58*, 85–88.

Hiraiwa, C., Han, D., Kuramitsu, A., Kuwabara, A., Takeuchi, H., Majima, M., et al. (2013). Chemical expansion and change in lattice constant of Y-doped BaZrO3 by hydration/dehydration reaction and final heat-treating temperature. *Journal of the American Ceramic Society, 96*, 879–884.

Huang, P.-N., & Petric, A. (1995). Electrical conduction of yttrium-doped strontium zirconate. *Journal of Materials Chemistry, 5*, 53–56.

Hui, R., Maric, R., Deces-Petit, C., Styles, E., Qu, W., Zhang, X., et al. (2006). Proton conduction in ceria-doped $Ba_2In_2O_5$ nanocrystalline ceramic at low temperature. *Journal of Power Sources, 161*, 40–46.

Huse, M., Norby, T., & Haugsrud, R. (2012). Effects of A and B site acceptor doping on hydration and proton mobility of $LaNbO_4$. *International Journal of Hydrogen Energy, 37*, 8004–8016.

Islam, M., Davies, R., Fisher, C., & Chadwick, A. (2001). Defects and protons in the CaZrO 3 perovskite and $Ba_2In_2O_5$ brownmillerite: Computer modelling and EXAFS studies. *Solid State Ionics, 145*, 333–338.

Ito, S., Watanabe, M., Saito, M., & Yamamura, H. (2012). Electrical properties of new brounmillerite-type $Ba_2In_{2-x}(Zn, Zr)_xO_5$ system. *Procedia Engineering, 36*, 68–73.

Ivanova, M. E., Meulenberg, W. A., Palisaitis, J., Sebold, D., Solís, C., Ziegner, M., et al. (2015). Functional properties of $La_{0.99}X_{0.01}Nb_{0.99}Al_{0.01}O_{4-\delta}$ and $La_{0.99}X_{0.01}Nb_{0.99}Ti_{0.01}O_{4-\delta}$ proton conductors where X is an alkaline earth cation. *Journal of the European Ceramic Society, 35*, 1239–1253.

Iwahara, H., Esaka, T., Uchida, H., & Maeda, N. (1981). Proton conduction in sintered oxides and its application to steam electrolysis for hydrogen production. *Solid State Ionics, 3*, 359–363.

Iwahara, H., Uchida, H., & Maeda, N. (1982). High temperature fuel and steam electrolysis cells using proton conductive solid electrolytes. *Journal of Power Sources, 7*, 293–301.

Iwahara, H., Uchida, H., & Tanaka, S. (1983). High temperature type proton conductor based on $SrCeO_3$ and its application to solid electrolyte fuel cells. *Solid State Ionics, 9*, 1021–1025.

Jankovic, J., Wilkinson, D. P., & Hui, R. (2012). Preparation and characterization of Ce-and La-doped $Ba_2In_2O_5$ as candidates for intermediate temperature (100–500 °C) solid proton conductors. *Journal of Power Sources, 201*, 49–58.

Kakinuma, K., Yamamura, H., Haneda, H., & Atake, T. (2001). Oxide-ion conductivity of $(Ba_{1-x}La_x)_2In_2O_{5+x}$ system based on brownmillerite structure. *Solid State Ionics, 140*, 301–306.

Kalyakin, A., Lyagaeva, J., Medvedev, D., Volkov, A., Demin, A., & Tsiakaras, P. (2016). Characterization of proton-conducting electrolyte based on $La_{0.9}Sr_{0.1}YO_{3-\delta}$ and its application in a hydrogen amperometric sensor. *Sensors and Actuators B: Chemical, 225*, 446–452.

Katahira, K., Kohchi, Y., Shimura, T., & Iwahara, H. (2000). Protonic conduction in Zr-substituted $BaCeO_3$. *Solid State Ionics, 138*, 91–98.

Kochetova, N., Spesivtseva, I., & Animitsa, I. (2013). Electrical properties of $Ba_2(In_{1-x}Al_x)_2O_5$ solid solutions. *Russian Journal of Electrochemistry, 49*, 176–180.

Kochetova, N., Animitsa, I., Medvedev, D., Demin, A., & Tsiakaras, P. (2016). Recent activity in the development of proton-conducting oxides for high-temperature applications. *RSC Advances, 6*, 73222–73268.

Kreuer, K.-D. (1996). Proton conductivity: Materials and applications. *Chemistry of Materials, 8*, 610–641.

Kreuer, K.-D., Fuchs, A., & Maier, J. (1995). HD isotope effect of proton conductivity and proton conduction mechanism in oxides. *Solid State Ionics, 77*, 157–162.

Kreuer, K.-D., Münch, W., Fuchs, A., Klock, U., & Maier, J. (2001). Proton conducting alkaline earth zirconates and titanates for high drain electrochemical applications. *Solid State Ionics, 145*, 295–306.

Kreuer, K. (1997). On the development of proton conducting materials for technological applications. *Solid State Ionics, 97*, 1–15.

Kreuer, K. (1999). Aspects of the formation and mobility of protonic charge carriers and the stability of perovskite-type oxides. *Solid State Ionics, 125*, 285–302.

Kreuer, K. (2003). Proton-conducting oxides. *Annual Review of Materials Research, 33*, 333–359.

Kröger, F. A. (1964). *The chemistry of imperfect crystals*. Amsterdam: North-Holland Publishing Company.

Krug, F., & Schober, T. (1996). The high-temperature proton conductor Ba3(Ca1.18Nb1.82) O9 − gd: Thermogravimetry of the water uptake. *Solid State Ionics, 92*, 297–302.

Kuzin, B., Beresnev, S., Bannykh, A., & Perfil'yev, M. (2000). Transport numbers of H^+ and O^{2-}in the electrochemical system (H_2 + H_2O), Me/$BaCe_{0.9}Nd_{0.1}O_{3-\alpha}$/Me,($H_2$ + H_2O). *Russian Journal of Electrochemistry, 36*, 424–430.

Kuzmin, A., Balakireva, V., Plaksin, S., & Gorelov, V. (2009). Total and hole conductivity in the $BaZr_{1-x}Y_xO_{3-\alpha}$ system (x = 0.02 − 0.20) in oxidizing atmosphere. *Russian Journal of Electrochemistry, 45*, 1351–1357.

Lagaeva, J., Medvedev, D., Demin, A., & Tsiakaras, P. (2015). Insights on thermal and transport features of $BaCe_{0.8-x}Zr_xY_{0.2}O_{3-\delta}$ proton-conducting materials. *Journal of Power Sources, 278*, 436–444.

Li, J., Luo, J.-L., Chuang, K. T., & Sanger, A. R. (2008). Chemical stability of Y-doped Ba (Ce, Zr)O_3 perovskites in H_2S-containing H_2. *Electrochimica Acta, 53*, 3701–3707.

Li, J., Yoon, H., & Wachsman, E. D. (2011). Hydrogen permeation through thin supported $SrCe_{0.7}Zr_{0.2}Eu_{0.1}O_{3-\delta}$ membranes; dependence of flux on defect equilibria and operating conditions. *Journal of Membrane Science, 381*, 126–131.

Lim, D.-K., Choi, M.-B., Lee, K.-T., Yoon, H.-S., Wachsman, E., & Song, S.-J. (2011). Non-monotonic conductivity relaxation of proton-conducting $BaCe_{0.85}Y_{0.15}O_{3-\delta}$ upon hydration and dehydration. *International Journal of Hydrogen Energy, 36*, 9367–9373.

Lin, B., Zhang, S., Zhang, L., Bi, L., Ding, H., Liu, X., et al. (2008). Prontonic ceramic membrane fuel cells with layered $GdBaCo_2O_{5+x}$ cathode prepared by gel-casting and suspension spray. *Journal of Power Sources, 177*, 330–333.

Liu, X.-M., Liu, Z.-G., Ouyang, J.-H., Gu, Y.-J., Xiang, J., & Yan, F.-Y. (2012). Structure and electrical conductivity of $BaCe_{0.7}In_{0.1}A_{0.2}O_{3-\delta}$(A = Gd, Y) ceramics. *Electrochimica Acta, 59*, 464–469.

Lyagaeva, Y. G., Medvedev, D., Demin, A., Tsiakaras, P., & Reznitskikh, O. (2015). Thermal expansion of materials in the barium cerate-zirconate system. *Physics of the Solid State, 57*, 285–289.

Lybye, D., Poulsen, F. W., & Mogensen, M. (2000). Conductivity of A-and B-site doped $LaAlO_3$, $LaGaO_3$, $LaScO_3$ and $LaInO_3$ perovskites. *Solid State Ionics, 128*, 91–103.

Münch, W., Seifert, G., Kreuer, K., & Maier, J. (1996). A quantum molecular dynamics study of proton conduction phenomena in BaCeO 3. *Solid State Ionics, 86*, 647–652.

Ma, G., Shimura, T., & Iwahara, H. (1998). Ionic conduction and nonstoichiometry in $Ba_xCe_{0.90}Y_{0.10}O_{3-\alpha}$. *Solid State Ionics, 110*, 103–110.

Magrasó, A., Fontaine, M.-L., Bredesen, R., Haugsrud, R., & Norby, T. (2014). Cathode compatibility, operation, and stability of $LaNbO_4$-based proton conducting fuel cells. *Solid State Ionics, 262*, 382–387.

Magrasó, A., Fontaine, M. L., Larring, Y., Bredesen, R., Syvertsen, G., Lein, H., et al. (2011). Development of proton conducting SOFCs based on $LaNbO_4$ electrolyte–status in Norway. *Fuel Cells, 11*, 17–25.

Mahato, N., Banerjee, A., Gupta, A., Omar, S., & Balani, K. (2015). Progress in material selection for solid oxide fuel cell technology: A review. *Progress in Materials Science, 72*, 141–337.

Mao, X., Yu, T., & Ma, G. (2015). Performance of cobalt-free double-perovskite $NdBaFe_{2-x}Mn_xO_{5+\delta}$ cathode materials for proton-conducting IT-SOFC. *Journal of Alloys and Compounds, 637*, 286–290.

Matsumoto, H., Kawasaki, Y., Ito, N., Enoki, M., & Ishihara, T. (2007). Relation between electrical conductivity and chemical stability of BaCeO3-based proton conductors with different trivalent dopants. *Electrochemical and Solid-State Letters, 10*, B77–B80.

Medvedev, D., Lyagaeva, J., Gorbova, E., Demin, A., & Tsiakaras, P. (2016). Advanced materials for SOFC application: Strategies for the development of highly conductive and stable solid oxide proton electrolytes. *Progress in Materials Science, 75*, 38–79.

Medvedev, D., Lyagaeva, J., Plaksin, S., Demin, A., & Tsiakaras, P. (2015a). Sulfur and carbon tolerance of $BaCeO_3$–$BaZrO_3$ proton-conducting materials. *Journal of Power Sources, 273*, 716–723.

Medvedev, D., Murashkina, A., & Demin, A. (2015b). Formation of dense electrolytes based on $BaCeO_3$ and $BaZrO_3$ for application in solid oxide fuel cells: The role of solid-state reactive sintering. *Review Journal of Chemistry, 5*, 193–214.

Medvedev, D., Murashkina, A., Pikalova, E., Demin, A., Podias, A., & Tsiakaras, P. (2014a). $BaCeO_3$: Materials development, properties and application. *Progress in Materials Science, 60*, 72–129.

Medvedev, D., Pikalova, E., Demin, A., Podias, A., Korzun, I., Antonov, B., & Tsiakaras, P. (2014b). Structural, thermomechanical and electrical properties of new (1 − x) $Ce_{0.8}Nd_{0.2}O_{2-\delta-x}BaCe_{0.8}Nd_{0.2}O_{3-\delta}$ composites. *Journal of Power Sources, 267*, 269–279.

Meng, X., Yang, N., Song, J., Tan, X., Ma, Z.-F., & Li, K. (2011). Synthesis and characterization of terbium doped barium cerates as a proton conducting SOFC electrolyte. *International Journal of Hydrogen Energy, 36*, 13067–13072.

Mitsui, A., Miyayama, M., & Yanagida, H. (1987). Evaluation of the activation energy for proton conduction in perovskite-type oxides. *Solid State Ionics, 22*, 213–217.

Mokkelbost, T., Lein, H. L., Vullum, P. E., Holmestad, R., Grande, T., & Einarsrud, M.-A. (2009). Thermal and mechanical properties of $LaNbO_4$-based ceramics. *Ceramics International, 35*, 2877–2883.

Nien, S., Hsu, C., Chang, C., & Hwang, B. (2011). Preparation of $BaZr_{0.1}Ce_{0.7}Y_{0.2}O_{3-\delta}$ based solid oxide fuel cells with anode functional layers by tape casting. *Fuel Cells, 11*, 178–183.

Nigara, Y., Yashiro, K., Kawada, T., & Mizusaki, J. (2003). Hydrogen permeability in $(CeO_2)_{0.9}(GdO_{1.5})$ = at high temperatures. *Solid State Ionics, 159*, 135–141.

Niwa, J., Suehiro, T., Kishi, K., Ikeda, S., & Maeda, M. (2003). Structure and electrical characteristics of Ce^{4+}-doped $Ba_2In_2O_5$. *Journal of Materials Science, 38*, 3791–3795.

Norby, T., Haugsrud, R., Marion, S., Einarsrud, M. A., Wiik, K., Andersen, O., et al. (2013). Proton conductors. Google Patents.

Norby, T., & Magrasó, A. (2015). On the development of proton ceramic fuel cells based on Ca-doped $LaNbO_4$ as electrolyte. *Journal of Power Sources, 282*, 28–33.

Nowick, A., & Du, Y. (1995). High-temperature protonic conductors with perovskite-related structures. *Solid State Ionics, 77*, 137–146.

Okuyama, Y., Kozai, T., Ikeda, S., Matsuka, M., Sakai, T., & Matsumoto, H. (2014). Incorporation and conduction of proton in Sr-doped $LaMO_3$ (M = Al, Sc, In, Yb, Y). *Electrochimica Acta, 125*, 443–449.

Okuyama, Y., Kozai, T., Sakai, T., Matsuka, M., & Matsumoto, H. (2013). Proton transport properties of $La_{0.9}M_{0.1}YbO_{3-\delta}$ (M = Ba, Sr, Ca, Mg). *Electrochimica Acta, 95*, 54–59.

Omata, T., Fuke, T., & Otsuka-Yao-Matsuo, S. (2008). Mixed dopant effect in $SrZrO_3$-based proton conductor. *Solid State Ionics, 179*, 1116–1119.

Omata, T., Ikeda, K., Tokashiki, R., & Otsuka-Yao-Matsuo, S. (2004). Proton solubility for $La_2Zr_2O_7$ with a pyrochlore structure doped with a series of alkaline-earth ions. *Solid State Ionics, 167*, 389–397.

Omata, T., Okuda, K., Tsugimoto, S., & Otsuka-Matsuo-Yao, S. (1997). Water and hydrogen evolution properties and protonic conducting behaviors of Ca^{2+}-doped $La_2Zr_2O_7$ with a pyrochlore structure. *Solid State Ionics, 104*, 249–258.

Osman, N., Talib, I. A., & Hamid, H. (2010). Effect of zirconium substitution on the phase formation and microstructure of BaCeO3. *Sains Malaysiana, 39*, 479–484.

Park, H. J. (2011). Electrical properties of the protonic conductor 1 mol% Y-doped $BaZrO_{3-\delta}$. *Journal of Solid State Electrochemistry, 15*, 2205–2211.

Park, K.-Y., Seo, Y., Kim, K. B., Song, S.-J., Park, B., & Park, J.-Y. (2015). Enhanced proton conductivity of yttrium-doped barium zirconate with sinterability in protonic ceramic fuel cells. *Journal of Alloys and Compounds, 639*, 435–444.

Pasierb, P., Wierzbicka, M., Komornicki, S., & Rekas, M. (2009). Electrochemical impedance spectroscopy of $BaCeO_3$ modified by Ti and Y. *Journal of Power Sources, 194*, 31–37.

Patakangas, J., Ma, Y., Jing, Y., & Lund, P. (2014). Review and analysis of characterization methods and ionic conductivities for low-temperature solid oxide fuel cells (LT-SOFC). *Journal of Power Sources, 263*, 315–331.

Pelosato, R., Cordaro, G., Stucchi, D., Cristiani, C., & Dotelli, G. (2015). Cobalt based layered perovskites as cathode material for intermediate temperature solid oxide fuel cells: A brief review. *Journal of Power Sources, 298*, 46–67.

Peng, C., Melnik, J., Luo, J.-L., Sanger, A. R., & Chuang, K. T. (2010). $BaZr_{0.8}Y_{0.2}O_{3-\delta}$ electrolyte with and without ZnO sintering aid: Preparation and characterization. *Solid State Ionics, 181*, 1372–1377.

Pergolesi, D., Fabbri, E., & Traversa, E. (2010). Chemically stable anode-supported solid oxide fuel cells based on Y-doped barium zirconate thin films having improved performance. *Electrochemistry Communications, 12*, 977–980.

Qiu, L.-G., & Wang, M.-Y. (2010). Ionic conduction and fuel cell performance of $Ba_{0.98}Ce_{0.8}Tm_{0.2}O_{3-\alpha}$ Ceramic. *Chinese Journal of Chemical Physics, 23*, 707–712.

Qiu, L. G., Ma, G. L., & Wen, D. J. (2005). Properties and application of ceramic $BaCe_{0.8}Ho_{0.2}O_{3-\alpha}$. *Chinese Journal of Chemistry, 23*, 1641–1645.

Radojković, A., Žunić, M., Savić, S., Branković, G., & Branković, Z. (2013). Chemical stability and electrical properties of Nb doped $BaCe_{0.9}Y_{0.1}O_{3-\delta}$ as a high temperature proton conducting electrolyte for IT-SOFC. *Ceramics International, 39*, 307–313.

Ranran, P., Yan, W., Lizhai, Y., & Zongqiang, M. (2006). Electrochemical properties of intermediate-temperature SOFCs based on proton conducting Sm-doped $BaCeO_3$ electrolyte thin film. *Solid State Ionics, 177*, 389–393.

Ricote, S., Bonanos, N., Manerbino, A., & Coors, W. (2012). Conductivity study of dense $BaCe_xZr_{(0.9-x)}Y_{0.1}O_{(3-\delta)}$ prepared by solid state reactive sintering at 1500 C. *International Journal of Hydrogen Energy, 37*, 7954–7961.

Rolle, A., Fafilek, G., & Vannier, R. (2008). Redox stability of $Ba_2In_2O_5$-doped compounds. *Solid State Ionics, 179*, 113–119.

Ruiz-Trejo, E., & Kilner, J. A. (2009). Possible proton conduction in $Ce_{0.9}Gd_{0.1}O_{2-\delta}$ nanoceramics. *Journal of Applied Electrochemistry, 39*, 523–528.

Sakai, T., Isa, K., Matsuka, M., Kozai, T., Okuyama, Y., Ishihara, T., et al. (2013). Electrochemical hydrogen pumps using Ba doped $LaYbO_3$ type proton conducting electrolyte. *International Journal of Hydrogen Energy, 38*, 6842–6847.

Sammes, N., Phillips, R., & Smirnova, A. (2004). Proton conductivity in stoichiometric and sub-stoichiometric yittrium doped SrCeO 3 ceramic electrolytes. *Journal of Power Sources, 134*, 153–159.

Sawant, P., Varma, S., Wani, B., & Bharadwaj, S. (2012). Synthesis, stability and conductivity of $BaCe_{0.8-x}Zr_xY_{0.2}O_{3-\delta}$ as electrolyte for proton conducting SOFC. *International Journal of Hydrogen Energy, 37*, 3848–3856.

Schober, T., & Bohn, H. (2000). Water vapor solubility and electrochemical characterization of the high temperature proton conductor $BaZr_{0.9}Y_{0.1}O_{2.95}$. *Solid State Ionics, 127*, 351–360.

Shao, Q., Ge, W., Lu, X., Chen, Y., Ding, Y., Lin, B., et al. (2015). A promising cathode for proton-conducting intermediate temperature solid oxide fuel cells: $Y_{0.8}Ca_{0.2}$ BaCo 4 O 7 + δ. *Ceramics International, 41*, 6687–6692.

Sharova, N., & Gorelov, V. (2003). Electroconductivity and ion transport in protonic solid electrolytes $BaCe_{0.85}R_{0.15}O_{3-\delta}$, where R is a rare-earth element. *Russian Journal of Electrochemistry, 39*, 461–466.

Sharova, N., & Gorelov, V. (2004). Effect of cation nonstoichiometry on the properties of solid electrolyte $Ba_xCe_{0.97}Nd_{0.03}O_{3-\delta}$($0.90 \leq x \leq 1.10$). *Russian Journal of Electrochemistry, 40*, 639–645.

Sharova, N., & Gorelov, V. (2005). Characteristics of Proton-Conducting Electrolytes $BaCe_{1-x}Nd_x O_{3-\delta}$ ($0 \leq x \leq 0.16$) in moist air. *Russian Journal of Electrochemistry, 41*, 1001–1007.

Shima, D., & Haile, S. (1997). The influence of cation non-stoichiometry on the properties of undoped and gadolinia-doped barium cerate. *Solid State Ionics, 97*, 443–455.

Shin, J. F., Orera, A., Apperley, D., & Slater, P. (2011). Oxyanion doping strategies to enhance the ionic conductivity in $Ba_2In_2O_5$. *Journal of Materials Chemistry, 21*, 874–879.

Shrivastava, U. N., Duncan, K. L., & Chung, J. (2012). Experimentally validated numerical modeling of Eu doped $SrCeO_3$ membrane for hydrogen separation. *International Journal of Hydrogen Energy, 37*, 15350–15358.

Solís, C., & Serra, J. M. (2011). Adjusting the conduction properties of $La_{0.995}Ca_{0.005}NbO_{4-\delta}$ by doping for proton conducting fuel cells electrode operation. *Solid State Ionics, 190*, 38–45.

Stroeva, A. Y., Gorelov, V., & Balakireva, V. (2010). Conductivity of $La_{1-x}Sr_xSc_{1-y}Mg_yO_{3-\alpha}$ ($x = y = 0.01$–0.20) in reducing atmosphere. *Russian Journal of Electrochemistry, 46*, 784–788.

Su, F., Xia, C., & Peng, R. (2015). Novel fluoride-doped barium cerate applied as stable electrolyte in proton conducting solid oxide fuel cells. *Journal of the European Ceramic Society, 35*, 3553–3558.

Sun, W., Fang, S., Yan, L., & Liu, W. (2012). Investigation on proton conductivity of $La_2Ce_2O_7$ in wet atmosphere: Dependence on water vapor partial pressure. *Fuel Cells, 12*, 457–463.

Sun, W., Liu, M., & Liu, W. (2013). Chemically stable yttrium and tin co-doped barium zirconate electrolyte for next generation high performance proton-conducting solid oxide fuel cells. *Advanced Energy Materials, 3*, 1041–1050.

Sun, W., Yan, L., Shi, Z., Zhu, Z., & Liu, W. (2010). Fabrication and performance of a proton-conducting solid oxide fuel cell based on a thin $BaZr_{0.8}Y_{0.2}O_{3-\delta}$ electrolyte membrane. *Journal of Power Sources, 195*, 4727–4730.

Sun, Z., Fabbri, E., Bi, L., & Traversa, E. (2011). Lowering grain boundary resistance of $BaZr_{0.8}Y_{0.2}O_{3-\delta}$ with $LiNO_3$ sintering-aid improves proton conductivity for fuel cell operation. *Physical Chemistry Chemical Physics, 13*, 7692–7700.

Syvertsen, G. E., Magrasó, A., Haugsrud, R., Einarsrud, M.-A., & Grande, T. (2012). The effect of cation non-stoichiometry in $LaNbO_4$ materials. *International Journal of Hydrogen Energy, 37*, 8017–8026.

Ta, T. Q., Tsuji, T., & Yamamura, Y. (2006). Thermal and electrical properties of $Ba_2In_2O_5$ substituted for In-site by rare earth elements. *Journal of Alloys and Compounds, 408*, 253–256.

Takahashi, T., & Iwahara, H. (1979). Protonic conduction in perovksite-type oxide solid solutions. Abstracts of papers of the American Chemical Society. AMER CHEMICAL SOC 1155 16TH ST, NW, WASHINGTON, DC 20036, 207-207.

Tao, Z., Zhu, Z., Wang, H., & Liu, W. (2010). A stable $BaCeO_3$-based proton conductor for intermediate-temperature solid oxide fuel cells. *Journal of Power Sources, 195*, 3481–3484.

Tarasova, N., & Animitsa, I. (2015). Protonic transport in oxyfluorides Ba_2InO_3F and $Ba_3In_2O_5F_2$ with Ruddlesden-Popper structure. *Solid State Ionics, 275*, 53–57.

Tarasova, N., Animitsa, I., Denisova, T., & Nevmyvako, R. (2015). The influence of fluorine doping on short-range structure in brownmillerite $Ba_{1.95}In_2O_{4.9}F_{0.1}$. *Solid State Ionics, 275*, 47–52.

Tolchard, J., & Grande, T. (2007). Physicochemical compatibility of $SrCeO_3$ with potential SOFC cathodes. *Journal of Solid State Chemistry, 180*, 2808–2815.

Tong, J., Clark, D., Hoban, M., & O'Hayre, R. (2010). Cost-effective solid-state reactive sintering method for high conductivity proton conducting yttrium-doped barium zirconium ceramics. *Solid State Ionics, 181*, 496–503.

Tsai, C.-L., Kopczyk, M., Smith, R., & Schmidt, V. (2010). Low temperature sintering of Ba $(Zr_{0.8-x}Ce_xY_{0.2})O_{3-\delta}$ using lithium fluoride additive. *Solid State Ionics, 181*, 1083–1090.

Tsipis, E. V., & Kharton, V. V. (2008). Electrode materials and reaction mechanisms in solid oxide fuel cells: A brief review. *Journal of Solid State Electrochemistry, 12*, 1367–1391.

Uchida, H., Maeda, N., & Iwahara, H. (1982). Steam concentration cell using a high temperature type proton conductive solid electrolyte. *Journal of Applied Electrochemistry, 12*, 645–651.

Uchida, H., Maeda, N., & Iwahara, H. (1983). Relation between proton and hole conduction in $SrCeO_3$-based solid electrolytes under water-containing atmospheres at high temperatures. *Solid State Ionics, 11*, 117–124.

Uchida, H., Yoshikawa, H., & Iwahara, H. (1989). Formation of protons in SrCeO 3-based proton conducting oxides. Part I. Gas evolution and absorption in doped $SrCeO_3$ at high temperature. *Solid State Ionics, 34*, 103–110.

Van de Grotthuss, C. J. T. (2006). Memoir on the decomposition of water and of the bodies that it holds in solution by means of galvanic electricity. *Biochimica et Biophysica Acta (BBA) -. Bioenergetics, 1757*, 871–875.

Virkar, A., & Maiti, H. (1985). Oxygen ion conduction in pure and yttria-doped barium cerate. *Journal of Power Sources, 14*, 295–303.

Vullum, F., Nitsche, F., Selbach, S. M., & Grande, T. (2008). Solid solubility and phase transitions in the system $LaNb_{1-x}Ta_xO_4$. *Journal of Solid State Chemistry, 181*, 2580–2585.

Wang, M.-Y., & Qiu, L.-G. (2008). Mixed Conduction in $BaCe_{0.8}Pr_{0.2}O_3$-Ceramic. *Chinese Journal of Chemical Physics, 21*, 286–290.

Wang, S., Zhao, F., Zhang, L., & Chen, F. (2012). Synthesis of $BaCe_{0.7}Zr_{0.1}Y_{0.1}Yb_{0.1}O_{3-\delta}$ proton conducting ceramic by a modified Pechini method. *Solid State Ionics, 213*, 29–35.

Wu, J., Li, L., Espinosa, W., & Haile, S. (2004). Defect chemistry and transport properties of $Ba_xCe_{0.85}M_{0.15}O_{3-\delta}$. *Journal of Materials Research, 19*, 2366–2376.

Wu, Z., & Liu, M. (1997). Stability of $BaCe_{0.8}Gd_{0.2}O_3$ in a H_2O-containing atmosphere at intermediate temperatures. *Journal of the Electrochemical Society, 144*, 2170–2175.

Xie, K., Yan, R., Chen, X., Wang, S., Jiang, Y., Liu, X., et al. (2009). A stable and easily sintering $BaCeO_3$-based proton-conductive electrolyte. *Journal of Alloys and Compounds, 473*, 323–329.

Xing, W., Dahl, P. I., Roaas, L. V., Fontaine, M.-L., Larring, Y., Henriksen, P. P., et al. (2015). Hydrogen permeability of $SrCe_{0.7}Zr_{0.25}Ln_{0.05}O_{3-\delta}$ membranes (Ln = Tm and Yb). *Journal of Membrane Science, 473*, 327–332.

Yamanaka, S., Fujikane, M., Hamaguchi, T., Muta, H., Oyama, T., Matsuda, T., et al. (2003). Thermophysical properties of $BaZrO_3$ and $BaCeO_3$. *Journal of Alloys and Compounds, 359*, 109–113.

Yamazaki, Y., Hernandez-Sanchez, R., & Haile, S. M. (2009). High total proton conductivity in large-grained yttrium-doped barium zirconate. *Chemistry of Materials, 21*, 2755–2762.

Yamazaki, Y., Hernandez-Sanchez, R., & Haile, S. M. (2010). Cation non-stoichiometry in yttrium-doped barium zirconate: Phase behavior, microstructure, and proton conductivity. *Journal of Materials Chemistry, 20*, 8158–8166.

Yokokawa, H., Horita, T., Sakai, N., Yamaji, K., Brito, M., Xiong, Y.-P., et al. (2004). Protons in ceria and their roles in SOFC electrode reactions from thermodynamic and SIMS analyses. *Solid State Ionics, 174*, 205–221.

Yoo, C.-Y., Yun, D. S., Joo, J. H., & Yu, J. H. (2015). The effects of NiO addition on the structure and transport properties of proton conducting $BaZr_{0.8}Y_{0.2}O_{3-\delta}$. *Journal of Alloys and Compounds, 621*, 263–267.

Yoshinaga, M., Yamaguchi, M., Furuya, T., Wang, S., & Hashimoto, T. (2004). The electrical conductivity and structural phase transitions of cation-substituted $Ba_2In_2O_5$. *Solid State Ionics, 169*, 9–13.

Zając, W., Hanc, E., Gorzkowska-Sobas, A., Świerczek, K., & Molenda, J. (2012). Nd-doped Ba (Ce, Zr)$O_{3-\delta}$ proton conductors for application in conversion of CO_2 into liquid fuels. *Solid State Ionics, 225*, 297–303.

Zhang, C., Zhao, H., & Zhai, S. (2011). Electrical conduction behavior of proton conductor $BaCe_{1-x}Sm_xO_{3-\delta}$ in the intermediate temperature range. *International Journal of Hydrogen Energy, 36*, 3649–3657.

Zhang, G., & Smyth, D. (1995a). Defects and transport of the brownmillerite oxides with high oxygen ion conductivity—$Ba_2In_2O_5$. *Solid State Ionics, 82*, 161–172.

Zhang, G., & Smyth, D. (1995b). Protonic conduction in $Ba_2In_2O_5$. *Solid State Ionics, 82*, 153–160.

Zhang, X., Qiu, Y. E., Jin, F., Guo, F., Song, Y., & Zhu, B. (2013). A highly active anode functional layer for solid oxide fuel cells based on proton-conducting electrolyte $BaZr_{0.1}Ce_{0.7}Y_{0.2}O_{3-\delta}$. *Journal of Power Sources, 241*, 654–659.

Interconnects for Solid Oxide Fuel Cells

Angeliki Brouzgou, Anatoly Demin and Panagiotis Tsiakaras

Abstract Interconnect materials serve an important role in solid oxide fuel cell (SOFC) technology, connecting the current collectors of each cell to either the next one or the electrical load. Up to date, two types of interconnects have been developed: (i) the ceramic and (ii) the metallic ones. The ceramic interconnects are oxides, which are very stable in oxidizing atmosphere, but their cost is high and they exhibit lower electrical conductivity in comparison with the metallic ones at the operating temperatures. The most studied ceramic interconnects are lanthanum and yttrium chromites, and perovskite p-type semiconductors. Currently, ceramic conductive materials are widely used as thin protective layers deposited on metallic interconnects. The metallic interconnects are cheaper than the ceramic ones, and they are used at lower operating temperatures. Compared with ceramics, they exhibit higher electronic conductivity, but they are not stable in oxidizing atmospheres. During the last decade, several solutions have been approached, predominating the surface modification of the metallic interconnects via protective oxide layers (ceramic one) deposition on them. This alternative approach increases the lifetime of metallic interconnects, especially under cathode conditions. Reactive element oxides, perovskites, spinels, and dual layers are the kind of coatings that have also been developed.

A. Brouzgou · P. Tsiakaras (✉)
Laboratory of Alternative Energy Conversion Systems, Department of Mechanical Engineering, School of Engineering, University of Thessaly, Pedion Areos, 383 34 Volos, Greece
e-mail: tsiak@mie.uth.gr

A. Demin · P. Tsiakaras
Laboratory of Electrochemical Devices Based on Solid Oxide Proton Electrolytes, Institute of High Temperature Electrochemistry, Ural branch of the Russian Academy of Sciences 20, Akademicheskaya St., 620990 Ekaterinburg, Russia

M. Boaro and A.S. Aricò (eds.), *Advances in Medium and High Temperature Solid Oxide Fuel Cell Technology*, CISM International Centre for Mechanical Sciences 574,
DOI 10.1007/978-3-319-46146-5_4

1 Introduction

Solid oxide fuel cell (SOFC) is an emerging power generation technology. The most attractive characteristics of SOFCs, among the other types of fuel cells, are their high efficiency, their flexibility in fuel choice, including carbon-based fuels, and their high-quality heat by-products (excess heat from SOFCs can be utilized as a second power source). However, from materials' standpoint, SOFC technology is still the most demanding one (Gannon et al. 2007; Zhang et al. 2015; Wu et al. 2015).

For practical SOFC applications, in order to reach the desired voltage output, several cells are stacked mainly in series and an interconnect is required for connecting the single cells (Fig. 1) (Zhu et al. 2003a, b). Tubular and flat plate or planar are the most studied SOFC designs. In any case, for an effective fuel cell stack operation, interconnect materials consist of an important key point (Stambouli et al. 2002).

In the last two decades, low operating temperatures (<800 °C) have been recognized as an essential point for lowering the cost and avoiding fast degradation of SOFCs. Nevertheless, even at low operating temperatures, the identification and fabrication of suitable interconnect materials are very challenging (Fergus 2005). Interconnects are exposed to both anode and cathode atmospheres (Fig. 1), and they play an important role for the appropriate SOFC operation (Chen et al. 2005;

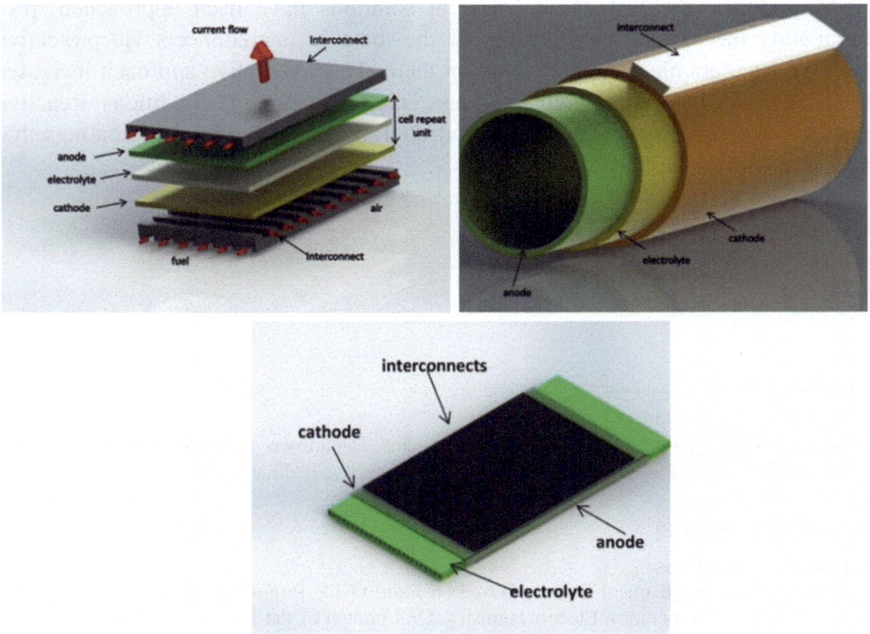

Fig. 1 Variable solid oxide fuel cell stack structures using interconnects

Fontana et al. 2007; Kurokawa et al. 2004; Gannon et al. 2004), demanding specific physicochemical and electrical properties, which are as follows:

1. *Chemical compatibility with the other SOFC components*: The interconnect material being in contact with both electrodes must be stable in both oxidizing and reducing atmospheres. Thus, any reaction or interdiffusion between interconnect and electrode materials should be excluded. In the opposite case, low conductive phases (mostly oxides) can be formed enhancing the electrical resistance and/or the mechanical degradation.
2. *High-temperature oxidation resistance in both anode and cathode atmospheres*: Poor oxidation resistance can lead to thermally grown of oxide scales and consequently to chemical and mechanical incompatibilities.
3. *High electrical and negligible ionic conductivity*: The acceptable area-specific resistance (ASR) level should be significantly lower than 100 mΩ cm^2.
4. *Thermal expansion compatibility with the other SOFC components*: The thermal expansion coefficient (TEC) value should match well with those of electrodes and electrolytes in order to avoid the generation of excessive stresses during the thermal cycling. In general, the most acceptable TEC values should be in the range of 9.9–10.6 $\times$ 10^{-6} K^{-1}.
5. *Adequate strength and creep resistance at elevated temperatures*: This is of special relevance to the planar SOFC where the interconnect serves as a structural support.
6. *Zero open porosity and sufficient mechanical strength*: Due to high operating temperatures and to avoid damages, good mechanical strength is necessary. Zero open porosity also helps avoiding reactant adsorption.
7. *Excellent imperviousness for oxygen and hydrogen*: It is to prevent direct combination of oxidant and fuel during fuel cell operation.
8. *Good thermal conductivity*: It is almost close to 5 Wm^{-1} K^{-1}.

Low-cost and easier methods of fabrication and inexpensive raw materials could be considered among the additional characteristics (Fergus 2005).

The following types of interconnects have been extensively studied and developed.

Ceramic interconnects (for operation at high temperatures: 900–1000 °C) are oxides stable in oxidizing atmospheres. However, compared with metallic interconnects, their cost is relatively high and they exhibit lower electrical conductivity at lower temperatures. They can be directly used as interconnect materials or can be deposited as a very thin protective layer between the metallic interconnect and the anode or the cathode. Doped lanthanum chromite ($LaCrO_3$) is the most widely investigated among the studied *ceramic interconnects* (Setz et al. 2015).

Metallic interconnects can be used only at intermediate temperatures (<800 °C), where the oxidation rate is quite low. They exhibit very good electrical conductivity, but in oxidizing atmospheres, they possess undesirable stability. Ferritic stainless steels (FSSs) that usually contain high chromium (Cr) percentage (10 and 26 wt%) are often used as metal interconnects. However, at high temperatures, Cr evaporates and tends to migrate to the three phase boundaries of cathode, promoting poisoning phenomena. Metallic interconnects usually require on the surface a protective coating against corrosion and SOFC's life time improvement. By lowering SOFC operation temperature ($\sim$600 °C or less), cheaper materials may be used (such as austenitic steels, which do not contain Cr) (Liu et al. 2009).

Coatings fall into three main categories: spinel oxide, perovskite, and reactive element oxide coatings. They are promising candidate materials for interconnects to be used in SOFC stacks. During long-term operation, material-related challenges may arise, including (i) surface instability (oxidation, spallation, and reactions with neighboring components), (ii) increased electrical resistance due to continuous oxide scale growth, and (iii) chromia-scale evaporation that may lead to cathode poisoning.

2 Ceramic Interconnects

As mentioned above, interconnects must be extremely stable, because they are exposed to both oxidizing and reducing sides of a solid oxide fuel cell. For this reason, for long-term operation, ceramics have been more successful than metals as interconnect materials. After many years of investigations, it has been concluded that chromites (more precisely $YCrO_3$ and $LaCrO_3$) are the most efficient ceramic interconnects, due to their acceptable stability under oxidizing conditions in a wide range of temperatures. However, in reducing atmospheres, a several number of oxides tend to lose oxygen, forming oxygen vacancies, and consequently dramatically decreasing their electrical conductivity (Singhal et al. 2003).

2.1 Lanthanum Chromite

At an early stage of the ceramic materials, research and development perovskite oxides have been considered among the best candidates. Doped $LaCrO_3$ belongs to the most promising interconnect materials as it exhibits acceptable conductivity at the operating temperatures and good stability in oxidizing and reducing atmospheres. More precisely, *lanthanum chromite* ($LaCrO_{3-\delta}$) has been found to be the most stable compound at low oxygen partial pressures, exhibiting a very high ($\sim$2450 °C) melting point (Hayashi et al. 1988). Moreover, $LaCrO_3$ exhibits good chemical compatibility with the stack components (sealant, electrolyte, and electrodes). However, at lower temperatures: (i) its not high conductivity, (ii) its mismatch TEC with the other components, and (iii) its difficulty to sinter to a high relative density, makes it insufficient for being used on its own. Additionally, $LaCrO_3$ exhibits low mechanical strength (Fergus 2004; Singhal et al. 2003).

Doping enhancement. Doping $LaCrO_3$ with alkaline earth elements (ions with lower valence-electron acceptors such as Ca, Mg, and Sr are substituted on either the La^{3+}or the Cr^{3+} sites) leads: (i) to an increase in p-type conductivity (conductivity due to small polaron hopping from room temperature at ~1400 °C and at oxygen pressures as low as 10^{-18} atm) and (ii) to a better matching of its TEC with the other SOFC components (Sakai et al. 2004). Among the most interesting properties of doped $LaCrO_3$ interconnects (especially those having the orthorhombic structure) are as follows: (i) high-melting points, (ii) good mechanical properties, and (iii) acceptable electrical conductivities at 1000 °C in oxidizing and reducing atmospheres (Fig. 2).

The effect of Sr, Ca, La, Y, Mg, Cu, Ni, Co, and other dopants on relevant properties for SOFC applications, including conductivity, chemical stability, thermal expansion and mechanical strength, is widely investigated in the last two decades.

Alkaline metals were conventionally doped at the La-site of $LaCrO_3$ and transition metals at the Cr-site. Ca and Sr are among the most used and preferable dopants exhibiting sufficient conductivity in fuel atmosphere that exceeds 1 S cm^{-1}. It was reported (Fergus 2004) that as the content of strontium and calcium increases, an increase in conductivity of lanthanum chromite at 1000 °C is observed. Ca has a more sound effect on electrical conductivity than Sr. In general,

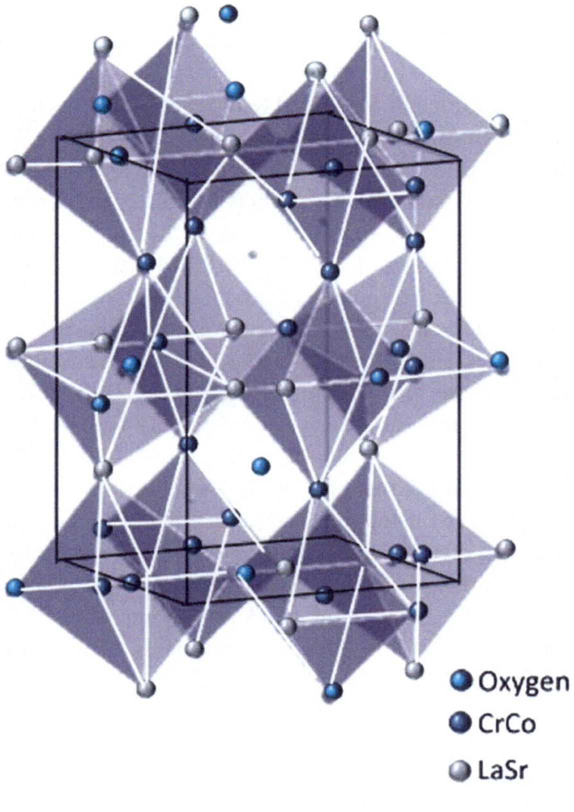

Fig. 2 Strontium and cobalt-doped lanthanum chromite orthorhombic unit cells

the electronic conductivity is enhanced as much closer is the radius of the introduced element to the radius of the host element, e.g., La in CrO_3.

It is noted that the extent of expansion due to oxygen loss is directly related to the oxygen vacancy concentration. Consequently, the substitution of divalent Ca and Sr with Y and La, respectively, induces the formation of small polarons as charge carriers, whereas the additional substitution of another element, e.g., Mn for Cr, results in the formation of a second charge carrier associated with Mn. As a consequence, TEC is improved (Singhal et al. 2003). Usually, when Sr is used, a second layer is formed when exposed in air, while when Ca is used only one single phase is formed. However, it has been stated that for Sr < 0.4, a stable single phase exists (Liu et al. 2000).

Mg, Cu, Ni, and Co have also been used as dopants for $LaCrO_3$. It was reported that the doping of $LaCrO_3$ with Mg leads to an increment of its conductivity, but not as much as when doped with Sr and Ca. The same results were observed for the doping of $LaCrO_3$ with Cu or Co (Fergus 2004). Mg, against to other alkaline earth elements, has much smaller radius and occupies Cr sites. TEC values of most of the above-mentioned materials are not acceptable.

Co-doping of chromites (Me-Sr-La chromite: Me = Co, Ca, etc.) can also enhance electrical conductivity. Very recently, Sr- and Co-doped $LaCrO_3$ materials with the composition of $La_{0.8}Sr_{0.2}Cr_{0.92}Co_{0.08}O_{3-\delta}$ (LSCC) were developed and tested (Setz et al. 2015). In this case, TEC values were high and close to 14.8×10^{-6} K^{-1}, while the electrical conductivity was close to 27.6 S cm^{-1} (at 950 °C in air).

Triple-doped $LaCrO_3$ led to more desirable results. Fe doping, at variable doping levels, in B-site of $La_{0.8}Ca_{0.2}Cr_{0.9}Co_{0.1}O_{3-\delta}$ (LCCC), resulted in a significant improvement in the electrical conductivity, about 67 S cm^{-1} (at 800 °C in air) and 71.9 S cm^{-1} (at 850 °C in air) (Fu et al. 2015). Up to date, this is the highest value reported in the literature. Indeed, as it can be distinguished in Fig. 3, the highest electrical conductivity is observed for Fe content close to 0.03. It is shown that for less or higher Fe content, electrical conductivity decreases. It was stated that at high oxygen activity, two compensatory mechanisms are activated: an ionic and an electronic one. The electronic/ionic ratio decreases as the Fe-doping level increases. The increase in Fe-doping level causes: (i) the formation of oxygen vacancy and (ii) the growth of the ionic conductivity contribution. Simultaneously, Fe doping has no significant effect on TEC value (11.47×10^{-6} K^{-1}) (Wei et al. 2014).

Fabrication and processing. An effort has been made to improve the properties of $LaCrO_3$ by using different fabrication methods. The main problem is related with the fact that Cr-containing oxides are difficult to sinter. More precisely, this problem derives from the vaporization of Cr-O species that leads to the enhancement of the evaporation–condensation mechanism of sintering. Consequently, the suppression of the densification causes powder coarsening. Thus, in order to achieve high densification level of $LaCrO_3$, many fabrication methods have been reported from the early stages, such as hot pressing (Singhal et al. 2003), co-sintering (Flandermeyer et al. 1988), electrochemical vapor deposition, plasma spraying, and laser ablation (Singhal 2000).

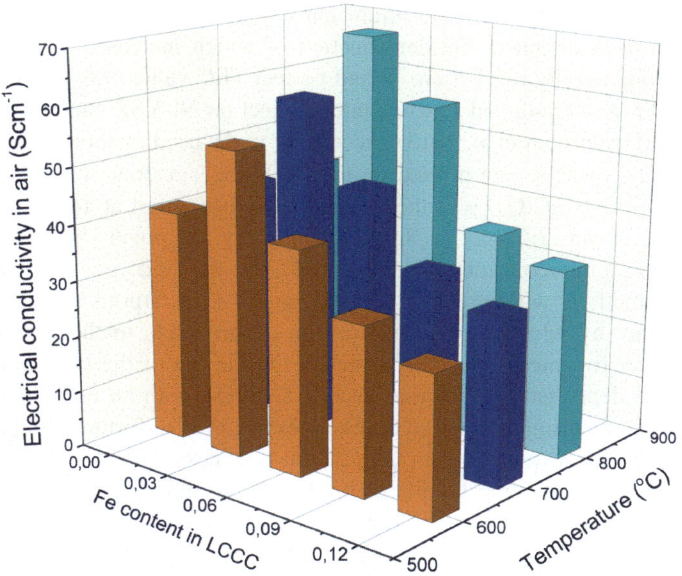

Fig. 3 Electrical conductivity for LCCCFe (Fe = 0–0.12) in air at 600, 700, and 800 °C (Fu et al. 2015)

The main disadvantages of the above-mentioned methods are mainly related with their high cost and the production of unstable interconnect materials. Up to 2006, the solid-state method has been the commonly used one: Lanthanum chromite was sintered at 1600 °C or above, presenting, however, no compatible TEC values with the common yttria zirconia electrolyte (Zhong 2006).

Today, novel preparation purposes are oriented toward: (i) stabilizing rhombohedral phase at room temperature, avoiding phase transition, and (ii) matching interconnects TEC to that of electrolyte in air atmosphere (Zhong 2006).

In this direction, relatively few research groups proposed cost-effective fabrication methods. In order to improve the properties of $LaCrO_3$, nano-sized La-doped $SrTiO_3$ powder was synthesized via Pechini method with citric acid. This powder was further used to form a thin interconnect layer on a porous Ni-YSZ anode support by the aid of a screen-printing process followed by co-sintering. With this preparation method, near-full densification ceramic materials were obtained, improving at a low extent the properties of $LaCrO_3$, as the interconnect material did not present any modification in reduction atmosphere (Park et al. 2012).

Very encouraging results, in terms of electrical conductivity and TEC values, were obtained by doping $LaCrO_3$ with Ca and Sr, after sintering them at temperatures lower than that for pure $LaCrO_3$ (Furtado et al. 2008). Furthermore, by adding Mg to single-doped LSC (lanthanum strontium chromite), the $La_{0.9}Sr_{0.10}Cr_{0.95}Mg_{0.05}O_3$ (LSCM) has been obtained, the densification of which increased, exhibiting both acceptable conductivity (22.1 S cm^{-1}) and perfect TEC value (9.8×10^{-6} K^{-1}).

A dense Sr- and Ca-doped $LaCrO_3$ interconnect on Ni-YSZ anode was recently investigated (Heidarpour et al. 2013). According to the used method, a base layer of $La_{0.8}Sr_{0.2}CrO_{3-\delta}$ was screen-printed on the NiO-YSZ substrate and a top layer of calcium chromite ($CaCrO_4$) was then coated and co-sintered at 1400 °C. Another dense single layer of $La_{1-x-y}Sr_xCa_yCrO_{3-\delta}$ by a novel bilayer $CaCrO_4$/ $La_{0.8}Sr_{0.2}CrO_{3-\delta}$ interconnect was prepared by the aid of a cost-effective co-sintering method with green NiO-YSZ anode (Heidarpour et al. 2012). In Fig. 4, the way by which the co-sintering procedure leads to the formation of a protective layer for interconnects is depicted. On the top of the interconnect layer ($LaSrCrO_3$) is deposited another top layer of $CaCrO_4$, with an initial thickness of 30 μm. During sintering, the top layer dissolves in $LaCrO_3$, forming finally a dense layer with few closed pores, of around 10 μm thickness. When a 50 μm $CaCrO_4$ top layer is deposited, some liquids of the initial Ca and Cr glycine–nitrate solutions, used during fabrication, sink down to the bottom of the deposited top layer ($CaCrO_4$) and they are mixed with the substrate, forming undesirable cracks.

Another approach for the fabrication of $La_{0.7}Ca_{0.3}CrO_3$, by reducing the sintering temperature using a combustion process for powder preparation, has also

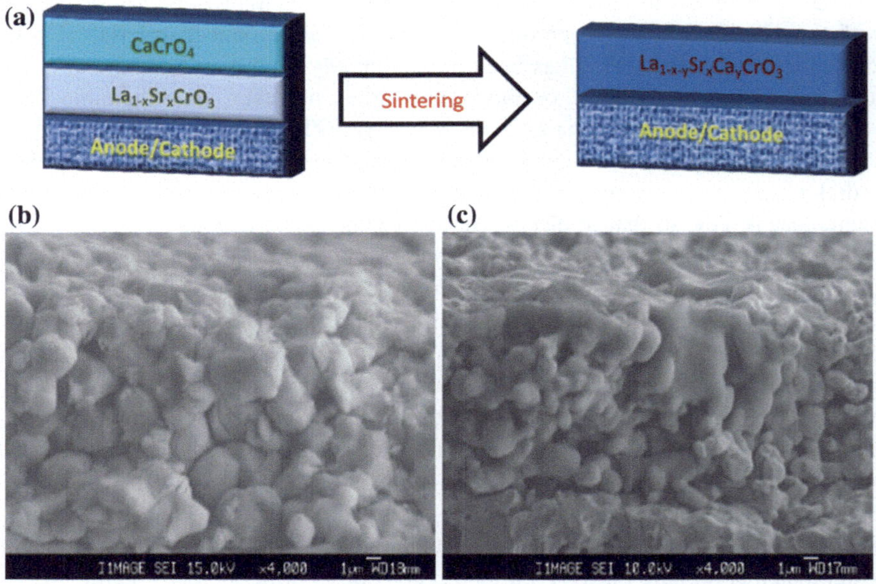

Fig. 4 Schematic representation of bilayer interconnect and anode before and after sintering (**a**). Cross-sectional SEM micrographs of LSC_{20}/CC with 30 and 50 μm $CaCrO_4$ top layer thickness, (**b**) and (**c**), respectively (Heidarpour et al. 2012)

been proposed (Nair et al. 2008). The conductivity of the prepared material was found to be close to 57 S cm^{-1} at 800 °C in air. The material maintained its single phase and a high density of 98 %. However, its low conductivity in fuel atmosphere and its chemical expansion still remain unsolved problems.

Despite the many years of research, there are still several problems to be solved such as (i) unmatched TEC with that of other components of the SOFC and (ii) insufficiently high conductivity.

Among the additional challenges that should be faced are also the following:

- their extremely inferior sintering behavior in air, which has been attributed to the formation of a thin layer of Cr_2O_3 at the interparticle necks during the initial stages of sintering and
- the reaction of $LaCrO_3$ with the electrolyte (e.g., YSZ) at high temperatures that causes the formation of a highly resistive lanthanum zirconate ($La_2Zr_2O_7$) phase, making impractical the cost-effective co-firing process (Zhu et al. 2003a, b).

Relatively few works concerning $LaCrO_3$ interconnects have been published in the last years, probably due to the appearance of the following new trends: the development of acceptor-doped yttrium chromite ceramic interconnects and the formation of very thin protective ceramic layers between the metallic interconnect and anode/cathode compartment (Zhu et al. 2003a, b).

2.2 *Yttrium Chromite*

$YCrO_3$, compared with $LaCrO_3$, is more stable under SOFC operational conditions, despite the fact that its melting point temperature is lower than that of $LaCrO_3$ (2300 °C against 2450 °C). At temperatures above 1000 °C, $YCrO_3$ starts to evaporate incongruently with the formation of an excess amount of Y_2O_3. Additionally, during the co-firing with SOFC's materials (usually at ~1400 °C), no other undesired chemical reaction-forming impure phases occurs (Wang et al. 2009a, b).

Among the investigated $YCrO_3$ interconnects, the one doped with Ca has been suggested for SOFC applications (Yoon et al. 2010). Ca-doped $YCrO_3$ (p-type conductor) conducts electricity via a small polaron hopping process. However, in order to avoid the formation of the secondary $CaCrO_4$ phase, Ca content should be kept constant (<25 %). This phase melts at 1228 °C during sintering, enhancing the densification (Wang et al. 2009a, b). Despite the fact that Ca-doped $YCrO_3$ exhibits lower electrical conductivity than $LaCrO_3$, its advantages such as higher mechanical strength (Paulik et al. 1999), higher stability (Armstrong et al. 1996), and better chemical compatibility with electrode's material/YSZ during co-sintering fabrication method (Wang et al. 2009a, b) make it more attractive.

In order to improve the electrical conductivity, many research groups used co-doping Ca-doped $YCrO_3$ with another element. For example, by adding Ni, the stability of $Y_{0.8}Ca_{0.2}Cr_{1-x}Ni_xO_{3\pm\delta}$ against reduction was improved; the oxygen vacancy formation was suppressed, resulting in the appearance of a significantly increased electrical conductivity of ~5.8 S cm^{-1}, in reducing environment at 900 °C. It was found that the content of Ni significantly affects the properties of the doped interconnect; when $x = 0.15$, the electrical conductivity reached its highest value of 34 S cm^{-1} (at 900 °C in air), with TEC = 11.5×10^{-6} K^{-1} (Yoon et al. 2010).

Zn was also used as a dopant element ($Y_{0.7}Ca_{0.3}Cr_{1-x}Zn_xO_{3-\delta}$), and when its content was $x = 0.10$, the highest electrical conductivity was measured (19 S cm^{-1} at 750 °C) and a TEC = 10.8×10^{-6} K^{-1}, a value close to that of YSZ electrolyte.

The doping with transition metals (Ni, Co, Cu), of Ca (20 at.%)-doped $YCrO_3$ ($Y_{0.8}Ca_{0.2}Cr_{0.85}Co_{0.1}Ni_{0.04}Cu_{0.01}O_3$) material, resulted in the closest match with the 8 mol% YSZ (TEC = 11.1×10^{-6} K^{-1} and 57 S cm^{-1} of electrical conductivity, at 900 °C in air atmosphere) (Yoon et al. 2011). Conclusively, the addition of Ni is responsible for the TEC improvement, while the addition of Co and Ni enhances the electrical conductivity. Moreover, Co improves the interconnect material toward stability in reducing atmospheres. Table 1 lists the thermal expansion coefficients and electrical conductivity of the above-discussed materials for interconnects.

As mentioned above, an acceptable TEC value falls in the range of 9.9–10.6 × 10^{-6} K^{-1} for the YSZ-based SOFCs and in the range of 8.0–9.8 × 10^{-6} K^{-1} for the SOFCs based on protonic electrolytes. However, as it is shown in Table 1, only the corresponding values of $La_{0.9}Sr_{0.10}Cr_{0.95}Mg_{0.05}O_3$ and $La_{0.803}Sr_{0.197}Cr_{0.92}Co_{0.08}O_{3-\delta}$ fall in the above range. The substitution of some

Table 1 Thermal expansion coefficient and electrical conductivity in air atmosphere of lanthanum and yttrium chromite interconnects

Material	Electrical conductivity in air (S cm^{-1})	TEC, (10^{-6} K^{-1})	T (°C)	Reference
$Y_{0.8}Ca_{0.2}Cr_{0.85}Co_{0.1}Ni_{0.04}Cu_{0.01}O_3$	57.0	11.1	900	Yoon et al. (2011)
$Y_{0.8}Ca_{0.2}Cr_{0.85}Ni_{0.15}O_{3-\delta}$	34.0	11.5	900	Yoon et al. (2010)
$La_{0.8}Sr_{0.2}Cr_{0.92}Co_{0.08}O_{3-\delta}$	27.6	14.8	1000	Setz et al. (2015)
$La_{0.9}Sr_{0.1}Cr_{0.95}Mg_{0.05}O_3$	22.1	9.8	1000	Furtado et al. (2008)
$La_{0.8}Sr_{0.2}Cr_{1-x}Fe_{0.5}O_{3-\delta}$	21.9	11.5	800	Wei et al. (2014)
$La_{0.803}Sr_{0.197}Cr_{0.92}Co_{0.08}O_{3-\delta}$	21.8	9.8	1000	Setz et al. (2015)
$Y_{0.7}Ca_{0.3}Cr_{0.9}Zn_{0.1}O_{3-\delta}$	19.0	10.8	750	Yoon et al. (2011)

elements improves TEC values, with their content to be a very sensitive factor. Moreover, it should be pointed out that their good electrical conductivity makes them appropriate materials for further investigation as interconnects.

3 Metallic Interconnects

Metallic interconnects have been set as the most suitable materials for improving the performance and cost-effectiveness of SOFCs, at operating temperatures <800 °C. They are characterized by relatively lower material and fabrication costs, easier and more complex shaping, and better electrical and thermal conductivity (in the range of 650–800 °C). During the last years, many alloys have been examined for the application as metallic interconnects.

Ferritic stainless steels are considered as the appropriate materials to be used as SOFC interconnects. In these materials, known as Fe–Cr alloys, the main alloying element is Cr, with contents typically between 10–26 %, although a higher Cr content of about 29 % is found in few specialized grades. However, the main drawback of these materials is that at high temperature, Cr evaporates and then it is deposited on the cathode surface of SOFC, damaging fuel cell's long-term stability.

Commercial AISI 430 ($\sim$16 wt% Cr) and Crofer 22 H ($\sim$22 wt% Cr), exhibiting TEC values closely matched to the most studied cell components, form a conductive double-layered oxide scale with a layer of loosely packed $(Mn, Cr)_3O_4$ on the top of dense Cr_2O_3 (Zhang et al. 2014a, b; Geng et al. 2006; Froitzheim et al. 2008; Rufner et al. 2008; Jian et al. 2006). However, they are lacking in sufficient oxidation resistance and electrical conductivity, causing cathode poisoning by Cr evaporated from the oxide scale layer. Improvements in surface modifications or coatings are often required for a long-term application with added fabrication complexity and extra cost (Hua et al. 2010, Zhang et al. 2011; Ajitdoss et al. 2013; Geng et al. 2012; Yang et al. 2012). Thus, the development of new alloys, especially with low-Cr content, is extremely desired.

An interesting research concerning commercial interconnect materials of Crofer 22 APU, Crofer 22H, Sanergy HT, ZMG232 G10, and E-Brite, exposed at 850 °C in a typical SOFC cathode-side environment (air + 3 % H_2O), was recently carried out (Sachitanand et al. 2013). All the investigated steels contained a Cr amount of about 22–26 % with balanced Fe. In all samples, an outer $(Cr,Mn)_3O_4$ spinel and an inner Cr_2O_3 corundum layer were formed. However, the spinel layer formed on the E-Brite steel was discontinuous in nature. It was observed that the steel with the higher Mn content (Crofer 22APU and Crofer 22H) leads to a lower Cr evaporation rate. Moreover, higher Mn content in the steel results in a thicker and Mn-rich spinel layer. It is believed that both factors seriously affect Cr evaporation rates.

Among the examined steels, E-Brite did not contain sufficient Mn for the formation of a continuous outer spinel layer. In this case, the Cr volatilization rate was found to be approximately four times higher than that for the other steels. It should be noted that from the examined commercial metallic interconnects, Crofer 22APU exhibited the highest mass gain, presenting no mass crackings. Different types of Crofer 22APU, such as K44M ferritic stainless steel, have also been investigated as possible interconnect materials (Piccardo et al. 2015).

Low-Cr alloy development. Ni-based and Cr-based alloys and FSSs have also been studied for their potential use as interconnect materials for solid oxide fuel cells (Fergus 2005). As mentioned above, TEC values of Ni-based alloys are considerably higher than those for the other SOFC components. Cr-based alloys are attractive when dispersed with stable oxides, but their fabrication is relatively expensive. Low-cost methods of deposition of Cr-based oxide dispersion-strengthened alloys on FSSs are the most attractive for the production of metallic interconnect materials.

In the absence of coatings, current metallic interconnects are not suitable for practical applications at <800 °C, since due to Cr-poisoning effect they may cause drastic performance degradation of the stack within its expected service lifetime (Zhu et al. 2003a, b). However, except for coatings' fabrication, another approach for overcoming poisoning effects from Cr evaporation is the development of low-Cr content materials.

Ardigo et al. (2015) examined the corrosion behavior and the electrical conductivity of a commercial stainless steel [K41X (AISI 441)] that contained 18 % of Cr in dual atmosphere (95 % O_2–5 % H_2O//10 %H_2–90 % H_2O, at 800 °C). They found that the alloy exhibits satisfactory oxidation resistance compared to the single atmosphere tests, as well as an acceptable ASR value of 100 mΩ cm^2, after 100 h at 800 °C. This ASR value can be attributed to the formation and the thickness of the oxide layer.

Zhang et al. (2014a, b) developed a novel low-Cr-containing Fe–Cr-Co alloy for intermediate temperature SOFCs. The formed oxide scale, after the exposure of the metallic interconnect at 750 °C for 1000 h in air, was lower than 1 μm, presenting a double-layered structure with dense $(Mn, Cr)_3O_4$ on the top of Cr_2O_3. In comparison with the commercial Crofer 22H alloy, Cr deposition in/around the lanthanum strontium manganite (LSM) electrode was remarkably reduced. The measured ASR was 14 mΩ cm^2. Additionally, the oxidation kinetics at 750 °C was found to obey the parabolic law, exhibiting a very low rate constant of 1.42×10^{-15} g^2 cm^{-4} s^{-1}. According to the authors, the significant enhancement of the oxidation resistance of Fe–Cr-Co alloy is due to the unique microstructure of the formed oxide scale.

Very recently, Jo et al. (2015) presented their results concerning a novel and low-cost ferritic stainless steel, named 460FC, being developed and fabricated by the conventional continuous casting and rolling process. The oxidation characteristics and high-temperature properties of 460FC were investigated at 800 °C and compared with those of Crofer 22. Although the as-prepared 460FC did not contain rare earth metals and did not limit the content of Si to extremely low level, it showed good electrical conductivity and low-Cr evaporation rate. More precisely, after oxidation at 800 °C for 500 h, the ASR was 12.5 and 13.1 mΩ cm^2 for the 460FC and the Crofer 22, respectively. The two oxidation rate constants were comparable with values of 1.20×10^{-10} g^2 cm^{-4} s^{-1} (460FC) and 1.12×10^{-10} g^2 cm^{-4} s^{-1} (Crofer 22).

Miguel-Pérez et al. (2012) compared Crofer 22APU (22 wt% Cr), Conicro 4023W188 (22 wt% Cr), and SS430 (18 wt% Cr) against the resistance to oxidation. After 100 h of exposure at 800 °C, the rate of oxidation was found to be 7.51×10^{-10}, 4.34×10^{-11}, and 1.12×10^{-7} g cm^{-4} s^{-1}, respectively; SS430 exhibited the lower resistance to oxidation.

Frangini et al. (2014) verified the feasibility of using low-Cr FSSs to improve the IT-SOFC's interconnect properties in terms of electrical conductivity and Cr evaporation resistance. They took the advantage of the low-Cr FSSs forming conductive and moisture-stable oxide layers of chromite spinels. According to their results for SS430, with 13.9 % Cr content after 2000 h of exposure at 800 °C, the Cr evaporation rate remained very low and stable at 1.9×10^{-14} g^2 cm^{-4} s^{-1}. The big difference reported for SS340 was mostly attributed in the different Cr contents.

Modification of Fe- and La/Cr-based alloys. According to the previous paragraphs, the behavior of metallic interconnects seems to be strongly dependent on the alloy composition, in particular on the percentage of Cr. It has been, however, reported that Fe–Cr interconnects with low-Cr content are susceptible to breakaway oxidation, due to hematite (Fe_2O_3)-phase nodular growth in the scale grown on the air side. In order to optimize Fe–Cr-based alloys with low-Cr content, the addition of Mn, Ti, Mo, Zr, La, and Y is considered as an alternative method to the known doping method. As an example (Hua et al. 2010), the addition of 1.05 wt% Mn, 0.52 wt% Ti, 2.09 wt% Mo, and other elements such as La, Y, and Zr into Fe–Cr alloy causes the formation of a multilayered structure with conductive Mn–Cr spinel in between the underneath Cr_2O_3 and the top Mn_2O_3 in the thermally grown oxide scale. For the alloy with the above composition, the oxidation rate constants obtained under both isothermal and cyclic oxidation condition were very low and within the range of 5.1×10^{-14} to 7.6×10^{-14} g^2 cm^{-4} s^{-1}, while the measured ASR value at 750 °C after 1000 h of oxidation was around 10 mΩ cm^2, lower than that measured for the conventional Fe–Cr stainless steels and comparable with that of the Ni-based alloys (Hua et al. 2010).

Ti and Nb have also been used to stabilize the ferritic structure (Wongpromrat et al. 2015). A typical dual-layer oxide scale, composed of a $(Mn,Cr)_3O_4$ spinel top layer and a Cr_2O_3-rich sublayer, was observed, with oxide nodules growing. After high-temperature oxidation of AISI 441 up to 24 h, nodules were observed on the top surface. Some of them were located above the substrate grain boundaries, while others were located above the heart of the grains. These nodules did not contain any iron and therefore were not sites of future catastrophic oxidation. The location of the nodules was strongly connected with the presence of niobium-containing phase(s) within the steel substrate (Wongpromrat et al. 2015).

The effect of Ti addition on the electrical conductivity and on Cr evaporation resistance was recently discussed (Seo et al. 2012). According to the published results, the addition of 1 wt% Ti promotes fast Cr_2O_3 growth due to the excess ionic defect in Cr_2O_3 matrix and the presence of a small amount of Ti in Cr_2O_3 matrix retards the oxidation rate by changing the oxidation kinetics. However, a high amount of Ti in Cr_2O_3 matrix is able to generate excess ionic defects, reducing the oxidation resistance. It was also reported that the co-addition of Ti and La enhances segregation of Ti near the scale/alloy interface; this interface segregation of Ti has beneficial effects, such as ASR and Cr evaporation reduction and FSS oxidation rate reduction. More precisely, the oxidation rate constant and the ASR value for the Cr_2O_3(La, Ti) sample, in ambient air at 800 °C, were found to be the optimal ones and equal to $0.58 \times 10^{-10} g^2\ mm^{-4}\ h^{-1}$ and 5.4 mΩ cm^2, respectively, while those for Crofer 22APU(La, Ti) were $1.62 \times 10^{-10} g^2\ mm^{-4}\ h^{-1}$ and 9.8 mΩ cm^2, respectively. After exposition in humid air for 100 h at 800 °C, Cr evaporation rate for Cr_2O_3(La, Ti) was $1.8 \times 10^{-7} g\ cm^{-2}\ h^{-1}$, while that for Crofer 22APU was $1.1 \times 10^{-7}\ g\ cm^{-2}\ h^{-1}$ and for the sample with high Ti (~1 wt%) and without La was $0.9 \times 10^{-7}\ g\ cm^{-2}\ h^{-1}$ (Seo et al. 2012).

Tungsten (W) is another element that has been added to Fe–Cr alloy. It was found that the high amount of W and low amount of Ti can reduce both the oxidation rate and the thickness of oxide scale and improve the ASR value. Among the investigated samples, the one with composition of Fe-19.8 Cr-0.44 Mn-0.18 C-0.23 Ti-3.98W exhibited at 800 °C the lowest Cr evaporation rate close to $3.0 \times 10^{-13}\ g^2\ cm^{-4}\ s^{-1}$ and an ASR value close to 8.5 mΩ cm^2 (Safikhani et al. 2014).

Ferritic stainless steel with composition of Fe-22Cr-0.5Mn (F) alloyed by (a third element) Nb (F-Nb) or by Mo (F-Mo) was evaluated (Yun et al. 2013) in a button cell configuration at 750 °C, in terms of degradation in ohmic resistance and cathodic polarization resistance; STS444 and Crofer 22APU were also evaluated for comparison reasons. F-Nb, F-Mo, and Crofer 22APU showed comparable ohmic resistance degradation rates 0.058, 0.066, and 0.074 Ω cm^2 k h^{-1}, respectively, and almost double of that measured on Pt under the same conditions. On the other hand, the ohmic resistance degradation rate for STS444 was found to be 0.134 Ω cm^2 kh^{-1}. Moreover, the degradation rates in the cathodic polarization resistance of the button cell determined by EIS after 400 h were 0.10, 0.06, 0.11, and 0.55 Ω cm^2 kh^{-1} for F-Nb, F-Mo, Crofer 22APU, and STS444, respectively. Finally, the F-Nb- or the F-Mo-alloyed steel showed similar degradation rates for the cathodic polarization resistance, expressed per unit of Cr deposition, with that of Crofer 22APU and close to 0.06 Ω cm^4 μg^{-1}, while for the STS444 was five times higher (0.30 Ω cm^4 μg^{-1}).

Among the major drawbacks of SOFCs, working at rather low temperatures (Sarda et al. 2013; Garcia-Fresnillo et al. 2014), is the tendency of the formation of the so called σ phase (σ-FeCr) at the interface between the nickel mesh and steel interconnect. The σ-FeCr phase is well known to be inherently brittle (Abiko 1997) and thus could deteriorate the nickel/steel interface, e.g., as a result of stresses initiated during thermal cycling. For this reason, the occurrence of this effect needs to be further investigated. Binary and ternary Fe–Cr alloys are advantageous for controlling the tendency of the σ-phase formation.

A binary Fe-30Cr and the corresponding ternary alloys containing Mn, Mo, or W (Fe-30Cr-2X) were studied by Niewolak et al. (2015), with respect to σ-phase formation at 650 °C. They found that the presence of W and especially of Mo promotes the formation of σ phase, while the presence of Ni results in accelerated formation of a σ-phase-rich layer at the ferrite/austenite interface, which is attributed to the inward diffusion of Ni.

It was also reported that the σ-phase-rich layer, which appears to act as an interdiffusion barrier, is thicker in the case of Fe–Cr and Fe–Cr-Mn alloys than for Mo- or W-rich ones. Thus, Ni has an important influence on the spinodal decomposition thermodynamics and kinetics in both Fe–Cr alloys and ferrite phases, reducing the fraction of Cr-rich phase in the equilibrium phase (Li et al. 2015; Hedström et al. 2013).

For the evaluation of the suitability of the Co-Cr-Mo alloys, as SOFC interconnects, the oxidation behavior of Co-29Cr-6Mo alloy with various Si concentrations (quaternary alloys) was investigated in air at 700–1000 °C (Tunthawiroon et al. 2015). It was found that the presence of alloyed Si improves the oxide scale adhesion by promoting the formation of a continuous Cr_2O_3 film. In the alloy with 1.0 wt% Si, the parabolic rate constant at 700 °C was 1.56×10^{-7}cm s$^{-1/2}$ and increased: (i) with decreasing Si concentration (1.93×10^{-7} for 0.5 wt% Si and 4.319×10^{-7} cm s$^{-1/2}$ for 0.1 wt% Si) and (ii) with increasing temperature (5.65×10^{-7} at 800 °C, 8.39×10^{-7} at 900 °C and 13.65×10^{-7} cm s$^{-1/2}$ at 1000 °C).

As known, SOFC performance degradation in terms of the protective oxide scale spalling and the subsequent destructive failure because of the attacks from the environment, primarily by oxygen, are highly concerned. Minimizing the thermal–mechanical stresses generated between interconnects and the adjacent ceramic solid electrolyte has become an important issue for further SOFC's performance enhancement. For this reason, the value of the TEC of the interconnect materials must be close to that of the ceramic solid electrolyte.

Molybdenum (Mo) is usually added to high-temperature alloys to increase high-temperature strength and creep resistance. It can be easily alloyed up to 4 wt% to stainless steel with commercial process. It has been reported that the addition of 0.1–2 wt% Mo into Fe-22Cr-0.5Mn steel reduces the oxidation rate and the ASR (Yun et al. 2012). It has also been reported that Mo addition increases the activation energy and suppresses the inward diffusion of oxygen, showing that the defect chemistry of oxide scale is altered. This causes the increase in oxidation resistance and electric conductivity. It was concluded that Mo addition does not significantly affect the evaporation of Cr. More precisely, it was found that the parabolic rate constant of the Mo-containing specimens (0, 0.1, 2, and 4 wt%) tested at 800 °C in ambient air for 300 h ranged $5.5–6 \times 10^{-15}$ g^2 mm^{-4} h^{-1} and increased to 9×10^{-15} g^2 mm^{-2} h^{-1}, after 300 h of exposition to air, when 4 wt% of Mo was added. Mo evaporation was not detected in 0.1 and 2 wt% Mo samples. However, in the sample with 4 wt% of Mo, after 100 h of oxidation, the observed Mo evaporation rate was 5.4×10^{-9} g cm^{-2} h^{-1} in humid air at 800 °C. The evaporation of volatile Mo species along with Cr reduced the stability of protective chromia, so that rapidly growing Fe-rich spinel was formed after 300 h of oxidation. The evaporation rate of Cr in all cases remained relatively high at 1–1.1×10^{-7} g cm^{-2} h^{-1}. With the addition of 0, 0.1, 2, and 4 wt% of Mo, the ASR values decreased from 35, 13, 12, to 7 mΩ cm^2, respectively (Yun et al. 2012).

Tensile and creep resistance of FSSs, used as small thickness interconnect, are considered as primary factors for the durable operation of planar SOFC system. Tensile and creep properties of a newly developed ferritic stainless steel (Crofer 22H) with additions of Nb and W for planar solid oxide fuel cell (pSOFC) interconnect were investigated at 25–800 °C (Chiu et al. 2012). These properties were compared with those of another ferritic stainless steel (Crofer 22 APU) without Nb or W. It was found that high-temperature tensile strength of the Crofer 22H steel with additions of Nb and W is superior to that of Crofer 22APU. This is due to a precipitation strengthening effect of $(Fe, Cr)_2(Nb, W)$ Laves phase. It was also found that the addition of Nb and W, in the Crofer 22H steel, enhances the creep resistance by reducing the creep strain rate (Chiu et al. 2012).

Doping technique showed positive effects for $LaCrO_3$, which is the most common base for SOFCs interconnects as it exhibits the highest conductivity among the lanthanide elements. It was reported that the addition of another element further enhances the conductivity and reduces the Cr evaporation rate. More precisely, the addition of Ca into $LaCrO_3$ coating ($LaCrO_3$-coated AISI 430) drastically lowered the oxidation rate and electrical resistance by approximately 2 and 6 times, respectively. The parabolic rate constant of the $La_{0.8}Ca_{0.2}CrO_3$, after 960 h of exposition at 800 °C in air, was found to be 6.94×10^{-14} g^2 cm^{-4} s^{-1}, compared with the 9.07×10^{-14} g^2 cm^{-4} s^{-1} exhibited by $La_{0.8}Sr_{0.2}CrO_3$ (Rashtchi et al. 2013).

As shown in Fig. 5, among the studied metallic interconnects (low-Cr-doped ferritic stainless steel), the Co-doped ferritic stainless steel with only 12.44 wt% Cr shows the lowest oxidation rate constant. According to the literature, its very good behavior can be attributed to the formation of a double-layer oxide scale with dense $(Mn, Cr)_3O_4$ spinel on the top of Cr_2O_3. It is believed that this dense outer $(Mn, Cr)_3O_4$ layer significantly enhances the resistance to oxidation at 750 °C in air and prevents the Cr volatilization from the thermally grown oxide scale and deposition in/around the electrode (Zhang et al. 2014a, b) (Table 2).

4 Coatings (Protective Ceramic Layers) for Metallic Interconnects

Metallic interconnects instead of the ceramic ones are mostly used today, as they exhibit acceptable conductivity at lower SOFC operating temperature and they are not cost-effective materials. Among the most promising metallic interconnects are FSSs (FSS with 10 and 26 wt% of Cr); for high operating temperatures are chromia-forming (developing Cr-oxide scales) materials. This is the reason FSS owns two major drawbacks: (i) the volatilization of undesired species of chromium oxyhydroxide $[CrO_2(OH)_2]$ and chromium oxides [Cr(IV) and/or Cr_2O_3] and

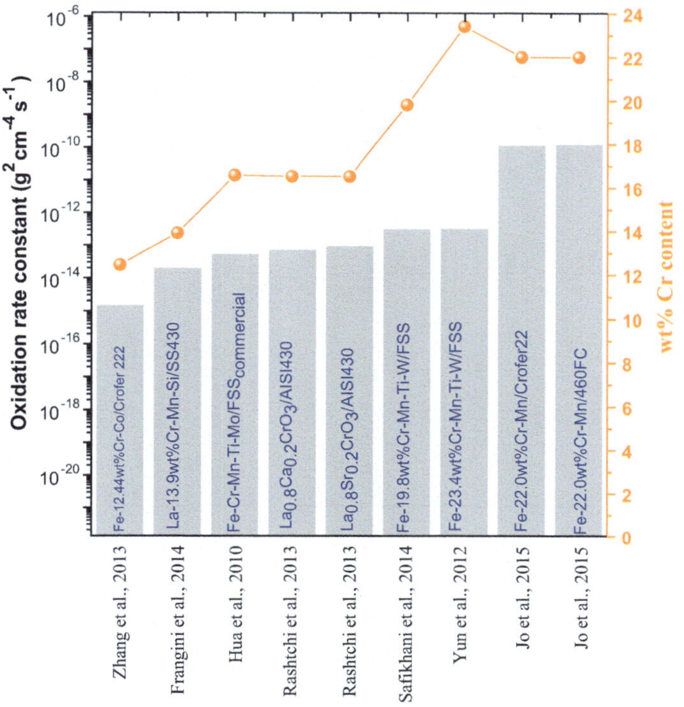

Fig. 5 Oxidation rate constant in terms of Cr content (wt%) for metallic interconnect materials (Zhang et al. 2014a, b; Frangini et al. 2014; Hua et al. 2010; Rashtchi et al. 2013; Safikhani et al. 2014; Yun et al. 2012; Jo et al. 2015)

(ii) the formation of oxide scales leading to the significant ohmic losses, mainly in water vapor-reducing atmospheres. The Cr volatile species are attributed to the reaction of chromium oxide with water and oxygen molecules (Kornely et al. 2011; Chen et al. 2011; Smeacetto et al. 2011).

As a consequence to Cr evaporation, the cathode is poisoned and fuel cell's performance is reduced. Therefore, in order to overcome the above-mentioned problems against corrosion, thin ceramic coatings are applied on the metallic interconnects (Fig. 6), reducing Cr volatilization and increasing, however, the ASR value. This is because the oxide scale formed on the metallic surface grows during SOFC's operation and exhibits lower electronic conductivity than bare stainless steels (Persson et al. 2012).

Interconnect coatings are considered as a potential solution for the reduction of the high-temperature corrosion issues. They are able to prevent the migration of Cr to the surface of interconnect and reduce its oxidation rate. A suitable coating must consist of a material with the appropriate characteristics; however, equally important is the process used for its deposition. Deposition process must enable coatings with good adhesion, low porosity, uniform thickness, good coverage, and at a low cost.

Table 2 Characteristics of some metallic interconnect materials. k_p is the parabolic rate constant, and ASR is the area-specific resistance

Alloy name	Content, wt%											k_p	ASR	References
	Cr	Mn	Si	La	Ni	Al	Nb	Co	Mo	Ti	W	$g^2\ cm^{-4}\ s^{-1}$	$\Omega\ cm^2$	
Crofer22 H (commercial)	20–24	0.3–0.8	0.1–0.6	–	–	–	–	–	–	–	–	5.29×10^{-15} (750 °C)	0.017 (800 °C)	Zhang et al. (2014a, b
Crofer22 APU (commercial)	22.1	0.41	0.01	0.08	–	–	–	–	–	–	–	3.1×10^{-14} (800 °C)	0.013 (800 °C)	Jo et al. (2015)
Crofer22 APU (commercial)	22.0	0.5	0.25	0.15	–	–	–	–	–	–	–	7.51×10^{-10} (800 °C)	–	Miguel-Pérez et al. (2012)
Crofer22 APU (commercial)	22.57	0.47	0.03	–	0.02	–	0.01	–	0.01	0.06	–	–	0.015	Chu et al. (2009)
SS430 (commercial)	18.0	0.45	0.4	–	0.20	–	–	–	–	–	–	1.12×10^{-7} (800 °C)	–	Miguel-Pérez et al. (2012)
SS405 (commercial)	13.9	0.40	0.44	–	0.24	0.25	–	–	–	–	–	1.9×10^{-14} (700 °C)	0.020 (700 °C)	Frangini et al. (2014)
SUS430 (commercial)	21.8	0.49	0.01	–	–	0.002	–	–	–	0.001	0.002	3.05×10^{-12} (800 °C)	0.0269	Safikhani et al. (2014)
ZMG232 (commercial)	23.57	0.56	0.16	–	0.3	–	0.01	–	0.01	–	–	–	0.022	Chu et al. (2009)
K41X (AISI 441) (commercial)	18.0	0.25	0.58	–	–	–	0.52	–	–	0.14	–	–	0.1 (800 °C)	Ardigo et al. (2015)
Low-Cr-containing FeCrCo	12.44	0.57	0.0016		–		–	9.68			–	1.42×10^{-15} (750 °C)	0.011 (800 °C)	Zhang et al. (2014a, b
460FC	22.3	<1.0	0.09	–	–		<0.5	–			–	3.3×10^{-14} (800 °C)	0.0125 (800 °C)	Jo et al. (2015)
Conicro 4023W188	22.0	0.75	0.3	0.1	22.0		–	bal.			–	4.34×10^{-11} (800 °C)	–	Miguel-Pérez et al. (2012)
Fe–Cr alloys	16.56	1.05	–	0.005	–	–	–	–	2.09	0.52	–	7.6×10^{-14} (750 °C)	0.068 (750 °C)	Hua et al. (2010)
Ti and Nb stabilized ferritic structure	17.83	0.24	0.60	–		0.006	0.55	–	0.010	0.13	–	8.8×10^{-7} (800 °C)	–	Wongpromrat et al. (2015)
Fe–Cr–W	19.8	0.44	0.01	–	–	0.002	–	–	–	0.52	2.0	3.05×10^{-13} (800 °C)	0.0085	Safikhani et al. (2014)

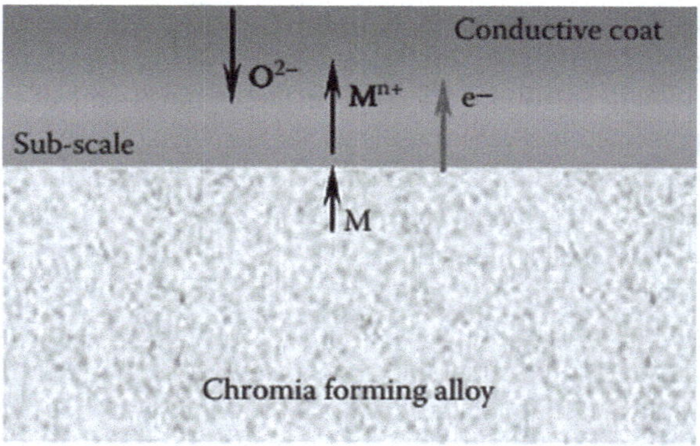

Fig. 6 Schematic of mass transport in a conductive oxide coating on a chromia-forming alloy (Fergus et al. 2008)

A protective ceramic-coating layer should mainly provide the following features (Park et al. 2013a, b; Wei et al. 2009; Wu et al. 2011; Zhang et al. 2011):

- High density,
- High stability,
- High electrical conductivity, and
- Intimidate contact with an interconnect substrate.

Coatings are usually oxides, owning the structure of fluorite, corundum, spinel, and perovskite classes. The most common of them are as follows: LSC, LSM, Co_3O_4, and $MnCo_2O_4$. Based on the literature observations (Persson et al. 2012), the interaction mechanisms between coating and forming oxides, in a metallic interconnect during oxidation, can be of the following types:

(a) Coatings that sinter/react with the growing oxide scale.
(b) Partial incorporation of the coating that reacts with the growing oxide scale, while main part of the coating remains porous.
(c) Incorporation of coating particles in the growing oxide scale.
(d) Elemental addition into the growing oxide scale from a coating that remains porous and is pushed in front of the growing oxide scale.

In the literature, two basic categories of protective ceramic layers were mainly investigated:

1. The single protective layers, which includes the following three subgroups of coatings: (a) spinel, (b) reactive elements oxides, and (c) perovskites, and
2. The dual layers, which have started being developed in the last decade.

Among the above-mentioned subgroups, *manganese cobaltite spinel* coatings have received great attention due to their high electrical conductivities at 800 °C, which can be as high as 60 S cm^{-1} (in $Mn_{1.5}Co_{1.5}O_4$ at 800 °C, in air) (Yang et al. 2007). $(Mn,Co)_3O_4$ is the most widely studied spinel; $(Cu,Mn)_3O_4$ is a potential candidate due to its high electrical conductivity (60 S cm^{-1}) (Hosseini et al. 2014) as well as its compatible, with ferritic stainless steel materials, TEC values (Shaigan et al. 2010).

A detailed classification of the coatings and of their preparation methods was presented in a recent review (Shaigan et al. 2010). According to the authors, ferritic stainless steel has become the standard material for SOFC interconnects, with some disadvantages that are tried to be overcome by the aid of coatings, surface treatments, and modifications.

Perovskites. Despite the fact that perovskite layers do not prevent the formation of chromia on the surface of the oxidized coating/metal composites, they may present: (i) considerable resistance to oxidation, (ii) desired electrical conductivity, and (iii) sufficient ability to absorb gaseous particles of chromium compounds during the oxidation of commercial ferritic steel used as the metallic interconnect in SOFCs.

Lanthanum strontium cobalt ferrite (LSCF) is the most investigated perovskite-type materials, used as cathodes in SOFC, for the oxygen reduction reaction (ORR). LSCF is a good electronic conductor, exhibiting high stability and high electrochemical activity for ORR at high temperatures. For these reasons, it has also been extensively studied as a ceramic protective layer for interconnect material. It was found that in LSCF films (Lee et al. 2010) posted with screen-printing method on Crofer 22APU (ferritic stainless steel with 22 wt% Cr, developed by Thyssen Krup), the ASR was remained close to 14 mΩ cm^2 for the first 40 h. However, from 40 to 300 h, the ASR increased up to 300 % ($\sim$54 mΩ cm^2). The change in ASR as a function of the oxidation time was attributed to the increase in the oxide scale thickness. Despite the relatively acceptable ASR, a mismatch of the TEC values was observed.

Moreover, $La_{0.6}Sr_{0.4}Co_{0.2}Fe_{0.8}O_3$ (LSCF) posted on Crofer 22APU via screen-printing method (Tsai et al. 2010), followed by sintering the screen-printed film at 1050 °C, exhibited an ASR value close to 54 mΩ cm^2 at 800 °C for 200 h. However, after 300-h operation, the ASR was further increased. Additional works showed the same behavior (Przybylski et al. 2014).

Except for the LSCF perovskite, lanthanum strontium manganites (LSM) have also been deposited on metallic interconnects, as perovskite ceramic protective layer. However, after heat treatment (such as calcination or annealing at elevated temperatures), cracks were formed on their surface, mostly due to the phase transformation from amorphous to crystalline and volume shrinkage of the LSM film (Da Conceição et al. 2013; Rashtchi et al. 2013). For example, applied LSM-coating films on Crofer 22APU and ZMG232 (Hitachi Metals, Japan), via screen painting and plasma sputtering methods, presented cracks formation. As a result, during the initial 300 h, the ASR value was doubled increasing from 30 to 60 mΩ cm^2 (Chu et al. 2009).

To avoid cracks formation, the preoxidation process is usually applied. The ASR of $La_{0.67}Sr_{0.33}MnO_{3-\delta}$-coated FSSs, Crofer 22APU, Crofer 22H, SS441, and ZMG232 L, due to changes in their microstructural and electrical properties (25-h preoxidation), was measured to be close to 2.24, 12.2,

2.3, and 6.8 mΩ cm^2, respectively, after 500 h of exposition in air atmosphere at 800 °C (Yang et al. 2012).

A solution to this problem is the formation of a highly dense coating that well adheres to the substrate, recommending the use of nanostructured powders in the form of spherical particles, the surfaces of which will constitute a 2D structure approximation (Brylewski et al. 2012). Moreover, it was found that a densely structured $(La,Sr)CrO_3$ thick film, deposited on a DIN 50049 ferritic stainless steel, showed reduced chromium evaporation rate. Under cathode environment, the ASR of the examined layers remained, after 300 h of operation, at about 90 mΩ cm^2 (Brylewski et al. 2012).

In order to overcome the phenomenon of cracking, the formation of a continuous protective/conductive layer was also suggested. Recently (Morán-Ruiz et al. 2015a, b), Fe-22Cr meshes were coated with $MnCo_{1.9}Fe_{0.1}O_4$ (MCF) and dipped into $LaNi_{0.6}Fe_{0.4}O_{3-\delta}$ (LNF). It was concluded that if the deposition of the protective coating is not enough to form a dense and continuous layer across the width of the mesh, the use of MCF spinel layer would not be enough to prevent the chromium migration. In this case, the ASR value was remained at 42.5 mΩ cm^2 for 400 min, at 800 °C in air.

LNF/MCF conductive/protective coatings have also been investigated with good results (Morán-Ruiz et al. 2014; Vidal et al. 2015). More precisely, the $LaNi_{0.6}Fe_{0.4}O_3$, prepared via a glycine–nitrate route method, exhibited a very close TEC value with that of interconnects materials. Moreover, $(La_{0.8}Sr_{0.2})_{0.95}Fe_{0.6}Mn_{0.3}Co_{0.1}O_3$ (LSFMC), $LaNi_{0.6}Fe_{0.4}O_{3-\delta}$ (LNF), and $LaNi_{0.6}Co_{0.4}O_{3-\delta}$ (LNC) were studied as contact materials between Crofer 22APU interconnect and a $La_{0.6}Sr_{0.4}FeO_3$ (LSF) cathode; LNC was found to be the most appropriate material. Under cathode conditions, the ASR value was very low and close to 6 mΩ cm^2 for 16 h and at 800 °C.

It has been reported that the LNC coating on AL441HP substrate maintained its ASR value steady at about 3 mΩ cm^2 for more than 500 h in 3 % H_2O humidified air and 800 °C (Seabaugh et al. 2012).

Lanthanum strontium manganese chromite-based materials (LSMCs) have also been investigated as ceramic coatings on FSSs (De Angelis Korb et al. 2013). However, the TEC values of the metal and layer were not compatible and Cr evaporation was observed. Nevertheless, it is necessary to further decrease the electrical resistance at the ceramics/metal interfaces for longer periods, i.e., thousands of hours. This is a fundamental issue for the proper operation of SOFCs, emphasizing also to the fabrication methods (Zhang et al. 2015; Huang et al. 2014).

The main drawback of the perovskite materials is that despite the fact that they lead to a decrease in the interfacial contact resistance, their sintering temperature is very high, leading to their compaction followed by Cr diffusion, especially at the cathode compartment.

Reactive element oxides and spinel coatings. The addition of Cr_2O_3-forming alloys even of small amounts of some elements, called reactive elements (La, Y, Nd, etc.), strongly enhances their resistance toward high-temperature oxidation. With respect to oxidation behavior, Co, Mn, Ti, Si, and Al are among the most important minor alloying additions in commercial ferritic steels, their concentration commonly being a few tenths of a percent (e.g., <0.005). For example, Mn is added to Crofer 22 APU to obtain the formation of an external spinel, which is expected to decrease the formation of volatile Cr species and to enhance the electrical conductivity of the oxide scale. Fontana et al. (2012) studied coatings composed of thin reactive element oxides made of La_2O_3 or Y_2O_3 deposited on Crofer 22APU. After 23,100 h in air, the measured ASR value was 410 mΩ cm^2, at 800 °C in air.

As shown in Fig. 7, the oxide scale, around 16 mm thick, is composed of a double-layered scale with a spinel oxide at the outer surface, covering a subscale of chromia. The YMn_2O_5 phase is almost completely dispersed across the oxide film, and only few precipitates are presented in the oxide surface.

Non-Cr-containing spinel coatings, such as $(Mn,Co)_3O_4$, are among the most promising barriers against Cr migration (Yang et al. 2005; Simner et al. 2005; Hoyt et al. 2012; Zhang et al. 2013; Kruk et al. 2013; Zhang et al. 2014a, b), preserving a relatively acceptable ASR value. It is worth to be noticed that the usual target of an ASR of each single SOFC component, in order to be sufficiently protective, is estimated to be close to 10 mΩ cm^2.

A dense, continuous, and crack-free (10–40-μm thickness) $Mn_{1.5}Co_{1.5}O_4$ spinel, coated on Crofer 22APU substrate by the aid of cathodic electrophoretic deposition (EPD) and followed by sintering at 800–1150 °C under different atmospheres, maintained stable for 2500 h of its ASR value to 20 mΩ cm^2, at 800 °C in air (Smeacetto et al. 2015).

The fabrication of interconnect protective layers requires the following: (i) synthesis of the desired phase, (ii) formation of the desired shape, and (iii) sintering to achieve the adequate strength and conductivity. For example, a mixture of $Mn_{1.5}Co_{1.5}O_4 + La_{0.60}Sr_{0.40}FeO_3$ was proved to be a good solution for the enhancement of the short-term performance of a SOFC stack, since $(MnCoO)_3O_4$ provides a protection layer for chromium evaporation and $La_{0.60}Sr_{0.40}FeO_3$ enhances the conductivity of the protective layer (Miguel-Pérez et al. 2013).

As it has already been mentioned, doping affects the properties of the materials. The effect of doping $(Mn,B)_3O_4$ materials at the B position, for the application as protective layers, was investigated by Pérez et al. (Miguel-Pérez et al. 2013). The chemical compatibility, Cr diffusion, and mechanical stability of metallic interconnects (Crofer 22APU, SS430, and Conicro 4023W188) and the assembled layers were examined. It was found that due to the reaction between the protective layer and the chromium species, the first of them exhibited the lowest ASR value of 0.5 mΩ cm^2 at 800 °C for 100 h in air (Miguel-Pérez et al. 2013). Moreover, incorporating copper dopant into $MnCoO_4$ ($Mn_{1.4}Co_{1.4}Cu_{0.2}O_4$ applied on Crofer 22APU), a very low ASR was achieved (<4 mΩ cm^2), which remained stable for 530 h at 800 °C in air (Chen et al. 2015).

In order to improve the SUS430 properties, a low-cost spinel coating of $MnCu_{0.5}Co_{1.5}O_4$ was deposited on it by the aid of the solgel method (Xiao et al. 2015). The measured ASR remained less than 4 mΩ cm^2 for 500 h, even though the coated alloy had undergone oxidation at 800 °C for 530 h in air and four thermal cycles from 800 °C at room temperature.

A resistivity of 37.4 mΩ cm^2 for 1000 h at 800 °C in air was observed in Crofer 22APU with $MnCo_{1.9}Fe_{0.1}O_4$ spinel as protective-coating layer and $La_{0.8}Sr_{0.2}Co_{0.75}Fe_{0.25}O_3$ perovskite as a contact layer (Montero et al. 2008).

Cu- and Ni-doped MnCo spinels were prepared via a slurry-coating process and subsequent heat treatment. The exhibited ASR for Cu-$Mn_{1.5}Co_{1.5}$ was found to be 17.6 mΩ cm^2 for 1000 h at 800 °C in air (Park et al. 2013a, b).

The behavior of AISI441 (SS441) coated with $Mn_{1.475}Co_{1.475}Ce_{0.05}O_4$ (Ce-MC) could be considered remarkable (Stevenson et al. 2013). Ce-MC consisted of a single-phase spinel, in which most of the Ce was actually presented in the form of finely dispersed ceria particles throughout the coating matrix, mainly because of the limited solubility of Ce. In this case, ASR was found to be close to 13 mΩ cm^2, for 9000 h at 800 °C in air.

Moreover, in Cu-Mn spinel, the fabrication of which depends on the Cu:Mn ratio, the higher copper content in addition with Cu-Mn spinel resulted in the formation of copper oxide (Chen et al. 2005; Bateni et al. 2007). $Cu_{1.3}Mn_{1.7}O_4$ thermally grown on AISI430 ferritic stainless steel showed an ASR value of 19.3 mΩ cm^2 for 500 h, at 800 °C in air, without allowing Cr migration (Hosseini et al. 2015). The ASR of the uncoated sample was measured to be 63.5 mΩ cm^2 after 500 h of oxidation ($\sim$3.3 times higher compared to the coated one). Furthermore, a good match between the coating material and the substrate was observed.

Coatings of the type Fe-Co-Ni have also been investigated (Geng et al. 2012). It was reported that the formed outer spinel layer of $(Fe,Co,Ni)_3O_4$ is able to suppress Cr outward migration and reduce the growth rate of the inner layer of Cr_2O_3; a stable surface oxide scale with ASR of 50 mΩ cm^2 (at 800 °C for 250 h in air) is formed and this value is lower than that of the bare steel.

Thus, except for the various metals doping into the spinel, many fabrication methods have been also suggested in the literature. According to Ou et al. (2014),

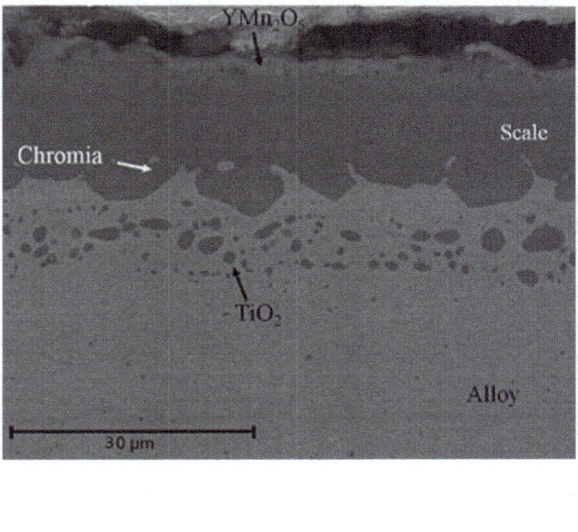

Fig. 7 SEM observations of Y_2O_3-coated Crofer22 APU after 23,100 h at 800 °C in air. Cross-sectional image. Reprinted from Fontana et al. (2012)

the dense reaction layer at the coating–alloy interface could be the key to improving the oxidation resistance of metallic interconnects.

Dual layers. The high sintering temperature is one of the main reasons of the development of ceramic layers on metallic interconnects. However, this has as a consequence the development of undesired layers and the increase in the ASR value. In order to override this drawback, dual-layer coatings have been developed and novel deposition methods have been investigated (Yang et al. 2015).

The comparison between the behavior of single and dual layer was studied in terms of ASR behavior (Da Conceição et al. 2013) by coating SS444 with one and two successive layers of $La_{0.7}Sr_{0.3}MnO_3$. The ASR value of the dual layer showed initially 30 mΩ cm^2 and reached a maximum of 40 mΩ cm^2, after 200 h at 800 °C in air. On the contrary, single-layer-coated SS444 maintained the ASR value at a very low level, 0.6 mΩ cm^2, for 200 h at 800 °C in air.

Two layers of LSC—10 wt% CGO ($Ce_{0.8}Gd_{0.2}O_{2-\delta}$) with thickness of ~30 and ~35 μm, respectively—maintained for 200 h of a low ASR value of about 22.8 mΩ cm^2 at 800 °C in air. More precisely, a cobalt layer has been deposited by electroplating on commercial SUS430 stainless steel. Then, a composite layer of LSC precursor and CGO was prepared on the cobalt coating by screen printing (Yang et al. 2015).

Sputtered Ni coating on ferritic stainless steel, obtained by means of magnetron sputtering method, was thermally converted into an electrically conductive outer layer of NiO with small amounts of $(Ni,Fe,Cr)_3O_4$, presenting an ASR of 40 mΩ cm^2 at 800 °C (Geng et al. 2012). Moreover, an ASR value of 25 mΩ cm^2 was measured in a dual layer of $Ni_{80}Cr_{20}/(La_{0.75}Sr_{0.25})_{0.95}MnO_3$ deposited on SUS430 via coating plasma spray, after 2800 operational hours at 800 °C (Wu et al. 2014).

In an effort to form a ceramic coating on metallic interconnect, a Fe-22Cr mesh was dipped into a ceramic ($LaNi_{0.6}M_{0.4}O_{3-\delta}$; M:Co, Fe) slurry, in order to form a metallic/ceramic material as contact coating for solid oxide fuel cells. Despite the efforts to form a dual layer, many cracks and high ASR values were observed (Morán-Ruiz et al. 2015a, b).

Crofer 22APU, Crofer 22H, E-Brite, and AL29-4C were coated with Co_3O_4 and studied in air + 1 % H_2O (Palcut et al. 2012). In Fig. 8, a schematic representation of the structure of the formed dual-layer metallic interconnect (Crofer 22APU) is depicted. The dense oxide scale consists of an inner chromium oxide layer followed by a thin chromium–cobalt oxide phase with the presence of some Mn and Fe. The inner coating layer, originally Co_3O_4, contains significant amounts of Cr and Mn (Palcut et al. 2012). The dual layer reduced the oxidation rate of the coated steels by 82–96 % in comparison with the uncoated one. More precisely, the oxidation rate values for the coated steels were measured to be in the following order: 0.89×10^{-14} (E-Brite) < 2.50×10^{-14} g^2 cm^{-4} s^{-1} (Al 49-4C) < 2.87×10^{-14} g^2 cm^{-4} s^{-1} (Crofer 22APU) < 5.45×10^{-14} g^2 cm^{-4} s^{-1} (Crofer 22H), while those for the uncoated ones were 18.9×10^{-14} (Crofer 22APU) < 25.0×10^{-14} (E-Brite) < 29.0×10^{-14} g^2 cm^{-4} s^{-1} (Al 49-4C) < 30.7×10^{-14} g^2 cm^{-4} s^{-1} (Crofer 22H).

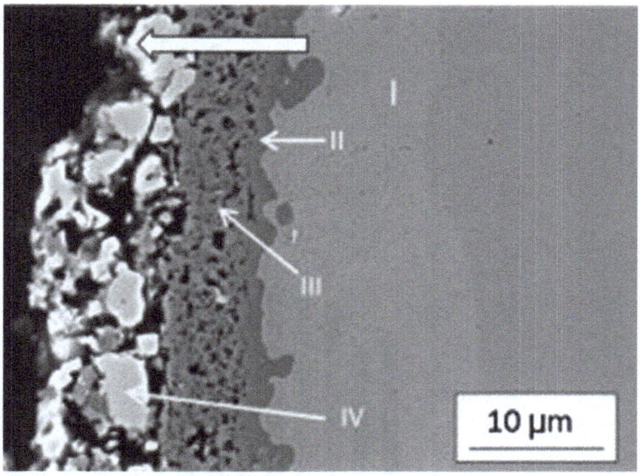

Fig. 8 Structure of a dual-layer metallic interconnect: (I) metal, (II) Cr_2O_3, (III) (Co, Mn, $Cr)_3O_4$, and (IV) $La_{0.85}Sr_{0.15}MnO_{3\text{-}\delta}$ (Palcut et al. 2012)

Very recently, Sanergy HT metallic interconnects have been investigated, presenting a very interesting behavior (Froitzheim et al. 2012). An oxide scale with an inner layer of Cr_2O_3 and a spinel layer on top were formed. In the case of the uncoated material, the spinel layer was of $(Cr,Mn)_3O_4$ type, while in the presence of a Co coating, the formation of $(Co,Mn,Fe)_3O_4$ was observed. The Cr evaporation measurements showed that despite the fact that the Co coating was very thin (640 nm), it effectively blocked Cr evaporation for at least 3000 h, without any sign of degradation. More precisely, for the uncoated interconnect, the evaporated amount of Cr at the first 1000 h was escalated above 2.0×10^{-3} kg m^{-2}, while for the coated interconnect at the first 1000 h was 0.5×10^{-3} kg m^{-2} and at 3000 h reached 1.0×10^{-3} kg m^{-2}. It should be noted that for the uncoated interconnect, the evaporation of Cr mass after 3000 h was higher than $\sim 10^{-2}$ kg m^{-2} (Froitzheim et al. 2012).

In Fig. 9, a comparison of several protective ceramic layers is presented, in terms of ASR and oxidation time.

Special interconnects and protective layers. Very recently, a novel interconnect, composed of Ag–glass materials, was investigated by Pi et al. (2013). Their purpose was to improve the density of the Ag interconnect-coating layers on anode-supported flat-tubular SOFCs, operated at an intermediate temperature (Fig. 10). As known, Ag exhibits extremely high electrical conductivity and stability in both oxidizing and reducing atmospheres, showing, however, very high vaporization rates even at intermediate temperatures. It was found that the sample of Ag + 10 wt% glass exhibited the lowest ASR, 2.8 mΩ cm^{-2} at 700 °C, and it is considered a very promising material for interconnect application.

Contrary to other SOFC types, in the case of planar–tubular one, it is difficult to sinter Ca- and Sr-doped $LaCrO_3$ interconnects with high densities, because the sintering procedure occurs via a gaseous Cr-O evaporation–condensation mechanism (Yokokawa et al. 1991). As an alternative, a dual-layer interconnect approach,

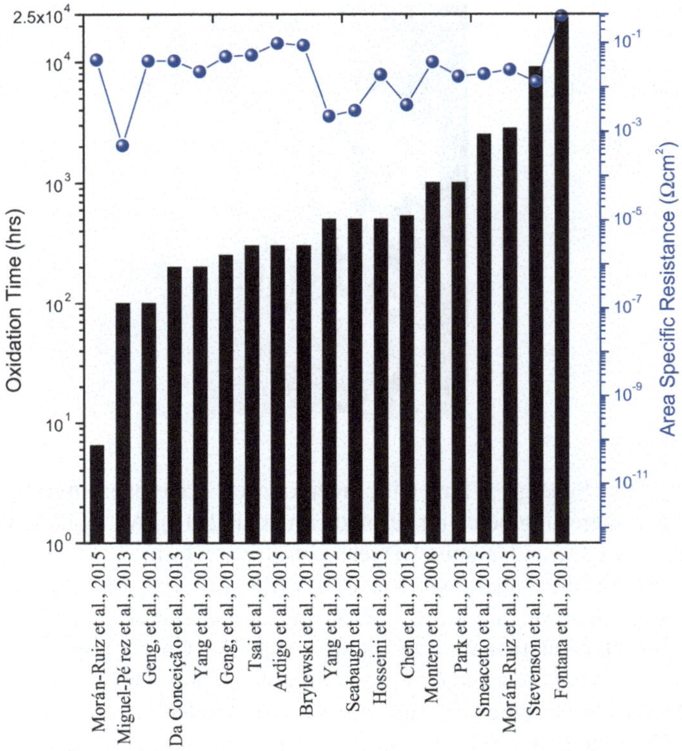

Fig. 9 Area-specific resistance and oxidation time for ceramic protective layers (Morán-Ruiz et al. 2015a, b; Miguel-Pérez et al. 2013; Geng et al. 2012; Da Conceição et al. 2013; Yang et al. 2015; Tsai et al. 2010; Ardigo et al. 2015; Brylewski et al. 2012; Yang et al. 2012; Seabaugh et al. 2012; Hosseini et al. 2015; Chen et al. 2015; Montero et al. 2008; Park et al. 2013a, b; Smeacetto et al. 2015; Stevenson et al. 2013; Fontana et al. 2012)

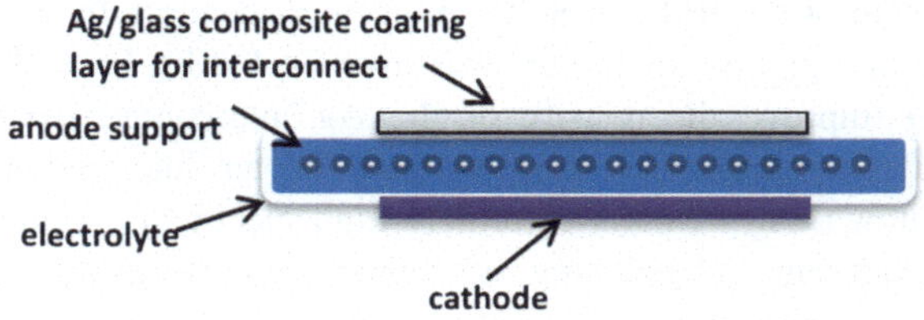

Fig. 10 Schematic diagram of the cross section of cathode/electrolyte/anode/Ag–glass interconnect (Pi et al. 2013)

consisting of an *n*-type-conducting layer on the anode side and a *p*-type-conducucing layer on the cathode side, was proposed (Gopalan 2005; Huang et al. 2006).

Among the *n*-type interconnects, $SrTiO_3$ perovskites doped with La or Y were found to be the most promising candidates for the use in dual-layer interconnects

(Marina et al. 2002; Wang et al. 2010). La-doped $SrTiO_3$ materials, synthesized by the Pechini method using citric acid, were tested by Park et al. (Park et al. 2012). The TEC value for $Sr_{0.7}La_{0.2}TiO_3$ was found to be close to 11.8×10^{-6} K^{-1}, similar to those of other SOFC components. The conductivity of $Sr_{0.8}La_{0.2}TiO_3$ and $Sr_{0.7}La_{0.2}TiO_3$ ranged between 9 and 18 S cm^{-1}, depending on the composition and the temperature, which exceeded the conductivity requirement for the SOFC interconnect (Park et al. 2012).

5 Concluding Remarks

The requirements of the materials for SOFC interconnects are the most demanding among the other cell components. Successful development of interconnect materials is vital for the large-scale commercialization of SOFCs technology.

The *ceramic interconnects* (oxides) are stable in oxidizing atmosphere and expensive materials, exhibiting lower electrical conductivity in comparison with their metallic analogs. The most studied ceramic interconnects are lanthanum and yttrium chromites with the latter to be more stable under SOFC operation conditions.

Among the available *metallic* materials, FSSs (Ferritic Stainless Steels) are good interconnect candidates because of their (i) high electrical conductivity oxide thin films formed on their surface, (ii) ease manufacturing, (iii) matched TEC, and (iv) their low cost; crofer22 APU is the best candidate due to its mass continuity under oxidizing conditions. However, under the SOFC operation temperatures, Cr evaporates from FSSs, and strongly damaging SOFCs' performance. Two of the major degradation mechanisms affecting SOFC stacks are directly linked to chromia formation of the metallic interconnect component: (i) higher stack ohmic resistance due to the formation of native chromium oxide (growing chromium oxide layer increases the ohmic resistance) and (ii) chromium poisoning of the SOFC cathode (chromium vapor diffuses through cracks or pores in the coating).

To avoid this drawback, *ceramic protective layers* (reactive elements, perovskites, spinels) deposited on the metallic interconnects have been developed.

The addition of *reactive elements* decreases the formation of volatile Cr species and enhances the electrical conductivity of the oxide scale. The change in the diffusion mechanism allows the corrosion rate to decrease and can significantly increase the scale adherence.

Perovskite structures have been widely studied, especially under cathode environment. Most of the studies report that $La_xSr_yFe_zO_3$-based coatings on ferritic alloys are not effective to prevent Cr outward diffusion. On the other hand, perovskite B-site cations (Ni, Fe, Co) diffuse, forming stable spinel with transition metals from interconnect.

Spinel oxides, in addition to the perovskites, can reduce oxidation and inhibit chromium volatility from FSSs. However, very long-term applications of these coatings still consists an issue. For this reason, Ce is added as reactive element in order to increase the oxide scale and coating adherence to the metal substrate.

Mn-Co spinels doped with various elements, such as Ni, Ti, Cu, and Fe, which exhibit higher electrical conductivity and a better densification.

Another examined solution is concerned with the development of dual layers (metallic–ceramic coatings). They can bridge the "differences" between interconnect and electrode, increasing the conductivity and adhesive ability of the interface, using low-temperature sintering process. Unfortunately, the relative low bulk conductivity of spinels and their imperfect thermal expansion match with the underlying stainless steel, which increases stack resistance. Cobalt physical vapor thin films (600-800 nm) deposited on ferritic stainless steel Sanergy HT™ have recently been shown to decrease dramatically the evaporation of Cr(VI) species, and with an additional 10 nm Ce layer under the Co coating, the oxidation resistance of the overall coating–alloy system is further improved.

Despite the progress of conductive/protective coatings, the stack of degradation rates is still higher than that required for SOFC commercialization.

Acknowledgments This chapter is prepared in the framework of the following project: the *Ministry of Education and Science of the Russian Federation* (Mega-grant contract No. 14. Z50.31.0001).

References

Abiko, K. (1997). The evolution of iron. *Physica Status Solidi A, 160*, 285–296.

Ajitdoss, L. C., Smeacetto, F., Bindi, M., Beretta, D., Salvo, M., & Ferraris, M. (2013). $Mn_{1.5}Co_{1.5}O_4$ protective coating on Crofer22APU produced by thermal co-evaporation for SOFCs. *Materials Letters, 95*, 82–85.

Ardigo, M. R., Popa, I., Combemale, L., Chevalier, S., Herbst, F., & Girardon, P. (2015). Dual atmosphere study of the K41X stainless steel for interconnect application in high temperature water vapour electrolysis. *International Journal of Hydrogen Energy, 40*, 5305–5312.

Armstrong, T. R., Stevenson, J. W., McCready, D. E., Paulik, S. W., & Raney, P. E. (1996). The effect of reducing environments on the stability of acceptor substituted yttrium chromite. *Solid State Ionics, 92*, 213–223.

Bateni, M. R., Wei, P., Deng, X., & Petric, A. (2007). Spinel coatings for UNS 430 stainless steel interconnects. *Surface and Coatings Technology, 201*, 4677–4684.

Brylewski, T., Dabek, J., Przybylski, K., Morgiel, J., & Rekas, M. (2012). Screen-printed (La, Sr) CrO_3 coatings on ferritic stainless steel interconnects for solid oxide fuel cells using nanopowders prepared by means of ultrasonic spray pyrolysis. *Journal of Power Sources, 208*, 86–95.

Chen, G., Xin, X., Luo, T., Liu, L., Zhou, Y., Yuan, C., et al. (2015). $Mn_{1.4}Co_{1.4}Cu_{0.2}O_4$ spinel protective coating on ferritic stainless steels for solid oxide fuel cell interconnect applications. *Journal of Power Sources, 278*, 230–234.

Chen, X., Hou, P. Y., Jacobson, C. P., Visco, S. J., & De Jonghe, L. C. (2005a). Protective coating on stainless steel interconnect for SOFCs: Oxidation kinetics and electrical properties. *Solid State Ionics, 176*, 425–433.

Chen, X., Huang, K., & Xu, X. B. (2005b). Automated design of a three-dimensional fishbone antenna using parallel genetic algorithm and NEC. *IEEE Antennas and Wireless Propagation Letters, 4*, 425–428.

Chen, X., Zhang, L., Liu, E., & Jiang, S. P. (2011). A fundamental study of chromium deposition and poisoning at $(La_{0.8}Sr_{0.2})0.95(Mn_{1-x}Co_x)O_{3\pm\delta}$ ($0.0 \leq x \leq 1.0$) cathodes of solid oxide fuel cells. *International Journal of Hydrogen Energy, 36*, 805–821.

Chiu, Y.-T., & Lin, C.-K. (2012). Effects of Nb and W additions on high-temperature creep properties of ferritic stainless steels for solid oxide fuel cell interconnect. *Journal of Power Sources, 198*, 149–157.

Chu, C.-L., Lee, J., Lee, T.-H., & Cheng, Y.-N. (2009). Oxidation behavior of metallic interconnect coated with La–Sr–Mn film by screen painting and plasma sputtering. *International Journal of Hydrogen Energy, 34*, 422–434.

Da Conceição, L., Dessemond, L., Djurado, E., & Souza, M. M. V. M. (2013). $La_{0.7}Sr_{0.3}MnO_3$-coated SS444 alloy by dip-coating process for metallic interconnect supported solid oxide fuel cells. *Journal of Power Sources, 241*, 159–167.

Da Conceição, L., & Souza, M. M. V. M. (2013). Synthesis of $La_{0.7}Sr_{0.3}MnO_3$ thin films supported on Fe–Cr alloy by sol–gel/dip-coating process: Evaluation of deposition parameters. *Thin Solid Films, 534*, 218–225.

De Angelis Korb, M., Savaris, I. D., Feistauer, E. E., Barreto, L. S., Heck, N. C., Müller, I. L., et al. (2013). Modification of the $La_{0.6}Sr_{0.4}CoO_3$ coating deposited on ferritic stainless steel by spray pyrolysis after oxidation in air at high temperature. *International Journal of Hydrogen Energy, 38*, 4760–4766.

Fergus, J., Hui, R., Li, X., Wilkinson, D. P., & Zhang, J. (2008). *Solid oxide fuel cells: Materials properties and performance*. Boca Raton: CRC press.

Fergus, J. W. (2004). Lanthanum chromite-based materials for solid oxide fuel cell interconnects. *Solid State Ionics, 171*, 1–15.

Fergus, J. W. (2005). Metallic interconnects for solid oxide fuel cells. *Materials Science and Engineering A, 397*, 271–283.

Flandermeyer, B. K., Poeppel, R. B., Dusek, J. T., & Anderson, H. U., (1988). *Sintering aid for lanthanum chromite refractories*. US Patent #4,749,632.

Fontana, S., Amendola, R., Chevalier, S., Piccardo, P., Caboche, G., Viviani, M., et al. (2007). Metallic interconnects for SOFC: Characterisation of corrosion resistance and conductivity evaluation at operating temperature of differently coated alloys. *Journal of Power Sources, 171*, 652–662.

Fontana, S., Chevalier, S., & Caboche, G. (2012). metallic interconnects for solid oxide fuel cell: Performance of reactive element oxide coating during 10, 20 and 30 months exposure. *Oxidation of Metals, 78*, 307–328.

Frangini, S., Masci, A., McPhail, S. J., Soccio, T., & Zaza, F. (2014). Degradation behavior of a commercial 13Cr ferritic stainless steel (SS405) exposed to an ambient air atmosphere for IT-SOFC interconnect applications. *Materials Chemistry and Physics, 144*, 491–497.

Froitzheim, J., Canovic, S., Nikumaa, M., Sachitanand, R., Johansson, L. G., & Svensson, J. E. (2012). Long term study of Cr evaporation and high temperature corrosion behaviour of Co coated ferritic steel for solid oxide fuel cell interconnects. *Journal of Power Sources, 220*, 217–227.

Froitzheim, J., Meier, G. H., Niewolak, L., Ennis, P. J., Hattendorf, H., Singheiser, L., et al. (2008). Development of high strength ferritic steel for interconnect application in SOFCs. *Journal of Power Sources, 178*, 163–173.

Fu, Y.-P., Wang, H.-C., Weng, C.-S., Hu, S.-H., & Liu, Y.-C. (2015). Characterizations of Fe doping on B-Site of $(La_{0.8}Ca_{0.2})(Cr_{0.9}Co_{0.1})O_{3-\delta}$ interconnect materials for SOFCs. *Journal of the American Ceramic Society, 98*(8), 2561.

Furtado, J. G. M., & Oliveira, R. N. (2008). Development of lanthanum chromites-based materials for solid oxide fuel cell interconnects. *Matéria (Rio de Janeiro), 13*, 147–153.

Gannon, P. E., Tripp, C. T., Knospe, A. K., Ramana, C. V., Deibert, M., Smith, R. J., et al. (2004). High-temperature oxidation resistance and surface electrical conductivity of stainless steels with filtered arc Cr–Al–N multilayer and/or superlattice coatings. *Surface and Coatings Technology, 188–189*, 55–61.

Gannon, P. E., & Montana State University. (2007). *Study of solid oxide fuel cell interconnects, protective coatings and advanced physical vapor deposition techniques.* Montana, United States of America: Montana State University.

Garcia-Fresnillo, L., Shemet, V., Chyrkin, A., de Haart, L., & Quadakkers, W. (2014). Long-term behaviour of solid oxide fuel cell interconnect materials in contact with Ni-mesh during exposure in simulated anode gas at 700 and 800° C. *Journal of Power Sources, 271*, 213–222.

Geng, S., Qi, S., Xiang, D., Zhu, S., & Wang, F. (2012a). Oxidation and electrical behavior of ferritic stainless steel interconnect with Fe–Co–Ni coating by electroplating. *Journal of Power Sources, 215*, 274–278.

Geng, S., Qi, S., Zhao, Q., Zhu, S., & Wang, F. (2012b). Electroplated Ni-Fe_2O_3 composite coating for solid oxide fuel cell interconnect application. *International Journal of Hydrogen Energy, 37*, 10850–10856.

Geng, S., Wang, Q., Wang, W., Zhu, S., & Wang, F. (2012c). Sputtered Ni coating on ferritic stainless steel for solid oxide fuel cell interconnect application. *International Journal of Hydrogen Energy, 37*, 916–920.

Geng, S., & Zhu, J. (2006). Promising alloys for intermediate-temperature solid oxide fuel cell interconnect application. *Journal of Power Sources, 160*, 1009–1016.

Gopalan, S. (2005). Bi-Layer pn junction interconnections for goal based solid oxide fuel cells, *Final Technical Report DOE, 1–21.*

Hayashi, S., Fukaya, K., & Saito, H. (1988). Sintering of lanthanum chromite doped with zinc or copper. *Journal of Materials Science Letters, 7*, 457–458.

Hedström, P., Huyan, F., Zhou, J., Wessman, S., Thuvander, M., & Odqvist, J. (2013). The 475 °C embrittlement in Fe-20Cr and Fe-20Cr-X (X = Ni, Cu, Mn) alloys studied by mechanical testing and atom probe tomography. *Materials Science and Engineering A, 574*, 123–129.

Heidarpour, A., Choi, G. M., Abbasi, M. H., & Saidi, A. (2012). A novel approach to co-sintering of doped lanthanum chromite interconnect on Ni–YSZ anode substrate for SOFC applications. *Journal of Alloys and Compounds, 512*, 156–159.

Heidarpour, A., Saidi, A., Abbasi, M. H., & Choi, G. M. (2013). In situ fabrication mechanism of a dense Sr and Ca doped lanthanum chromite interconnect on Ni-YSZ anode of a solid oxide fuel cell during co-sintering. *Ceramics International, 39*, 1821–1826.

Hosseini, N., Abbasi, M. H., Karimzadeh, F., & Choi, G. M. (2015). Development of $Cu_{1.3}Mn_{1.7}O_4$ spinel coating on ferritic stainless steel for solid oxide fuel cell interconnects. *Journal of Power Sources, 273*, 1073–1083.

Hosseini, N., Karimzadeh, F., Abbasi, M. H., & Choi, G. M. (2014). Microstructural characterization and electrical conductivity of $Cu_xMn_{3-x}O_4$ ($0.9 \leq x \leq 1.3$) spinels produced by optimized glycine–nitrate combustion and mechanical milling processes. *Ceramics International, 40*, 12219–12226.

Hoyt, K. O., Gannon, P. E., White, P., Tortop, R., Ellingwood, B. J., & Khoshuei, H. (2012). Oxidation behavior of (Co, Mn)3O_4 coatings on preoxidized stainless steel for solid oxide fuel cell interconnects. *International Journal of Hydrogen Energy, 37*, 518–529.

Hua, B., Pu, J., Lu, F., Zhang, J., Chi, B., & Jian, L. (2010a). Development of a Fe–Cr alloy for interconnect application in intermediate temperature solid oxide fuel cells. *Journal of Power Sources, 195*, 2782–2788.

Hua, B., Zhang, W., Wu, J., Pu, J., Chi, B., & Jian, L. (2010b). A promising $NiCo_2O_4$ protective coating for metallic interconnects of solid oxide fuel cells. *Journal of Power Sources, 195*, 7375–7379.

Huang, J.-J., Fu, Y.-P., Wang, J.-Y., Cheng, Y.-N., Lee, S., & Hsu, J.-C. (2014). Characterization of Fe–Cr alloy metallic interconnects coated with LSMO using the aerosol deposition process. *Materials Research Bulletin, 51*, 63–68.

Huang, W., & Gopalan, S. (2006). Bi-layer structures as solid oxide fuel cell interconnections. *Solid State Ionics, 177*, 347–350.

Jian, P., Jian, L., Bing, H., & Xie, G. (2006). Oxidation kinetics and phase evolution of a Fe–16Cr alloy in simulated SOFC cathode atmosphere. *Journal of Power Sources, 158*, 354–360.

Jo, K. H., Kim, J. H., Kim, K. M., Lee, I.-S., & Kim, S.-J. (2015). Development of a new cost effective Fe–Cr ferritic stainless steel for SOFC interconnect. *International Journal of Hydrogen Energy. 40*(30), 9523.

Kornely, M., Neumann, A., Menzler, N. H., Leonide, A., Weber, A., & Ivers-Tiffée, E. (2011). Degradation of anode supported cell (ASC) performance by Cr-poisoning. *Journal of Power Sources, 196*, 7203–7208.

Kruk, A., Stygar, M., & Brylewski, T. (2013). Mn–Co spinel protective–conductive coating on AL453 ferritic stainless steel for IT-SOFC interconnect applications. *Journal of Solid State Electrochemistry, 17*, 993–1003.

Kurokawa, H., Kawamura, K., & Maruyama, T. (2004). Oxidation behavior of Fe–16Cr alloy interconnect for SOFC under hydrogen potential gradient. *Solid State Ionics, 168*, 13–21.

Lee, S., Chu, C.-L., Tsai, M.-J., & Lee, J. (2010). High temperature oxidation behavior of interconnect coated with LSCF and LSM for solid oxide fuel cell by screen printing. *Applied Surface Science, 256*, 1817–1824.

Li, S., Wang, Y., & Wang, X. (2015). Effects of Ni content on the microstructures, mechanical properties and thermal aging embrittlement behaviors of Fe–20Cr–xNi alloys. *Materials Science and Engineering A, 639*, 640–646.

Liu, W. N., Sun, X., Stephens, E., & Khaleel, M. A. (2009). Life prediction of coated and uncoated metallic interconnect for solid oxide fuel cell applications. *Journal of Power Sources, 189*, 1044–1050.

Liu, X., Su, W., Lu, Z., Liu, J., Pei, L., Liu, W., et al. (2000). Mixed valence state and electrical conductivity of $La_{1-x}Sr_xCrO_3$. *Journal of Alloys and Compounds, 305*, 21–23.

Marina, O. A., Canfield, N. L., & Stevenson, J. W. (2002). Thermal, electrical, and electrocatalytical properties of lanthanum-doped strontium titanate. *Solid State Ionics, 149*, 21–28.

Miguel-Pérez, V., Martínez-Amesti, A., Nó, M. L., Larrañaga, A., & Arriortua, M. I. (2012). Oxide scale formation on different metallic interconnects for solid oxide fuel cells. *Corrosion Science, 60*, 38–49.

Miguel-Pérez, V., Martínez-Amesti, A., Nó, M. L., Larrañaga, A., & Arriortua, M. I. (2013). The effect of doping (Mn, B)3O4 materials as protective layers in different metallic interconnects for solid oxide fuel cells. *Journal of Power Sources, 243*, 419–430.

Montero, X., Tietz, F., Sebold, D., Buchkremer, H. P., Ringuede, A., Cassir, M., et al. (2008). $MnCo_{1.9}Fe_{0.1}O_4$ spinel protection layer on commercial ferritic steels for interconnect applications in solid oxide fuel cells. *Journal of Power Sources, 184*, 172–179.

Morán-Ruiz, A., Vidal, K., Laguna-Bercero, M. Á., Larrañaga, A., & Arriortua, M. I. (2014). Effects of using $(La_{0.8}Sr_{0.2})_{0.95}Fe_{0.6}Mn_{0.3}Co_{0.1}O_3$ (LSFMC), $LaNi_{0.6}Fe_{0.4}O_{3-\delta}$ (LNF) and $LaNi_{0.6}Co_{0.4}O_{3-\delta}$ (LNC) as contact materials on solid oxide fuel cells. *Journal of Power Sources, 248*, 1067–1076.

Morán-Ruiz, A., Vidal, K., Larrañaga, A., Porras-Vázquez, J. M., Slater, P. R., & Arriortua, M. I. (2015a). Evaluation of using protective/conductive coating on Fe-22Cr mesh as a composite cathode contact material for intermediate solid oxide fuel cells. *International Journal of Hydrogen Energy, 40*, 4804–4818.

Morán-Ruiz, A., Vidal, K., Larrañaga, A., Porras-Vázquez, J. M., Slater, P. R., & Arriortua, M. I. (2015b). Laser machining of $LaNi_{0.6}M_{0.4}O_{3-\delta}$ (M: Co, Fe) dip-coated on a Fe–22Cr mesh material to obtain a new contact coating for SOFC: Interaction between Crofer22APU interconnect and $La_{0.6}Sr_{0.4}FeO_3$ cathode. *International Journal of Hydrogen Energy, 40*, 8407–8418.

Nair, S. R., Purohit, R. D., Tyagi, A. K., Sinha, P. K., & Sharma, B. P. (2008). Low-temperature sintering of $La(Ca)CrO_3$ powder prepared through the combustion process. *Journal of the American Ceramic Society, 91*, 88–91.

Niewolak, L., Garcia-Fresnillo, L., Meier, G. H., & Quadakkers, W. J. (2015). Sigma-phase formation in high chromium ferritic steels at 650 °C. *Journal of Alloys and Compounds, 638*, 405–418.

Ou, D. R., & Cheng, M. (2014). Effect of pre-oxidation on the oxidation resistance of spinel-coated Fe–Cr ferritic alloy for solid oxide fuel cell applications. *Journal of Power Sources, 247*, 84–89.

Palcut, M., Mikkelsen, L., Neufeld, K., Chen, M., Knibbe, R., & Hendriksen, P. V. (2012). Efficient dual layer interconnect coating for high temperature electrochemical devices. *International Journal of Hydrogen Energy, 37*, 14501–14510.

Park, B.-K., Lee, J.-W., Lee, S.-B., Lim, T.-H., Park, S.-J., Park, C.-O., et al. (2013). Cu- and Ni-doped $Mn_{1.5}Co_{1.5}O_4$ spinel coatings on metallic interconnects for solid oxide fuel cells. *International Journal of Hydrogen Energy, 38*, 12043–12050.

Park, B.-K., Lee, J.-W., Lee, S.-B., Lim, T.-H., Park, S.-J., Song, R.-H., et al. (2012). La-doped $SrTiO_3$ interconnect materials for anode-supported flat-tubular solid oxide fuel cells. *International Journal of Hydrogen Energy, 37*, 4319–4327.

Park, B. K., Lee, J. W., Lee, S. B., Lim, T. H., Park, S. J., Park, C. O., et al. (2013). Cu- and Ni-doped Mn1.5Co1.5O4 spinel coatings on metallic interconnects for solid oxide fuel cells. *International Journal of Hydrogen Energy*, 38, 12043–12050.

Paulik, S. W., Baskaran, S., & Armstrong, T. R. (1999). Mechanical properties of calcium-substituted yttrium chromite. *Journal of Materials Science Letters, 18*, 819–822.

Persson, Å. H., Mikkelsen, L., Hendriksen, P. V., & Somers, M. A. J. (2012). Interaction mechanisms between slurry coatings and solid oxide fuel cell interconnect alloys during high temperature oxidation. *Journal of Alloys and Compounds, 521*, 16–29.

Pi, S. H., Lee, S. B., Song, R. H., Lee, J. W., Lim, T. H., Park, S. J., et al. (2013). Novel Ag–glass composite interconnect materials for anode-supported flat-tubular solid oxide fuel cells operated at an intermediate temperature. *Fuel Cells, 13*, 392–397.

Piccardo, P., Anelli, S., Bongiorno, V., Spotorno, R., Repetto, L., & Girardon, P. (2015). K44M ferritic stainless steel as possible interconnect material for SOFC stack operating at 600 °C: Characterization of the oxidation behaviour at early working stages. *International Journal of Hydrogen Energy, 40*, 3726–3738.

Przybylski, K., Brylewski, T., Durda, E., Gawel, R., & Kruk, A. (2014). Oxidation properties of the Crofer 22 APU steel coated with $La_{0.6}Sr_{0.4}Co_{0.2}Fe_{0.8}O_3$ for IT-SOFC interconnect applications. *Journal of Thermal Analysis and Calorimetry, 116*, 825–834.

Rashtchi, H., Sani, M. A. F., & Dayaghi, A. M. (2013). Effect of Sr and Ca dopants on oxidation and electrical properties of lanthanum chromite-coated AISI 430 stainless steel for solid oxide fuel cell interconnect application. *Ceramics International, 39*, 8123–8131.

Rufner, J., Gannon, P., White, P., Deibert, M., Teintze, S., Smith, R., et al. (2008). Oxidation behavior of stainless steel 430 and 441 at 800 °C in single (air/air) and dual atmosphere (air/hydrogen) exposures. *International Journal of Hydrogen Energy, 33*, 1392–1398.

Sachitanand, R., Sattari, M., Svensson, J.-E., & Froitzheim, J. (2013). Evaluation of the oxidation and Cr evaporation properties of selected FeCr alloys used as SOFC interconnects. *International Journal of Hydrogen Energy, 38*, 15328–15334.

Safikhani, A., & Aminfard, M. (2014). Effect of W and Ti addition on oxidation behavior and area-specific resistance of Fe–22Cr–0.5Mn ferritic stainless steel for SOFCs interconnect. *International Journal of Hydrogen Energy, 39*, 2286–2296.

Sakai, N., Yokokawa, H., Horita, T., & Yamaji, K. (2004). Lanthanum chromite-based interconnects as key materials for SOFC stack development. *International Journal of Applied Ceramic Technology, 1*, 23–30.

Sarda, V., Auvinen, S., Shemet, V., Quadakkers, W. J., Pihlatie, M., Kiviaho, J., et al. (2013). Long term resistivity behavior of SOFC interconnect/Ni-Mesh/anode interfaces. *ECS Transactions, 57*, 2279–2288.

Seabaugh, M. M., Ibanez, S., & Swartz, S. L. (2012). Protective coatings for metal alloys and methods incorporating the same. Google Patents.

Seo, H. S., Yun, D. W., & Kim, K. Y. (2012). Effect of Ti addition on the electric and ionic property of the oxide scale formed on the ferritic stainless steel for SOFC interconnect. *International Journal of Hydrogen Energy, 37*, 16151–16160.

Setz, L. F. G., Santacruz, I., León-Reina, L., De la Torre, A. G., Aranda, M. A. G., Mello-Castanho, S. R. H., et al. (2015). Strontium and cobalt doped-lanthanum chromite: Characterisation of synthesised powders and sintered materials. *Ceramics International, 41*, 1177–1187.

Shaigan, N., Qu, W., Ivey, D. G., & Chen, W. (2010). A review of recent progress in coatings, surface modifications and alloy developments for solid oxide fuel cell ferritic stainless steel interconnects. *Journal of Power Sources, 195*, 1529–1542.

Simner, S. P., Anderson, M. D., Xia, G. G., Yang, Z., & Stevenson, J. W. (2005). Long-term SOFC stability with coated ferritic stainless steel interconnect. Ceramic Engineering and Science Proceedings, 83–90.

Singhal, S. C. (2000). Advances in solid oxide fuel cell technology. *Solid State Ionics, 135*, 305–313.

Singhal, S. C., & Kendall, K. (2003). *High-temperature solid oxide fuel cells: Fundamentals, design and applications: Fundamentals, design and applications*. Oxford: Elsevier Science.

Smeacetto, F., De Miranda, A., Cabanas Polo, S., Molin, S., Boccaccini, D., et al. (2015). Electrophoretic deposition of $Mn_{1.5}Co_{1.5}O_4$ on metallic interconnect and interaction with glass-ceramic sealant for solid oxide fuel cells application. *Journal of Power Sources, 280*, 379–386.

Smeacetto, F., Salvo, M., Leone, P., Santarelli, M., & Ferraris, M. (2011). Performance and testing of joined Crofer22APU-glass-ceramic sealant-anode supported cell in SOFC relevant conditions. *Materials Letters, 65*, 1048–1052.

Stambouli, A. B., & Traversa, E. (2002). Solid oxide fuel cells (SOFCs): A review of an environmentally clean and efficient source of energy. *Renewable and Sustainable Energy Reviews, 6*, 433–455.

Stevenson, J. W., Yang, Z. G., Xia, G. G., Nie, Z., & Templeton, J. D. (2013). Long-term oxidation behavior of spinel-coated ferritic stainless steel for solid oxide fuel cell interconnect applications. *Journal of Power Sources, 231*, 256–263.

Tsai, M.-J., Chu, C.-L., & Lee, S. (2010). $La_{0.6}Sr_{0.4}Co_{0.2}Fe_{0.8}O_3$ protective coatings for solid oxide fuel cell interconnect deposited by screen printing. *Journal of Alloys and Compounds, 489*, 576–581.

Tunthawiroon, P., Li, Y., Tang, N., Koizumi, Y., & Chiba, A. (2015). Effects of alloyed Si on the oxidation behaviour of Co–29Cr–6Mo alloy for solid-oxide fuel cell interconnects. *Corrosion Science, 95*, 88–99.

Vidal, K., Morán-Ruiz, A., Larrañaga, A., Porras-Vázquez, J. M., Slater, P. R., & Arriortua, M. I. (2015). Characterization of $LaNi_{0.6}Fe_{0.4}O_3$ perovskite synthesized by glycine-nitrate combustion method. *Solid State Ionics, 269*, 24–29.

Wang, S., Lin, B., Chen, Y., Liu, X., & Meng, G. (2009a). Evaluation of simple, easily sintered $La_{0.7}Ca_{0.3}Cr_{0.97}O_{3-\delta}$ perovskite oxide as novel interconnect material for solid oxide fuel cells. *Journal of Alloys and Compounds, 479*, 764–768.

Wang, S., Lin, B., Dong, Y., Fang, D., Ding, H., Liu, X., et al. (2009b). Stable, easily sintered Ca–Zn-doped $YCrO_3$ as novel interconnect materials for co-fired yttrium-stabilized zirconia-based solid oxide fuel cells. *Journal of Power Sources, 188*, 483–488.

Wang, Z., Mori, M., & Itoh, T. (2010). Thermal expansion properties of $Sr1 - xLaxTiO_3$ ($0 \leq x \leq 0.3$) perovskites in oxidizing and reducing atmospheres. *Journal of the Electrochemical Society, 157*, B1783–B1789.

Wei, T., Liu, X., Yuan, C., Gao, Q., Xin, X., & Wang, S. (2014). A modified liquid-phase-assisted sintering mechanism for $La_{0.8}Sr_{0.2}Cr_{1-x}Fe_xO_{3-\delta}$—A high density, redox-stable perovskite interconnect for solid oxide fuel cells. *Journal of Power Sources, 250*, 152–159.

Wei, W., Chen, W., & Ivey, D. G. (2009). Oxidation resistance and electrical properties of anodically electrodeposited Mn–Co oxide coatings for solid oxide fuel cell interconnect applications. *Journal of Power Sources, 186*, 428–434.

Wongpromrat, W., Parry, V., Charlot, F., Crisci, A., Latu-Romain, L., Chandra-ambhorn, W., et al. (2015). Possible connection between nodule development and presence of niobium and/or titanium during short time thermal oxidation of AISI 441 stainless steel in wet atmosphere. *Materials at High Temperatures, 32*, 22–27.

Wu, J., Gemmen, R. S., Manivannan, A., & Liu, X. (2011). Investigation of Mn/Co coated T441 alloy as SOFC interconnect by on-cell tests. *International Journal of Hydrogen Energy, 36*, 4525–4529.

Wu, W., Guan, W., Wang, G., Liu, W., Zhang, Q., Chen, T., et al. (2014). Evaluation of $Ni_{80}Cr_{20}$/ $(La_{0.75}Sr_{0.25})_{0.95}MnO_3$ dual layer coating on SUS 430 stainless steel used as metallic interconnect for solid oxide fuel cells. *International Journal of Hydrogen Energy, 39*, 996–1004.

Wu, W., Guan, W., & Wang, W. (2015). Contribution of properties of composite cathode and cathode/electrolyte interface to cell performance in a planar solid oxide fuel cell stack. *Journal of Power Sources, 279*, 540–548.

Xiao, J., Zhang, W., Xiong, C., Chi, B., Pu, J., & Jian, L. (2015). Oxidation of $MnCu_{0.5}Co_{1.5}O_4$ spinel coated SUS430 alloy interconnect in anode and cathode atmospheres for intermediate temperature solid oxide fuel cell. *International Journal of Hydrogen Energy, 40*, 1868–1876.

Yang, P., Liu, C.-K., Wu, J.-Y., Shong, W.-J., Lee, R.-Y., & Sung, C.-C. (2012). Effects of pre-oxidation on the microstructural and electrical properties of La0.67Sr0.33MnO3 − δ coated ferritic stainless steels. *Journal of Power Sources, 213*, 63–68.

Yang, X., Tu, H., & Yu, Q. (2015). Fabrication of Co_3O_4 and $La_{0.6}Sr_{0.4}CoO_{3-\delta}$–$Ce_{0.8}Gd_{0.2}O_{2-\delta}$ dual layer coatings on SUS430 steel by in-situ phase formation for solid oxide fuel cell interconnects. *International Journal of Hydrogen Energy, 40*, 607–614.

Yang, Z., Xia, G., Simner, S. P., & Stevenson, J. W. (2005). Thermal growth and performance of manganese cobaltite spinel protection layers on ferritic stainless steel SOFC interconnects. *Journal of the Electrochemical Society, 152*, A1896–A1901.

Yang, Z., Xia, G. G., Li, X. H., & Stevenson, J. W. (2007). (Mn, Co)3O4 spinel coatings on ferritic stainless steels for SOFC interconnect applications. *International Journal of Hydrogen Energy, 32*, 3648–3654.

Yokokawa, H., Sakai, N., Kawada, T., & Dokiya, M. (1991). Chemical thermodynamic considerations in sintering of $LaCrO_3$-based perovskites. *Journal of the Electrochemical Society, 138*, 1018–1027.

Yoon, K. J., Cramer, C. N., Stevenson, J. W., & Marina, O. A. (2010a). Advanced ceramic interconnect material for solid oxide fuel cells: Electrical and thermal properties of calcium- and nickel-doped yttrium chromites. *Journal of Power Sources, 195*, 7587–7593.

Yoon, K. J., Cramer, C. N., Thomsen, E. C., Coyle, C. A., Coffey, G. W., & Marina, O. A. (2010b). Calcium- and cobalt-doped yttrium chromites as an interconnect material for solid oxide fuel cells. *Journal of the Electrochemical Society, 157*, B856–B861.

Yoon, K. J., Stevenson, J. W., & Marina, O. A. (2011). High performance ceramic interconnect material for solid oxide fuel cells (SOFCs): Ca- and transition metal-doped yttrium chromite. *Journal of Power Sources, 196*, 8531–8538.

Yun, D. W., Seo, H. S., Jun, J. H., & Kim, K. Y. (2013). Evaluation of Nb- or Mo-alloyed ferritic stainless steel as SOFC interconnect by using button cells. *International Journal of Hydrogen Energy, 38*, 1560–1570.

Yun, D. W., Seo, H. S., Jun, J. H., Lee, J. M., & Kim, K. Y. (2012). Molybdenum effect on oxidation resistance and electric conduction of ferritic stainless steel for SOFC interconnect. *International Journal of Hydrogen Energy, 37*, 10328–10336.

Zhang, H., Wu, J., Liu, X., & Baker, A. (2013). Studies on elements diffusion of Mn/Co coated ferritic stainless steel for solid oxide fuel cell interconnects application. *International Journal of Hydrogen Energy, 38*, 5075–5083.

Zhang, W., Pu, J., Chi, B., & Jian, L. (2011). $NiMn_2O_4$ spinel as an alternative coating material for metallic interconnects of intermediate temperature solid oxide fuel cells. *Journal of Power Sources, 196*, 5591–5594.

Zhang, W., Yan, D., Yang, J., Chen, J., Chi, B., Pu, J., et al. (2014a). A novel low Cr-containing Fe–Cr–Co alloy for metallic interconnects in planar intermediate temperature solid oxide fuel cells. *Journal of Power Sources, 271,* 25–31.

Zhang, Y., Javed, A., Zhou, M., Liang, S., & Xiao, P. (2014b). Fabrication of Mn–Co spinel coatings on Crofer 22 APU stainless steel by electrophoretic deposition for interconnect applications in solid oxide fuel cells. *International Journal of Applied Ceramic Technology, 11,* 332–341.

Zhang, Z., Yue, D., Yang, G., Chen, J., Zheng, Y., Miao, H., et al. (2015). Three-dimensional CFD modeling of transport phenomena in multi-channel anode-supported planar SOFCs. *International Journal of Heat and Mass Transfer, 84,* 942–954.

Zhong, Z. (2006). Stoichiometric lanthanum chromite based ceramic interconnects with low sintering temperature. *Solid State Ionics, 177,* 757–764.

Zhu, W. Z., & Deevi, S. C. (2003a). Development of interconnect materials for solid oxide fuel cells. *Materials Science and Engineering A, 348,* 227–243.

Zhu, W. Z., & Deevi, S. C. (2003b). Opportunity of metallic interconnects for solid oxide fuel cells: A status on contact resistance. *Materials Research Bulletin, 38,* 957–972.

Catalysts and Processes in Solid Oxide Fuel Cells

Alfonsina Pappacena, Luca Bardini and Marta Boaro

Abstract This chapter describes the requirements necessary for the development of suitable solid oxide fuel cell electrodes. The materials and the chemical and electrochemical processes involved in the anode compartment are then specifically considered. Particular emphasis is given to the advances that have been achieved and the necessary improvements that are still needed in order to operate at intermediate temperature with renewable resources or light hydrocarbons. Moreover, the relationship between catalysis and electrocatalysis and the issues correlated with an integrated or direct oxidation will be introduced.

1 Electrode Properties

By definition, the electrodes in a fuel cell are the place where the redox processes take place: the oxygen reduction at the cathode and the H_2 oxidation at the anode. A prerogative of solid oxide fuel cells (SOFCs) is the possibility to use other fuels such as CO, CH_4, alcohols, and even solid carbon besides the standard H_2.

An SOFC electrode's main property is its capability to catalyze the proper electrochemical reaction and to provide pathways for the released electrons to reach the current collectors. To be suitable for this kind of application, a material must possess the following characteristics: (i) mixed electronic conductivity (EC) and ionic conductivity (IC), (ii) good stability at SOFC operating conditions (e.g. stable at high temperature and reducing/oxidizing atmosphere, tolerant to impurities, and

A. Pappacena · L. Bardini · M. Boaro (✉)
Polytechnic Department of Engineering and Architecture, University of Udine, Via del Cotonificio, 108, 33100 Udine, Italy
e-mail: marta.boaro@uniud.it

M. Boaro and A.S. Aricò (eds.), *Advances in Medium and High Temperature Solid Oxide Fuel Cell Technology*, CISM International Centre for Mechanical Sciences 574,
DOI 10.1007/978-3-319-46146-5_5

insensitive to redox regime), (iii) compatibility with the electrolyte and interconnect materials, and (iv) electrocatalytic activity toward oxidation reactions (1, 2 at the anode side) and oxygen reduction (3, at the cathode side)

$$H_2 + 1/2\,O^{2-}_{(l)} \leftrightarrow H_2O + 2e^- \tag{1}$$

$$C_nH_{2m} + (2n + m)O^{2-} \leftrightarrow nCO_2 + mH_2O + (4n + 2m)e^- \tag{2}$$

$$O_2 + 4e^- \leftrightarrow 2O^{2-} \tag{3}$$

Moreover, electrode's microstructure must be designed so as to allow transport of products and reactants (porosity >30 vol.%) and, in specific configurations, to work as a support.

From Eqs. (1) to (3), it is evident that an electrochemical reaction can take place only if gas-reactant molecules, electrons, and oxygen anions coexist. If the conduction is purely electric in the electrode and ionic in the electrolyte, the three phases come into simultaneous contact at the interface between electrolyte and electrode. This zone is denominated "three-phase boundary" (TPB), and maximizing its length is crucial for the optimization of cell performance. This can be achieved by employing *met*al and *cer*amic composites (*cermets*) where metallic catalysts and an ionic conductor are mixed in the proper amount to reach the reciprocal percolation. In this way, all the metal will be "connected" and allows electrons to flow from the reaction sites to the external circuit and there will be "fins" that transfer ions from the electrolyte to the electrode (see Fig. 1). Alternatively, a two-dimensional development

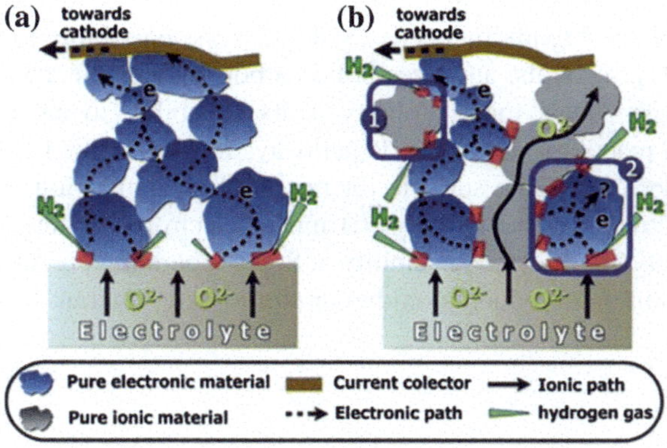

Fig. 1 Representations of the triple-phase boundary (TPB) zones (in *red color*) in **a** a pure electronic conductor in contact with an electrolyte, showing the movement of oxygen ions in the electrolyte toward the TPB, which is limited to the electrolyte–electrode interface, and the movement of the electrons, produced in the TPB, toward the current collector. **b** In a cermet material (with a metal and a pure ionic conductor), some potential TPB places may be deactivated because either there is no oxygen pathway toward the TPB zones (*1*), or the electrons (*2*) cannot reach the current collector [reprinted with permission from The Royal Society of Chemistry @ 2010, (Ruiz-Morales et al. 2010)]

of TPB can be achieved through the use of ionic and electronic mixed conductors (MIEC) (Fig. 2).

In any case, there are many possible combinations for the preparation of composites as it is shown in Fig. 3g–j. Figure 3 summarizes also the functionalities that electrodes should have to operate effectively (vertices) and the classes of materials used for electrode assembly (Irvine et al. 2016). The picture evidences that it is difficult to find a perfect material that alone embodies all the desired characteristics. In this light, double perovskites, i.e., oxygen-deficient layered perovskites, seem promising. For example, a good electroactivity and excellent redox stability with tolerance to coke and sulfur contamination from hydrocarbon fuels are demonstrated by (Sengodan et al. 2015) for the catalyst $PrBaMn_2O_{5+\delta}$.

The requirements for electrode materials become more and more stringent with the development of SOFC technology which aims to operate at intermediate temperature and with complex fuels (hydrocarbons, alcohols, biogas). This challenges scientists to investigate new materials with a higher activity toward the oxygen reduction and toward the direct oxidation of carbon-based fuels (Zhao et al. 2013; Brett et al. 2008 and Ge et al. 2012).

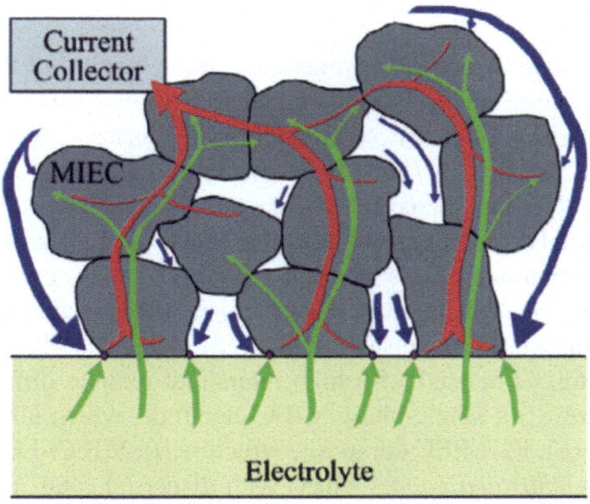

Fig. 2 Schematic diagram of charge and mass transport within and on the surface of a single-phase mixed conducting porous electrode: *green path* = flux of holes or electrons, *red path* = flux of oxygen vacancies/O^{2-} ions, and *blue path* = reactant gases or surface adsorbates. In this case, the electrochemical reaction can occur on the surfaces of MIEC exposed to gas [adapted from (Liu et al. 2011), © Elsevier Ltd. Open access under CC-BY-NC-ND license]

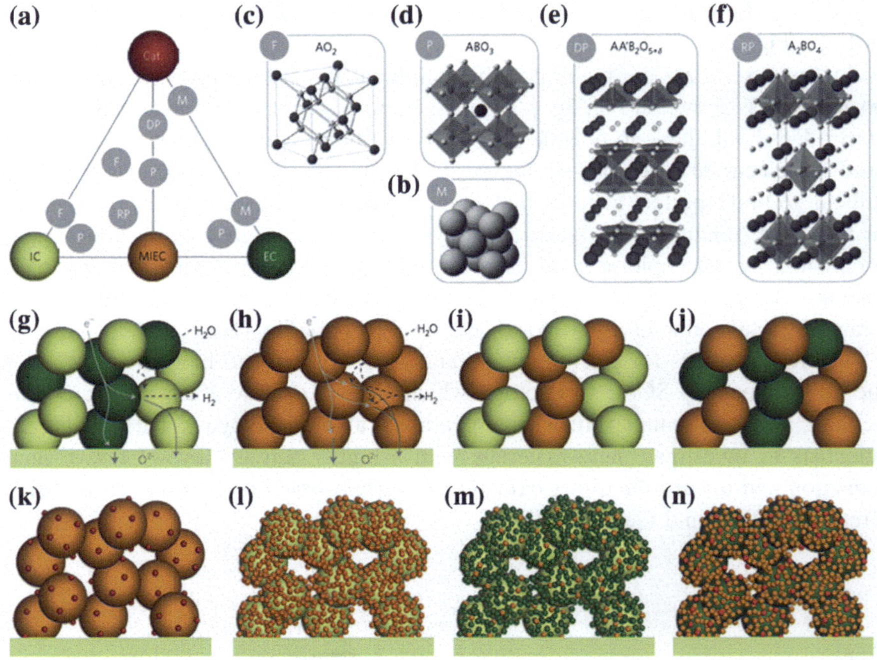

Fig. 3 SOFC electrode materials and microstructure: On the vertices of the triangle the three key functionalities that an electrode should exhibit and classes of materials employed to deliver these functionalities (*green sphere*): *IC* Ionic conduction, *EC* Electronic conduction, *Cat* Catalytic activity for the desirable reaction, *MIEC* Mixed electronic and ionic conduction; M = crystal structure of a metal (as Ni) that behaves as catalyst and electronic conductors, F = fluorite structures which show catalytic and ionic functions, P = crystal structure of perovskite, which shows a wide range of functionalities, but often not excels in catalytic activity, DP = double perovskite, which is an oxygen-deficient layered structure, which results almost an ideal electrode since most of the functionalities result optimized; RP = Ruddlesden–Popper phases, a perovskite-related structure, associated with high interstitial oxygen diffusion rates. *From g to j type of possible composites*: (*g*); single-phase MIEC electrode, which also illustrates a 2 PB fuel electrode in SOEC mode (*h*); IC–MIEC composite structure (*i*); MIEC–EC composite structure (*j*); *From k to n examples of infiltrated anode*: MIEC with dispersed catalyst particles (*k*); IC coated with a percolating layer of MIEC (*l*); IC coated with a percolating layer of EC and MIEC (*m*); EC coated with a percolating layer of MIEC and dispersed catalyst (*n*). Reprinted with permission from Macmillan Publishers Ltd: (Nature Energy) (Irvine et al. 2016)

2 Electrode Materials

2.1 Cathodes

A cathode material must catalyze the dissociation and reduction of oxygen (ORR, reaction 3). The kinetics of this reaction is complex and comprises many steps. Figure 4 outlines some of the mechanisms suggested in the literature to be important in determining the reaction rate (Adler 2004).

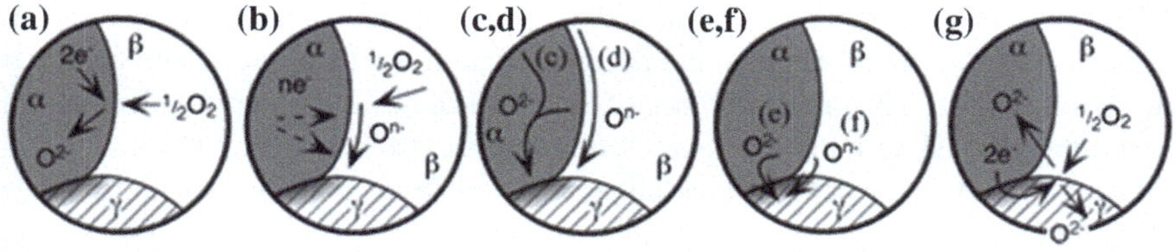

Fig. 4 Some mechanisms thought to govern oxygen reduction in SOFC cathodes. Phases a, *b*, and γ refer to the electronic phase, gas phase, and ionic phase, respectively: **a** Incorporation of oxygen into the bulk of the electronic phase (if mixed conducting); **b** adsorption and/or partial reduction of oxygen on the surface of the electronic phase; **c** bulk or **d** surface transport of O^{2-} or O^{n-}, respectively, to the α/γ interface, **e** electrochemical charge transfer of O^{2-} or **f** combinations of O^{n-} and e-, respectively, across the α/γ interface, and **g** rates of one or more of these mechanisms wherein the electrolyte itself is active for generation and transport of electroactive oxygen species. Reprinted with permission from American Chemical Soc. (Adler 2004)

Oxygen molecules are generally thought to adsorb somewhere onto one or more solid surface(s), where they undergo catalytic and/or electrocatalytic reduction steps to form partially reduced ionic/atomic species. Before, after, or between partial reduction steps, these species must be transported along surfaces, interfaces, or inside the bulk of the electrode material(s) to the electrolyte, where they are fully and formally incorporated as electrolytic O^{2-}. Due to this complex kinetics, the cathode is responsible for a large portion of cell voltage losses. If, how, and where any of these processes happen and what step(s) are rate-determining for a particular electrode depend on the investigated material as well as its microstructure and the conditions of testing.

Many cathode materials belong to the perovskite family (Jun et al. 2016). Perovskite with the general formula ABO_3 consists of corner-shared BO_6 octahedra together with A cations at the corner of the unit cell (Fig. 5).

Fig. 5 Outline of the ideal cubic perovskite structure ABO_3 [from web https://www.kek.tp/en/NewsRoom/Realese/20140516150000/]

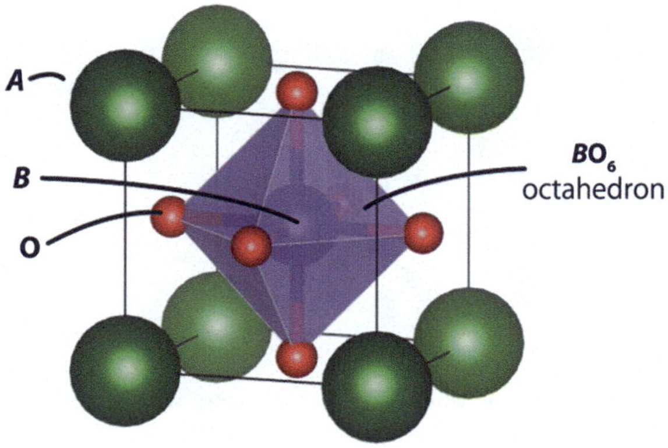

The perovskite structure can "iron out" considerable lattice mismatches between the A–O and B–O bond lengths and accommodate more than one A site and/or B site cation species. The great tolerance of lattice vacancies and intergrowth structure extend the perovskite family to layered or complex structures (Fig. 6).

Generally speaking, perovskites allow first-row transition metal cations to be introduced into B site and rare earth elements into A site. The doping allows us to modify the properties of the oxide. First-row transition metal cations usually exhibit multivalence in function of oxygen pressure, so a proper choice of B dopants can

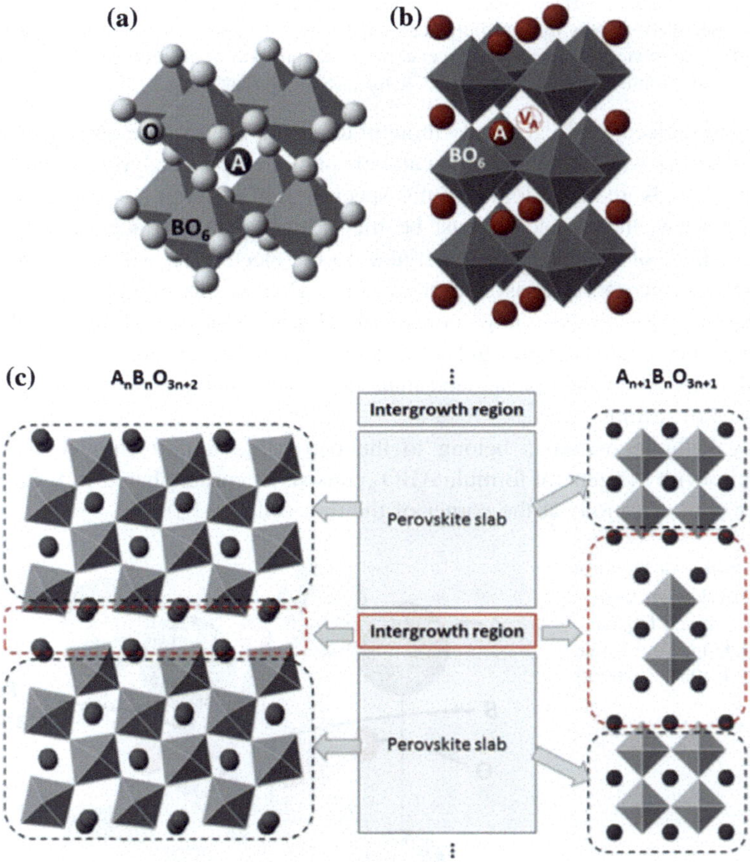

Fig. 6 Accommodation of no stoichiometry with respect to the ideal perovskite structure: **a** the ideal perovskite structure, **b** deficiency through vacancies denoted by *red hollow spheres*, **c** excess (A/B>1 and /or O/B>3) through intergrowths (marked with red dotted lines). In the case where only oxygen excess is present, the intergrowth comprises of the region where perovskite slabs are offset in such way to allow accommodation of extra oxygen ions (e.g., the $A_nB_nO_{3n+2}$ series $La_xSr_{1-x}TiO_{3+x/2}$), A-site super-stoichiometry is incorporated by an intergrowing of perovskite slabs with other crystal structures such as rock salt lattice, leading to the well-known Ruddlesden–Popper structure (e.g., to the Ruddlesden–Popper phases, $A_{n+1}B_nO_{3n+1}$) [reprinted with permission from Macmillan Publishers Ltd: (Nature Chem) (Neagu et al. 2013)]

modify electronic conductivity and catalytic properties of a given perovskite family (Goodenough 2004). Doping in A affects the concentration of vacancies and the ionic and electronic conductivity of materials.

The most popular material used for cathodes is strontium-doped lanthanum manganite, $La_{1-x}Sr_xMnO_3$ (LSM) for fuel cells operating at 800–1000 °C (Jiang 2008). In this temperature range, LSM has high electronic conductivity (200 S/cm at 900 °C) but is a very poor ionic conductor (10^{-8} S/cm). It follows that, the oxygen reduction is restricted to the electrode/electrolyte interface, and the rate-determining step of reaction is the oxygen surface exchange. To increase the three-phase boundary, LSM is generally combined with an ionic conductor such as yttrium-doped zirconia (YSZ).

In order to develop cathodes for intermediate- and low-temperature applications, the research has been recently focused on materials with an increased oxygen exchange rate and a high ionic conductivity (Sun et al. 2010). Alternative perovskite materials such as $La_{1-x}Sr_xFeO_3$ (LSF), $La_{1-x}Sr_xCoO_3$ (LSC), and $La_{1-x}Sr_xFe_{1-y}Co_yO_3$ (LSCF) are also being investigated. These compounds generally show a higher ionic conductivity (0.4 S/cm at 800 °C) in comparison with LSM and reasonably good performance. However, they suffer from a poor chemical stability due to solid-state reactions with the YSZ electrolyte during the manufacturing process. A strategy to prevent such an undesired interaction is the deposition of a thin protective layer of gadolinium- or samarium-doped ceria (GDC, SDC) (Adijanto et al. 2001).

Another way to tackle the problem is to prepare the cathode via impregnation (see further in this chapter) of the suitable catalyst into the porous electrolyte structure, which requires lower sintering temperatures compared to the conventional manufacturing methods (Vohs and Gorte 2009).

An interesting option is the use of more chemically stable perovskite materials that also offer higher ionic and electronic conductivity at lower temperatures. It was found, for example, that $Pr_{0.8}Sr_{0.2}FeO_3$ showed no solid-state reactions with YSZ and an area-specific polarization resistance of 0.204 and 0.164 Ω cm^2, respectively, at 800 and 850 °C (Piao et al. 2007). Another interesting material for its good oxygen surface exchange and diffusion properties is $Ba_{0.5}Sr_{0.5}Co_{0.6}Fe_{0.4}O_{3-\delta}$ (BSCF). Shao and Haile (2004) demonstrated for the first time that the area-specific resistances of this material are 0.055–0.071 and 0.51–0.61 Ω cm^2 at 600 and 500 °C, respectively. Despite its excellent electrocatalytic activity, the use of BSCF can be problematic since its thermal expansion coefficient (TEC) (20×10^{-6} K^{-1} between 50 and 1000 °C) does not match the TEC of the YSZ electrolyte (Wei et al. 2006). Moreover, BSCF, as most of the perovskite oxides containing alkaline earth elements, is susceptible to CO_2 poisoning, especially at temperatures below 550 °C where poisoning effects are irreversible (Yan et al. 2006).

An alternative strategy retaining some of the characteristics of perovskite is the development of parent structures that would promote ionic and electronic conductivity. Several A_2BO_4- structures of the Ruddlesden–Popper type have been proposed as SOFC cathodes, including those A = La, Sr, Ba, Pr, Nd, and M = Ni, Cu, Co Fe (Skinner et al. 2013). Among the phases first investigated, there are the

$La_2NiO_{4+\delta}$ family of compounds (Yoo et al. 2012). They are predominantly *p*-type semiconductors with a conductivity around of 100 S/cm. These phases consist of alternating layers of perovskite structure ABO_3 and rock salt structure AO along the crystallographic c direction (Fig. 7).

In these compounds, ionic transport involves mobility of interstitial oxygen ions in the rock salt-type layers and vacancies in the perovskite layers, which resulted in a low polarization resistance (Tsipis and Kharton 2008). Increasing the number of perovskite blocks in the RP phases leads to an increase in the B site valences, which leads to changes in the electronic conductivity with significant increases for $n = 2$ or $n = 3$ (Takahashi et al. 2010). To date, there are no reports on the ionic conductivity in these higher-order phases, and hence, there is a considerable room for future developments of such RP materials, incorporated either as single cathodes or as electronic conductors to enhance current collection.

Further information can be found in the cited literature (Sun et al. 2010; Tsipis and Kharton 2008; Skinner et al. 2013).

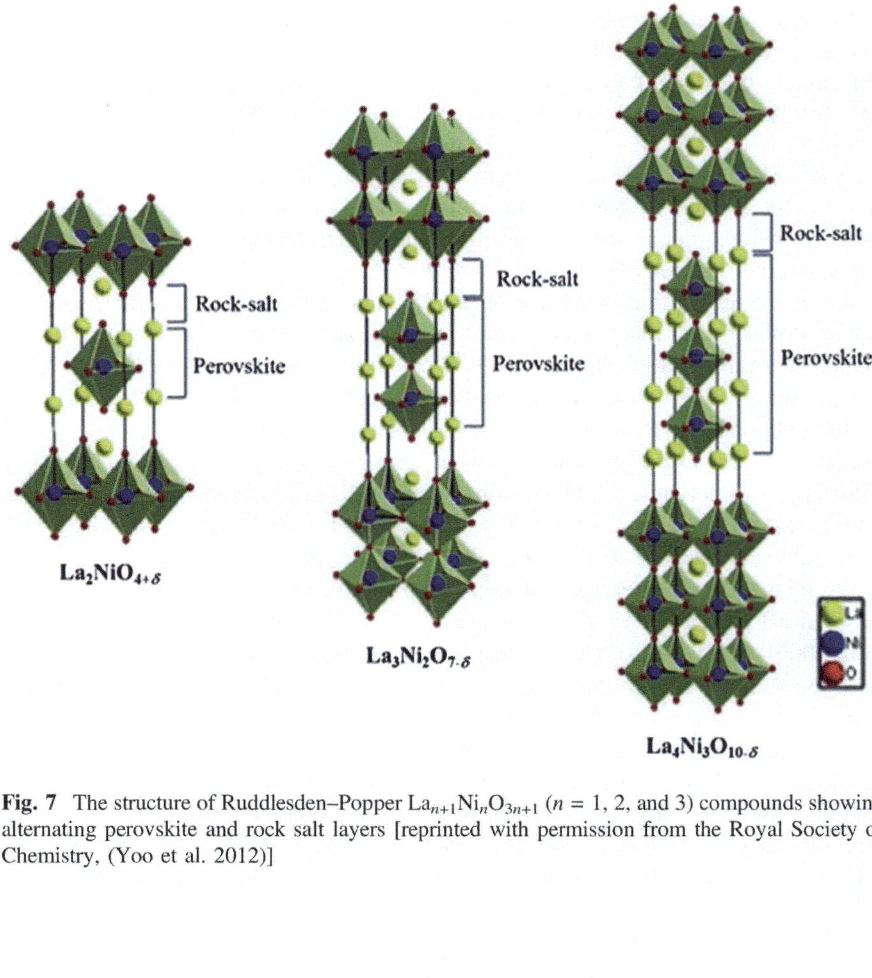

Fig. 7 The structure of Ruddlesden–Popper $La_{n+1}Ni_nO_{3n+1}$ ($n = 1$, 2, and 3) compounds showing alternating perovskite and rock salt layers [reprinted with permission from the Royal Society of Chemistry, (Yoo et al. 2012)]

2.2 *Anodes*

Materials suitable for anode applications have to satisfy the general requirements listed in the first paragraph and catalyze the electro-oxidation of fuels (reaction 1, 2). Depending on the anode configuration, materials may also be selective and promote the internal reforming (see below).

Hydrogen and hydrocarbon oxidation (HOR) can occur following different mechanisms, depending on the microstructure and composition of the anode and on the operating conditions. Figure 8 shows the possible mechanisms of H_2 electro-oxidation in a traditional Ni-based anode.

The reader can refer to several reviews for a detailed description of anode materials and their relative configurations (Atkinson et al. 2004; Gorte and Vohs 2009; Mahato et al. 2015). Here, we highlight the main differences among the materials focusing on their use directly with hydrocarbons and biofuels (Ge el al. 2012).

Ni-Based Anodes In the case of YSZ-based fuel cells, the most studied and used anode is a cermet of nickel and YSZ. For ceria-based electrolytes, nickel is generally mixed with a ceria-doped oxide [gadolinium- or samarium-doped ceria (GDC, SDC)].

In a typical SOFC, NiO and YSZ are mixed and cosintered together at high temperature and the NiO is reduced in situ via a careful procedure of reduction. The reduction of NiO usually occurs at high temperature (800 °C) and takes several hours to complete, it leads to a porous anode owing to the different density of Ni and NiO.

Both the Ni and YSZ phases have convenient properties to make a suitable anode. Ni metal has excellent catalytic properties toward the hydrogen dissociation reaction at SOFC operating temperatures and high electrical conductivity (10^5–10^4 S/cm). Moreover, NiO has a melting point of 1955 °C that allows us to cosinter NiO and YSZ at high temperature (1500 °C) without the risk of a harmful

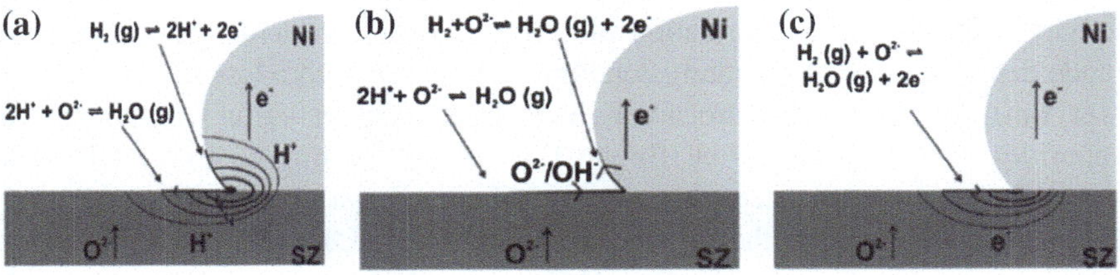

Fig. 8 Schematic representations of the probable mechanisms of anode processes proposed by various research groups: **a** H_2 adsorption on the Ni and formation of H^+ ions. The latter migrate to the site for water formation either along the surfaces or through bulk Ni- and bulk-stabilized ZrO_2 (SZ). **b** Migration of O^{2-} or OH^- from the SZ to the Ni along the respective surfaces resulting in the formation of water at Ni sites. **c** Water formation occurs at stabilized zirconia surface and electron transport to the Ni is possible along the surface or via the SZ. Reprinted with permission from 2015 Elsevier Ltd (Mahato et al. 2015)

interaction between the two. YSZ is a strong material, easy to form in comparison with other electrolyte materials, and relatively inexpensive. When mixed with NiO, it modifies the expansion thermal coefficient of the latter and provides ion channels for carrying O^{2-} into the electrode. This, in turn, facilitates the formation of an optimal interface and the extension of TPB.

As both YSZ and Ni play an important role in anode performance, the balance between the amounts of the two phases is critical to optimize cell performance (Lee et al. 2002). The Ni-YSZ percolation threshold is at about 30 % (v/v) of Ni with a conductivity of 10 S/cm, and an amount over 50 % (v/v) permits to reach higher conductivity (1000 S/cm). In general, the volume ratio of Ni to YSZ phases varies from 35:65 to 55:45. Besides composition, characteristics of starting oxide powders , such as the reciprocal particle size and particle shape, have a profound effect on the SOFC performance; thus, they can be important variables for the engineering of an adequate anode microstructure (Fig. 9). In any case, the optimum particle size and shape depend on the deposition techniques used and on the processing steps (Jiang et al. 2000; Primdahl and Mogensen 1997; Mahato et al. 2015).

Taking into account how many factors can influence the performance of a Ni-YSZ cermet, not surprising is the difficulty to reproduce and compare cermet electrode data. Most of the results were obtained in the temperature range of 800–1000 °C, and the polarization resistance values were found to vary by a factor from 10 to 10^2, especially when extrapolated to low-temperature regions. Figure 10 highlights some of the factors that affect the performance of a Ni-YSZ electrode.

The major problem with Ni-YSZ anodes is their redox instability (Busawon et al. 2008) which is associated with the re-expansion of Ni upon its reoxidation. Reoxidation can occur near the seal in the event of a leak or during shutdown procedure because Ni easily oxidizes when exposed to high H_2O:H_2 ratio at low temperature. Moreover, Ni is sensitive to poisoning. For example, common impurities in starting materials, such as Na_2O and SiO_2, may segregate as a glassy sodium silicate phase at the anode/electrolyte interface reducing performance in few hundred hours, especially if impurities' concentration is higher than 100 ppm.

Moreover, besides their high manufacturing flexibility and high performance, Ni-YSZ systems are not ideal to work directly with hydrocarbons. The cracking activity of Ni causes the formation of graphitic fibers or tar deposition into the anode depending on the operating conditions of the cell with a loss of performance over time. In addition, hydrocarbon fuels often contain sulfur compounds as impurities which poison nickel (Bøgild Hansen and Rostrup-Nielsen 2010). The nickel deactivation can occur through two different ways depending on the operating temperature and on the sulfur concentration inside the fuel. At high temperature, the dissociative adsorption of H_2S on metal surface with the irreversible formation of bulk nickel sulfide is strongly favored even with low concentrations of pollutant. At low temperature, the degradation is reversible and due to surface sulfides' deposition. The tolerance toward sulfur poisoning has been recently revised (Da Silva and Heck 2015; Gong et al. 2007; Yang et al. 2012), and the reader can refer to the cited references for more details.

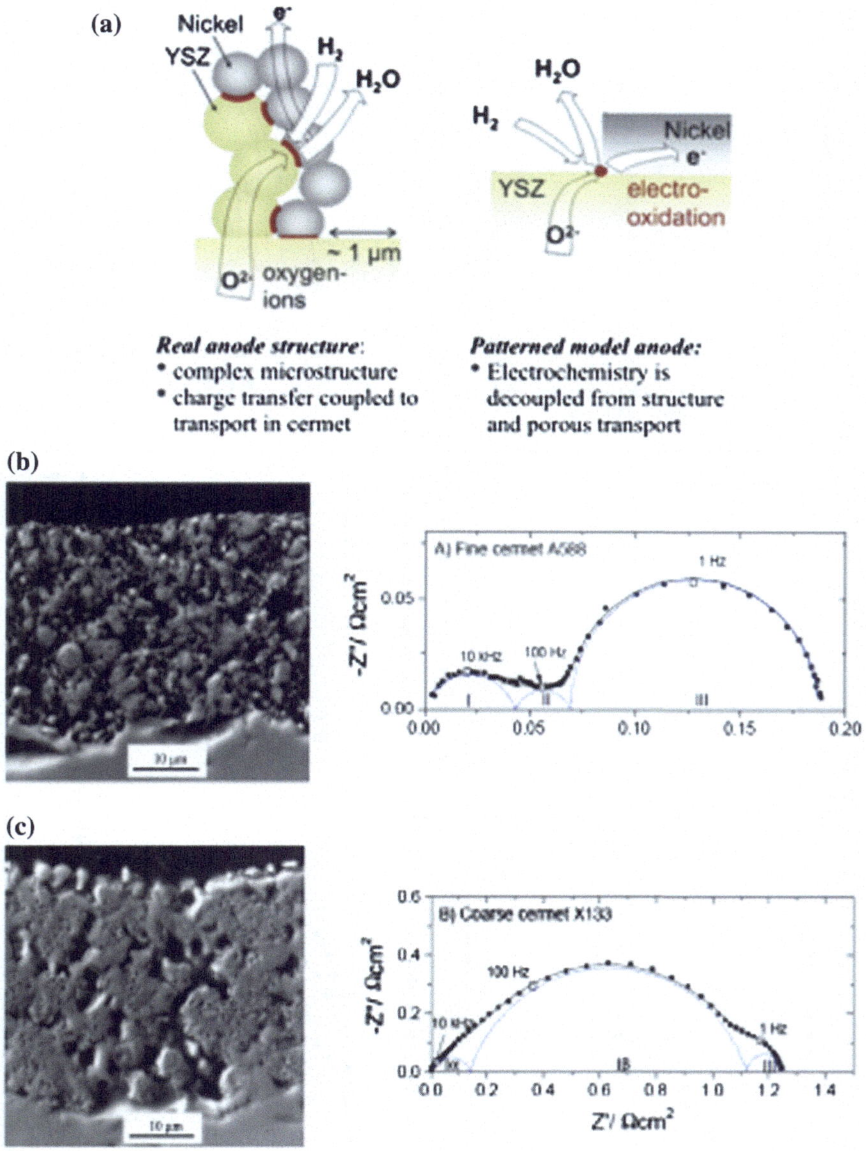

Fig. 9 **a** Schematic showing how a cermet structure offers extended TPB length, and so how the TPB can be modified or affected by the reciprocal size of Ni and YSZ particles. **b** Fine cermet Ni/YSZ anode prepared by spray painting with Ni/YSZ volume ratio 40/60 and the related impedance data. **c** Coarse cermet anodes prepared by screen printing and the related impedance data. Adapted (Mahato et al. 2015) by permission from © 2015, Elsevier Ltd

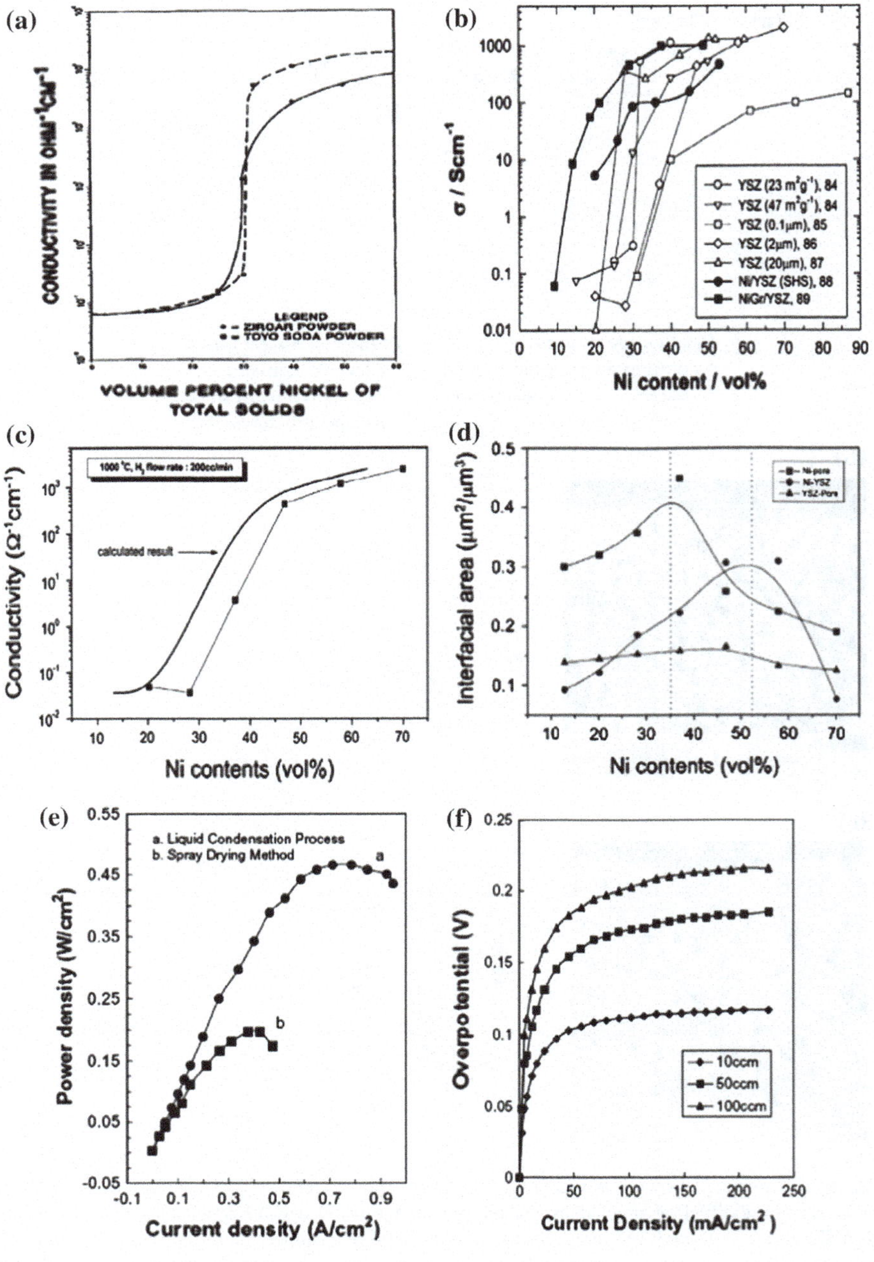

Fig. 10 (continued)

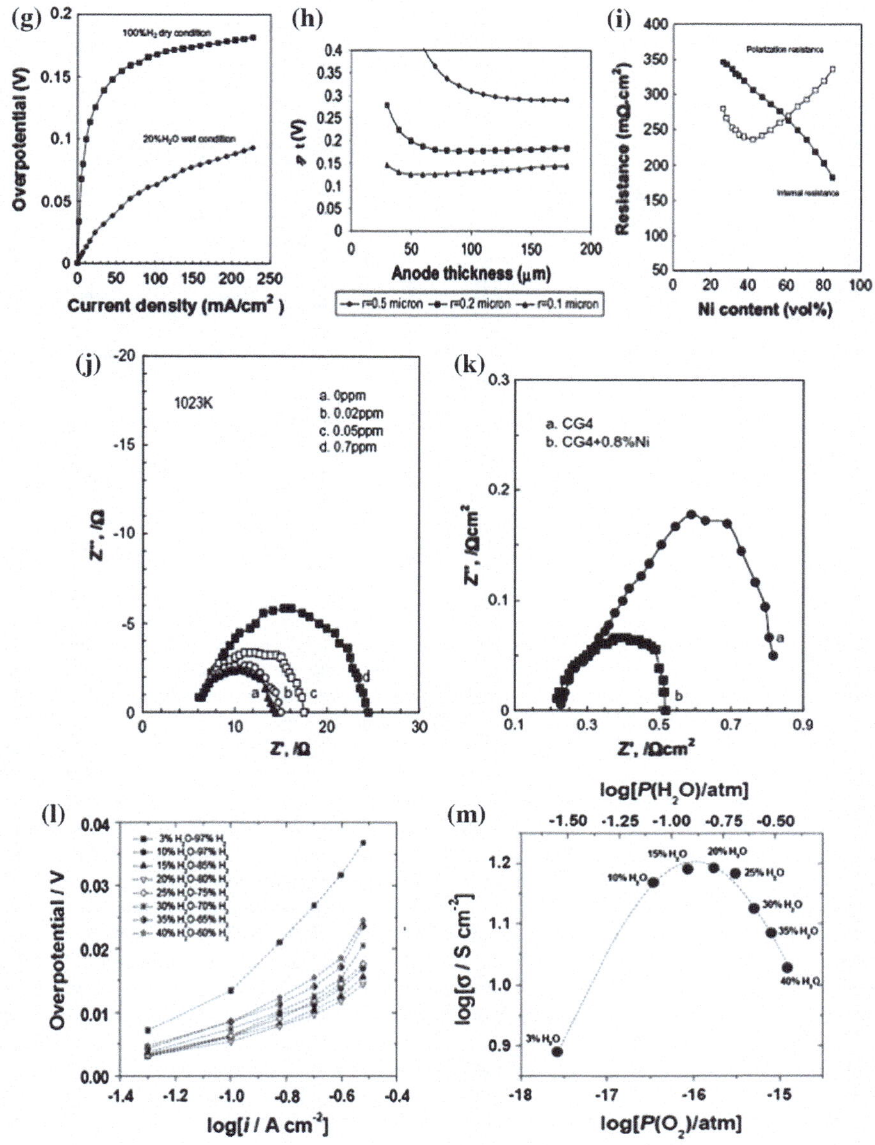

Fig. 10 (continued)

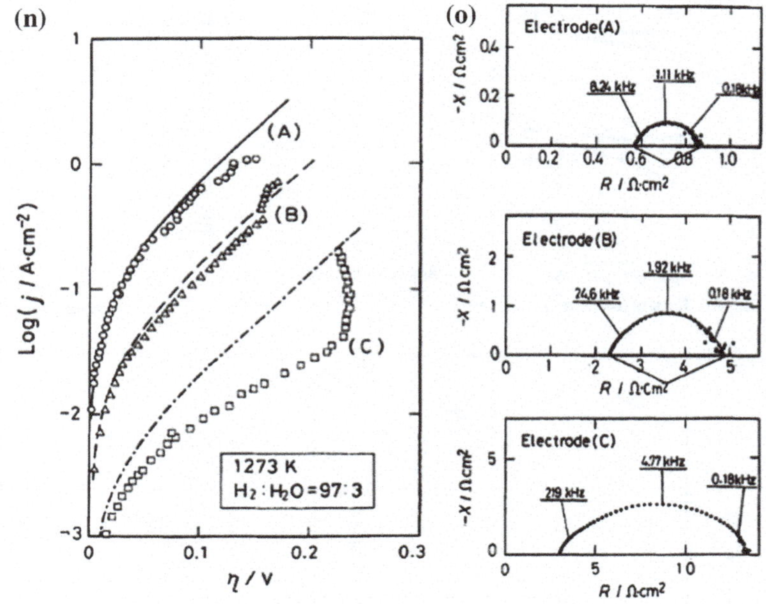

Fig. 10 **a** Conductivity of Ni–YSZ cermets versus *the volume percent of Ni fabricated with zirconia powders from two different sources*, viz. Toya Soda and Zircar from Ref. (Vohs and Gorte 2009), **b** conductivity of Ni–YSZ cermet anodes as a function of *Ni content* measured at 1000 °C in humidified H_2 from Ref. (Vohs and Gorte 2009), **c** variation of electrical content as a function of Ni content from Ref. (Wang et al. 2005), **d** variation of *interfacial area of Ni pore*, Ni–YSZ, and YSZ pore from Ref. (Wang et al. 2005), **e** comparison of the power density output of the unit *cell fabricated using two different approaches* from Ref. (Wang and Lu 1996), **f** dependence of overpotential on the current density of Ni–YSZ cermet anode with 40 vol.% of Ni measured at *various gas flow rates* from Ref. (Wei et al. 2006), **g** dependence of overpotential on the current density of Ni–YSZ cermet anode with 40 vol.% of Ni measured in *dry and wet conditions* at 850 ° C from Ref. (Wei et al. 2006), **h** influence of *particle size and anode thickness* on the polarization behavior for H_2 oxidation (hydrogen humidified at 50 °C, or H_2:H_2O = 87.6:12.4) from Ref. (Xiang et al. 2012), **i** effect of *Ni content* on the polarization and internal resistance of Ni–YSZ anode tested in 200 ml/min of H_2 as anode gas and 1000 ml min^{-1} of oxygen as cathode gas from Ref. (Wang and Lu 1996), **j** AC impedance spectra of SOFC with Ni–YSZ cermet anode measured after equilibration in *humidified H_2 mixed with several concentrations of H_2S* from Ref. (Xuan et al. 2009), **k** AC impedance spectra of CG4 electrode fed with humidified H_2 showing dramatic suppression of the semicircle in the low-frequency region as a result of *minor Ni addition* from Ref. (Yakabe et al. 2000), **l** the overpotential of Ni–YSZ cermet (50:50 vol.%) anode *as a function of current density* at 1000 °C from Ref. (Yan et al. 2006), **m** interfacial conductivity of Ni–YSZ cermet (50:50 vol.%) anode *as a function of oxygen partial pressure* measured at 1000 °C under OCP; anode gas used: x %H_2O–(100 − x) %H_2 from Ref. (Yan et al. 2006), **n** polarization behavior of Ni–YSZ anode prepared with constant nickel content (40 vol.%), at a constant precalcination temperature (1400 °C) and at *various baking temperatures* (*A*) 1500 °C, (*B*) 1400 ° C, and (*C*) 1200 °C, and (o) impedance behavior of Ni–YSZ cermet anodes prepared at a constant precalcination temperature (1400 °C) and at various baking temperatures (*A*) 1500 °C, (*B*) 1400 ° C, and (*C*) 1200 °C from Ref. (Medford et al. 2015). Reprinted with permission from 2015 © Elsevier (Mahato et al. 2015). For the cited reference, see *source*

To overcome the drawbacks inherent in using a Ni-YSZ electrode directly with hydrocarbon, two strategies are adopted: The first implies a modification of Ni reactivity, and the second consists of using different materials as constituents of the anode. A different approach is to proceed to an external or an internal reforming of hydrocarbons to obtain syngas which is in turn directly converted into electricity at the anode (see below). Figure 11 summarizes the possible methods to modify nickel-based anodes.

A common practice (Fig. 11, *approach a*) to avoid carbon deposition is to add large amounts of steam, with a steam–carbon (S/C) ratio higher than 2; however, it is possible to obtain the same result with almost dry methane by optimizing temperature and applying high current (Zhan et al. 2006). In case of CH_4, a temperature window between 550 and 650 °C, where carbon would not be thermodynamically stable, was individuated. Thermal cracking of CH_4 to carbon and H_2 was inhibited below 650 °C. Above 550 °C, the disproportion of CO to carbon and CO_2 was shifted to the CO side (Lin et al. 2005).

Another way (Fig. 11, *approach b*) is to incorporate carbon-resistant species, such as ceria and noble metal particles, into the Ni cermet matrix. Effective metals are Ru, Pd, Co, Sn, and Cu. Metals can form alloys with Ni (i.e., NiCu or NiCo alloys) (An et al. 2011) or interact with it (i.e., Sn) (Nikolla et al. 2007), mitigating the Ni-cracking activity. On the other hand, metals themselves can be good

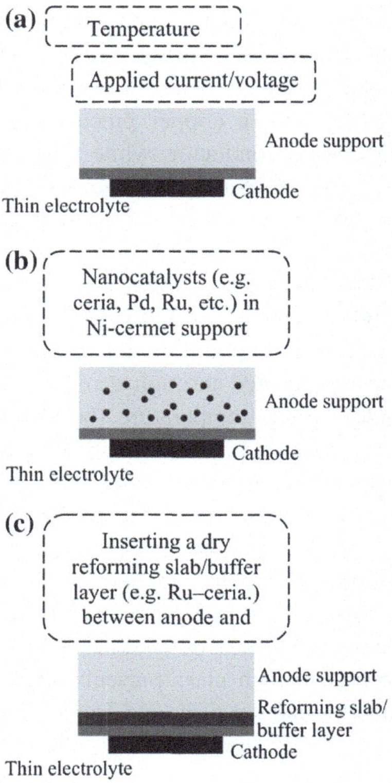

Fig. 11 The three strategies used in Ni cermet-based SOFC anodes for direct hydrocarbon utilization. Reprinted with permission from © 2012 Wiley-VCH Verlag GmbH & Co. KGaA Weinheim (Ge et al. 2012)

catalysts for the reforming and oxidation of hydrocarbons (Fig. 11, *approach c*). Ceria and ceria-based oxygen ion conductors are effective in carbon removal, thanks to their ionic and electronic mixed properties under reducing atmospheres. The latter with or without a noble metal can be deposited as a porous protective slab above the Ni electrode. It is believed that such a layer may work even as a diffusion layer favoring the preferential diffusion of H_2 at the electrolyte/electrode interface. All these strategies have been demonstrated effective in laboratory scale, but the scale-up to industrial applications remains challenging. Moreover, for each of them, there are drawbacks. An O^{2-} flux from the electrolyte may contribute to the carbon removal not beyond the TPB region; therefore, the first approach cannot protect the inactive region of Ni-based cermet anodes from carbon coking. The effectiveness of the second method can be reduced by the rapid coarsening and sintering of the added catalyst. In the last approach, the insertion of a thick dry reforming layer increases the complexity of the system and, usually, the ohmic resistance of the electrode. These strategies are not mutually exclusive, and their synergic use deserves further investigations.

Alternative Anodes Alternatively to a Ni cermet, it is possible to develop cermet with metals that are poor catalysts of cracking. To be suitable, the metals need to be stable at high operating temperature, not prone to oxidation if exposed to steam and eventually affordable. Copper seems to be a good candidate since it does not catalyze carbon formation and is stable at the oxygen partial pressures typically encountered in SOFC anodes. However, the melting temperature (T_{mp}) of the related oxides (CuO and Cu_2O) is around 1200–1300 °C, which makes inapplicable the conventional firing procedure to obtain a cermet. Copper-based cermets are usually prepared via impregnation of a porous oxide skeleton (ZrO_2- or CeO_2-based oxide) with a copper precursor salt solution. In this system, copper works as electronic conductor, while a co-catalyst, such as CeO_2 or/and noble metals (Pd, Ru, Pt), is requested to achieve good performance (McIntosh and Gorte 2004). Copper–CeO_2 anode allows the use of heavy hydrocarbons, and it is quite resistant to sulfur poisoning. Nevertheless, the long-term stability of the microstructure of Cu-based cermet anode is of great concern especially at operating temperature higher than 800 °C because of copper sintering (T_{mp} of Cu = 1064 °C). It is interesting to point out that recently, it has been found these kinds of electrodes can be rejuvenated through a redox cycle. Oxidation of the aged electrode in air at intermediate temperature (600–700 °C) causes the redispersion of copper if the co-catalyst is a ceria–zirconia mixed oxide (CZ). In addition, it is also observed that copper nanoparticles could play a role in the mechanism of oxygen transport due to a strong interaction with the CZ (Boaro et al. 2014).

The development of materials and systems in alternative to the Ni-based electrode has brought several researchers to investigate also many other classes of oxides. These are generally classified on the basis of their crystalline structure, namely fluorites, perovskites, double perovskites, pyrochlores, rutiles, and tungsten bronzes. Each class presents advantages and disadvantages to be used as SOFC anode material (Tao and Irvine 2004).

Early attempts to identify an alternative anode material consisted in doping the zirconia and ceria lattice with ions of transition metals and rare earth elements, respectively. In any case, the electronic conductivity and catalytic activity of the so-modified material remain too low for anode applications.

Recently, great attention has been given to the development of perovskites and double perovskites. As discussed before, the doping at the site B can modify the catalytic properties of the perovskites. Table 1 lists some of the possible metal to be included in a perovskite structure, highlighting those that would be expected to allow a decrease in coordination number in the perovskite lattice and hence afford catalytic behavior. The elements stable in the reducing conditions of an SOFC anode are indicated on the bottom line.

The use of these species produces a dimensionally stable anode that does not change significantly between fully oxidized and fully reduced anode. For operation with hydrocarbon fuels, the ability to oxide the anode, without damaging it, is particularly useful, since periodic oxidation cycles will remove impurities brought in with the fuel or carbon deposits formed by gas-phase reactions.

Promising electrochemical performances have been obtained with $La_{0.75}Sr_{0.25}Cr_{0.5}Mn_{0.5}O_3$ (LSCM), which showed excellent redox stability, good conductivity, and good electrode kinetics. A particularly important property of LSCM is its catalytic activity for direct oxidation of hydrocarbons, which is more significant than its reforming activity (Ge et al. 2012; Gorte and Vohs 2011).

Figure 12 shows a qualitative classification of these alternative materials in terms of electrical conductivity, electrocatalytic activity, redox stability, and fuel flexibility.

Titanates show higher electronic conductivity when materials are sintered in reducing atmosphere. $La_xSr_{1-x}TiO_3$ ($x = 0.2$, 0.3, and 0.4) sintered in H_2 shows initial conductivity of the order of 400–5000 S cm^{-1} and redox stability up to 1200 °C. $La_{0.4}Sr_{0.4}TiO_3$ does not react with YSZ and has a very close thermal expansion coefficient (TEC) with YSZ (Burnat et al. 2012). For these characteristics, LST has been often used as anode support, anode scaffold, and anode barrier layer. The catalytic properties of titanates are modest, but can be improved by doping B sites with Mn, Sc, or Ga (Ge et al. 2012 and reference therein). Mn- and Ga-codoped LSTs were proven to be excellent SOFC anode materials for the direct use of hydrocarbons, showing high OCV (1.25–1.4 V) in wet methane at 950 °C (Ruiz Morales et al. 2006).

Table 1 Transition metal elements in a perovskite framework, relative coordination number MO_x, and stability in fuel atmosphere

	Sc	Ti	V	Cr	Mn	Fe	Co	Ni	Cu	Zn
MO_5			V^V		✓	✓	✓	✓	Cu^{II}	
MO_4tet			V^V		?		?	?		✓
MO_4Sq								?	Cu^{II}	
Fuel	✓	✓	✓	✓	✓	?				

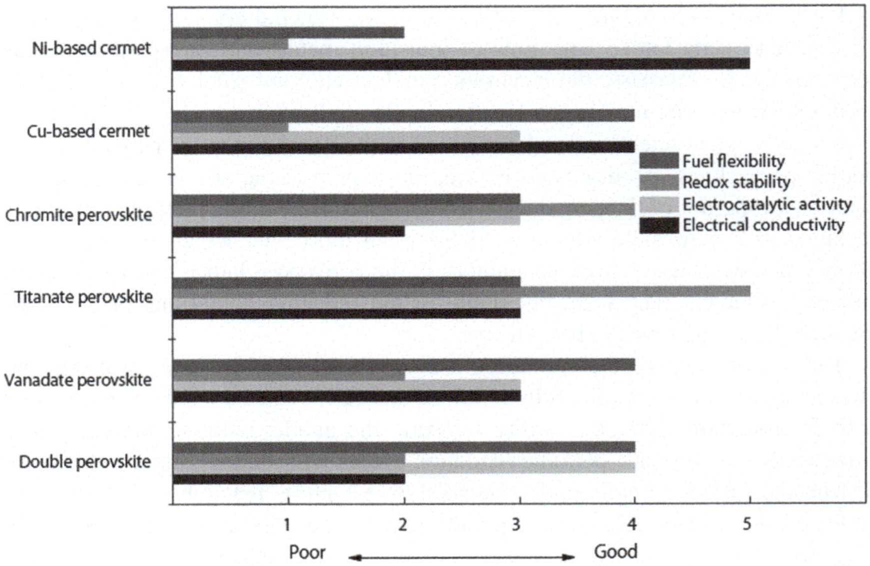

Fig. 12 Qualitative indexing of different SOFC anode materials in terms of electrical conductivity, electrocatalytic activity, redox stability, and fuel flexibility. Reprinted with permission from © Wiley-VCH Verlag GmbH & Co. KGaA, Weinheim (Ge et al. 2012)

Vanadate perovskites as $La_{1-x}Sr_xVO_3$ (LSV) are good electronic conductor, active toward hydrocarbon oxidation, and thermally stable in reducing condition; moreover, they show a high resistance to carbon deposition and sulfur poisoning. In particular, LSV anodes underwent continuous improvement rather than being poisoned, in H_2S-laden coal gas and biogas up to 500 h (Ge et al. 2011).

To summarize, with respect to a conventional Ni-based electrode, all these alternative materials generally show higher redox stability and higher tolerance to carbon formation and sulfur poisoning, albeit they show lower conductivity and more sluggish reaction kinetics. Generally, good performance requires operating temperature higher than 800 °C, in opposite to the necessity of developing materials active below 700 °C.

It is worth noting that some of these materials can operate also as cathode catalysts, and for example, $La_{0.75}Sr_{0.25}Cr_{0.5}Mn_{0.5}O_{3-\delta}$ (LSCM) (Bastidas et al. 2006; Ruiz-Morales et al. 2006, 2007), $La_4Sr_8Ti_{12-x}Fe_xO_{38-\delta}$ (LSTF) (Canales-Vázquez et al. 2007), and CeO_2 (Ruiz-Morales et al. 2008) have been demonstrated to be good candidates for symmetric cell. This can simplify a lot SOFC manufacturing as shown in Fig. 13.

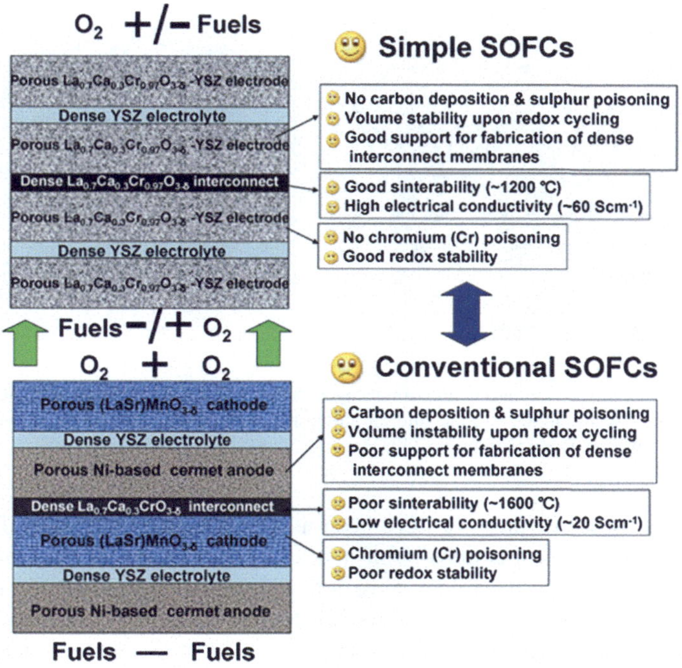

Fig. 13 A schematic illustration of a simple SOFC based on the use of chromite oxide as anode and cathode. Reprinted with permission from Elsevier B.V (Lin et al. 2010)

3 Engineering Electrodes: Development of Nanostructured Interfaces

It is well-known that to ensure high power densities with target of 1 W/cm^2 at 0.7 V, the total area-specific resistance ($ASR = ASR_{electrolyte} + ASR_{anode} + ASR_{cathode}$) of a cell requires to be of 0.3 Ω cm^2 (0.1 Ω cm^2 for each component) (Steel and Heinzel 2001). This means that the conductivity of 1 mm-thick electrode must be around 1 S cm^{-1}.

Keeping in mind this, an approach to optimize mechanical properties and performance consists in splitting the electrode into two layers: a thick conductive layer engineered to have high porosity to facilitate the diffusion of reactants and a high conductivity to decrease ohmic losses, and a thin (5–10 μm) functional layer near the electrolyte, designed to maximize the TPB extension, with only enough conductivity and porosity to minimize potential drops across its thickness (see Fig. 14) (Gross et al. 2007).

With this strategy, materials in each layer can be optimized independently for their specific function, thus allowing greater flexibility in electrode design.

The functional layer where the faradaic processes take place is often considered the "true electrode" since cell performance is largely determined by its properties at

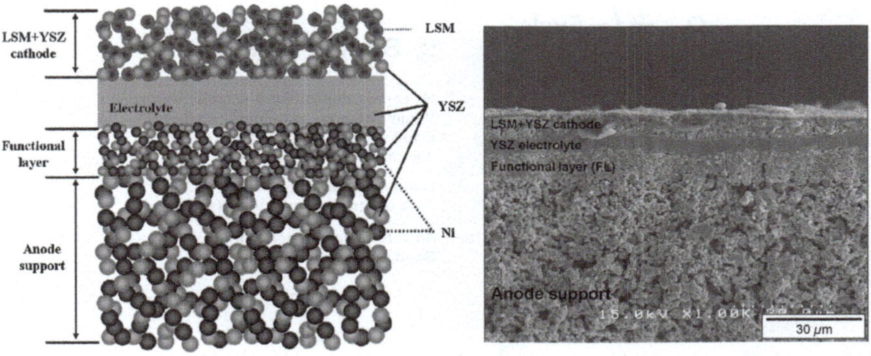

Fig. 14 Concept of the additional functional layer. Schematic of anode-supported SOFC single cell with a tape-casting anode support and an anode functional layer. Reprinted with permission from © 2007 Elsevier B.V (Kim et al. 2007)

nanoscale. As proof of this, we have been assisting to an evolution of functional electrode layer's nanostructures with the development of SOFC technology. Therefore, single-phase electrodes have left the path to composite electrodes, to impregnated skeletons, and nowadays to locally exsolved structures. In this evolvement, the main goal is the optimization of cell performance and durability by carefully designing and controlling the interface architecture. Taking into account that electrolyte/electrode interface can be also modified under operating conditions, there are many approaches that can be used to achieve such purpose (selection and combination of specific materials, their treatment, etc.). Infiltration and in situ exsolution are among the methods most investigated.

3.1 Preparation of Electrodes by Infiltration (Impregnation)

A useful way to modify an electrode is to deposit another active component into a porous skeleton through impregnation. This technique is often used to prepare supported catalysts and offers several advantages when used for the preparation of electrodes. There are excellent reviews to engineer both cathodes and anodes with this approach (Gorte and Vohs 2009; Liu et al. 2013; Jiang 2012; Jiang et al. 2010; Ding et al. 2014). The usual procedure to prepare electrodes via impregnation is the use of suspensions or precursor solutions of catalytically active materials to wet and cover the surface of a porous scaffold previously cosintered with the electrolyte. The final microstructure is obtained through a suitable thermal treatment and often after reduction in situ (in case of anode), achieving a good distribution and high surface areas for the impregnated components. The infiltrated catalytic nanoparticles can form discrete distribution or a thin and continuous network on the surface of the porous scaffold. Key parameters for a good deposition are (i) the concentration of solution, (ii) the type of solvent, and (iii) the heating rate used to dry and to decompose the salt. From our experience, the best results are often obtained with diluted solution and slow heating rate (Boaro 2014; Jung et al. 2006).

Figure 15 represents the process (Fig. 15a–b) and the different distributions (Fig. 15c–d) that can be obtained.

Porosity higher than 70 % is necessary to allow the introduction of enough active material on the scaffold, which must be mechanically strong. Catalyst loading is generally around 10–20 wt% of total electrode weight using non-noble metal and up to 5 % when the catalyst is a precious metal.

In general, it would be very difficult for concentrated precursor solutions to penetrate and infiltrate uniformly into micro and nanopores of the scaffold by capillary force and multiple infiltration–calcination steps are required, particularly in the case of a thick anode/cathode substrate. The application of vacuum was demonstrated beneficial to expediting the infiltration process, which despite that remains a long procedure (Jiang et al. 2002).

To obtain infiltrated electrodes, there are five possible impregnation's ways: (i) nanoparticle dispersion, (ii) metal–salt precipitation, (iii) metal–salt plus additive precipitation, (iv) molten salt decomposition and (v) molten salt plus surfactant decomposition (Sholklapper et al. 2008).

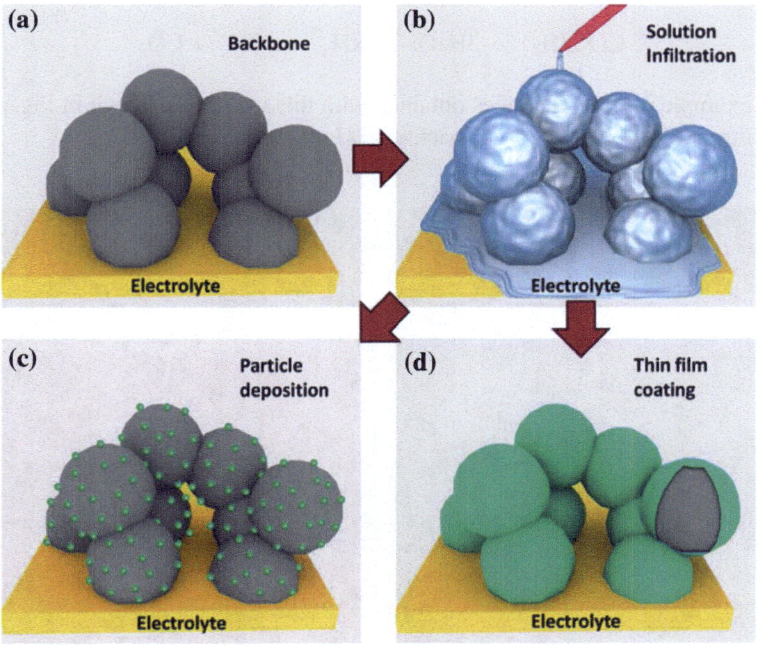

Fig. 15 A schematic of a typical infiltration process: **a** an as-fired electrode backbone; **b** a process of solution drops entering into the electrode backbone; two typical morphologies of infiltrated electrode after thermal treatment: **c** particle deposition and **d** thin-film coating. Reprinted with permission from © 2014, The Royal Society of Chemistry, (Ding et al. 2014)

The most straightforward one is to disperse catalyst nanoparticles in a solvent, then to impregnate the suspension into the electrode, and finally to calcine the resultant product. This method allows for a high level of control in the catalysts, since synthesis can be done before and in a controlled environment. A limitation is that the nanoparticle agglomeration may be difficult to control. Additionally, the electrode itself acts as a filter preventing the deposition of catalysts deep inside the pores which clog, thus leading to gas starvation. To avoid this inconvenience, the catalyst can be deposited from a solution of salt precursors, such as nitrates. As example, the involved reaction could be the decomposition of metal nitrates:

$$Ce^{3+} + NO_3^- \leftrightarrow Ce(NO_3)_3 \xrightarrow{\Delta} CeO_2 \tag{4}$$

This method is not suitable to prepare complex phases such as perovskites although it has been often adopted for the impregnation of a single component (Lu et al. 2006). The method does not give uniform deposition due to thermal gradients of the heating source used to dry the solution.

More uniform distribution can be achieved using urea as a precipitant agent (Jung et al. 2006). In this case, the nucleation of the catalyst is driven by the precipitation with urea which decomposes following the reaction below releasing OH^- and CO_3^- as precipitating ligands (Qin et al. 2015).

$$CO(NH_2)_2 + 3H_2O \xrightarrow{\Delta} 2NH_4^+ + 2OH^- + CO_2 \tag{5}$$

An example of the differences obtained with this method is shown in Fig. 16 for the infiltration of CuO in a YSZ backbone (Jung et al. 2006).

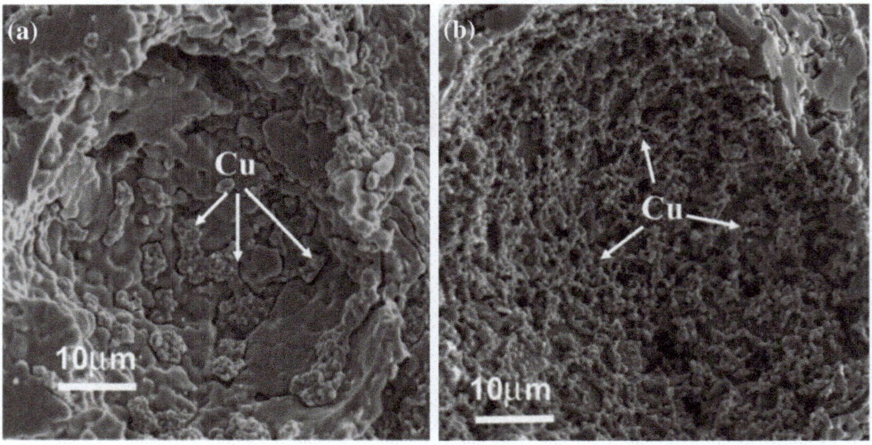

Fig. 16 SEM images of porous YSZ blocks after impregnation of CeO_2 and Cu (13 vol.% Cu, 5 vol.% CeO_2) and subsequent reduction in H_2 for 2 h at 973 K. The Cu was added using the **a** nitrate only and **b** nitrate/urea methods. Reprinted with permission from © 2006, Elsevier VB (Jung et al. 2006)

These images show large differences in the morphology of the Cu deposits produced by the two different impregnation methods. Impregnation using only aqueous copper nitrate produced relatively large patches of copper agglomerates, while a much more continuous Cu layer was produced when urea was added to the impregnation solution. In a conventional impregnation from copper nitrate solution, the morphology of the initial copper deposits is controlled by the drying process. The size and shape of the Cu deposits therefore depend on the size of the water droplets during drying. When using solutions containing urea, the following reactions take place upon heating to 80 °C.

$$CO(NH_2)_2 + 3H_2O \rightarrow 2NH_4 + 2OH^- + CO_2(g) \quad (6)$$

$$Cu^{2+} + 2OH^- \rightarrow Cu(OH)_2(s) \quad (7)$$

The $Cu(OH)_2$ nucleation occurs prior to evaporation, and its deposition remains homogenous and not affected by the drying process. Thus, following reduction, it was possible to obtain a more even distribution of Cu in the YSZ matrix.

The urea-based precursor method allows the uniform deposition of the catalyst on the walls of the pores avoiding their clogging. However each impregnation step requires 2 h, and many steps are necessary to obtain contiguous coatings. The method is thus best suited for producing dispersed catalysts and is less practical when contiguous coatings are required. Such a pattern is necessary when the purpose is to increase electronic or ionic electrode's conductivity.

Taking into account that many salt precursors have a low melting point or a high solubility in water, the impregnation of the porous ceramic skeleton can be made using molten salt precursors or highly concentrated salt precursor solutions. Using this approach, it is possible to accelerate the process of deposition; however, the approach is restricted to simple compounds that melt at low temperature and are thermally stable above 100 °C. Additionally, the catalyst distribution is often random and not uniform. The temperature gradient between precursors and substrate is the leading cause of poor impregnations since it induces viscosity variations; thus, it must be reduced for an optimal deposition.

A higher versatility in the compositions of infiltrated compunds is given from the use of surfactants such as Triton X-100. Surfactants can coordinate the metal ions favoring the formation of complex phases. Also, surfactants' addition improves the wettability of the substrate, allowing a homogeneous deposition of the composites/catalysts. The wettability can be further improved with a combined use of surfactants and ethanol (Lou et al. 2010).

The advantages and disadvantage of the above methods are summarized in Table 2.

Other approaches to infiltrate active/or conductive phases into the electrode backbone can be properly adapted from the conventional methods to prepare filters and structured porous reactors such as sol-gel, incipient wetness, and solution combustion synthesis (Boaro et al. 2014; Gorte and Vohs 2009). In these methods,

Table 2 Impregnation methods and their properties discussed in this paper

Impregnation method					
	Nanoparticle dispersion	Metal salt precipitation	Metal salt–urea precipitation	Molten salt	Molten salt plus surfactant
Infiltrant properties	(*) Low loading	(*) Low loading	(*) Low loading	(*) High loading	(*) High loading
	(+) Fast	(+) Fast	(+) Preferential deposition	(+) Fast	Non-preferential deposition
	(+) Complex phases possible	(–) Only simple phases possible	(+) Complex phases possible	(–) Only simple phases possible	(+) Fast
	(–) Non-uniform deposition	(–) Non-uniform deposition	(–) Slow	(–) random deposition	(+) Complex phases possible
	(–) Electrode acts as filter	–	–	–	(+) Continuous in minimum steps
	(–) Limited to highly porous electrodes	–	–	–	–

Neutral, positive, and negative properties are denoted by *, +, or –, respectively. Reprinted with permission from Wiley-VCH Verlag GmbH & Co. KGaA, Weinheim (Sholklapper et al. 2008)

the addition of a complexant agent such as citrate acid or glycine is crucial to get homogeneous compounds.

The first anode developed via infiltration was a YSZ porous backbone impregnated with CeO_2 (as catalyst component) and copper (as conductive phase) designed in order to develop a coking-resistant electrode (Gorte and Vohs 2009). The technique was then adopted to modify the conventional nickel cermet with the aim of improving its cooking and sulfur tolerance and redox properties (Busawon et al. 2008). Typical impregnates to these ends are copper, iron, ruthenium, palladium, and (doped) ceria.

Regarding ceramic anodes, infiltration is the way often used to introduce and to enhance catalytic properties. Figure 17 shows the performance of different $(La_{0.75}Sr_{0.25})(Cr_{0.5}Mn_{0.5})O_3$ (LSCM/YSZ) composite anodes in the H_2 oxidation reaction (Kim et al. 2009) when impregnated with 0.46 % of Pd.

LSCM/YSZ electrodes were prepared by traditional method (a) or through impregnation of LSCM in the scaffold of YSZ (b). The apparently significant differences in the promotion effect of infiltrated Pd are due to the different morphology of LSCM. After reduction, the infiltrated LSCM layer is broken into very small particles and is highly porous (see Fig. 17c), while LSCM prepared with a traditional method shows larger and compact grains. Pd particles can be therefore distributed differently, remaining isolated on the surface in the conventional electrode while penetrating at the interface between YSZ and the infiltrated LSCM in the LSCM-impregnated electrode.

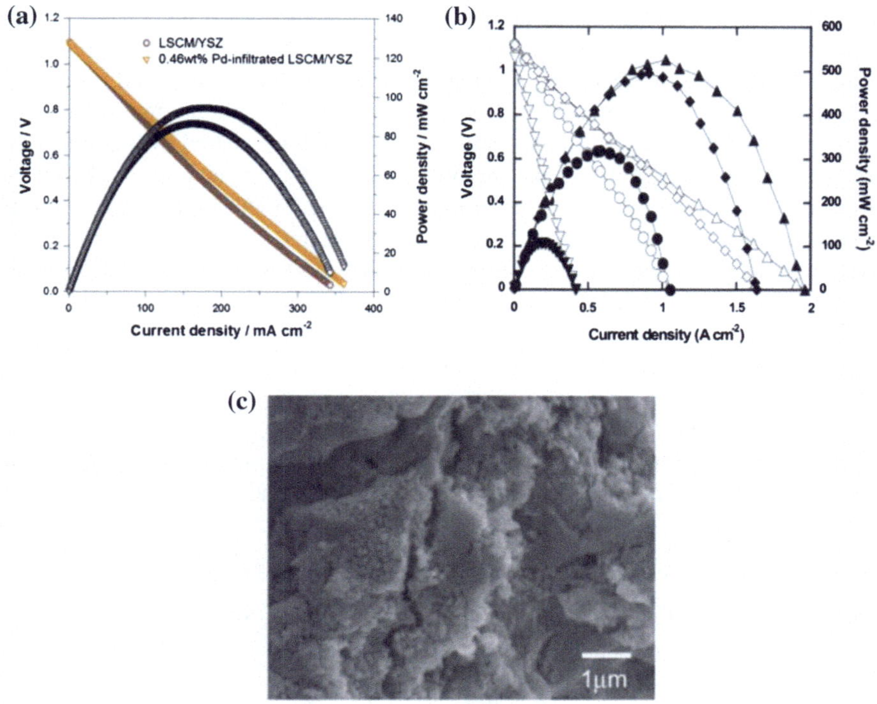

Fig. 17 Cell performance in humidified H_2 (3 % H_2O) of **a** conventional LSCM/YSZ composite anodes with and without Pd infiltration at 800 °C; the infiltrated Pd loading was 0.46 wt%; **b** infiltrated LSCM/YSZ composite anodes at 700 °C: (*inverted triangles*) indicate the anode infiltrated only with LSCM; (*circles*) indicate the LSCM/YSZ anode with 5 wt% ceria; (*rhombuses*) indicate the LSCM/YSZ anode with 0.5 wt% Pd, and (*triangles*) indicate the LSCM/YSZ anode with 5wt% ceria and 0.5 wt% Pd; **c** SEM micrograph of an infiltrated LSCM/YSZ composite anode after reduction in humidified H_2 at 800 °C for 4 h. Reprinted with permission from © Elsevier Ltd, 2012 (Jiang 2012)

It is clear that the morphology and microstructure of scaffold could affect the distribution and thus the catalytic activity of the infiltrated nanoparticles.

The infiltration method has been greatly utilized also to improve cathode properties or to prepare alternative cathodes (Ding et al. 2014; Jiang et al. 2010 and reference therein). Infiltration of $La_{0.8}Sr_{0.2}MnO_3$ (LSM) into a porous electrolyte backbone has enabled much improved performance as compared with conventionally processed LSM-YSZ-based electrodes. This because, via impregnation LSM can be dispersed in small nanoparticles, that leads to an increase of the number of sites available for oxygen reduction. It has been also shown that the conductivity of the infiltrated LSM-YSZ composite cathodes increases at very low LSM loading with a threshold of 20 % for an acceptable conductivity (according to the percolation theory for the composites, the minimum value is 30 %) (Vohs and Gorte 2009; He et al. 2004). Moreover, since it requires only low temperatures to synthesize and calcine the catalysts, infiltration resulted an expedient approach for the preparation of low/intermediate-temperature cathodes based on cobaltites $La_{1-x}Sr_xCoO_{3-\delta}$ (LSC) and ferrites $La_{1-x}Sr_xCo_{1-y}Fe_yO_{3-\delta}$. These latter are more active than the LSM since they possess a higher ionic conductivity, but would react with YSZ at the temperatures necessary for sintering with conventional methods.

Some of the most promising nanostructured electrodes prepared by impregnation are reported in Table 3. Here, the promotion factor, f_p is a measure of the effectiveness of this technique compared to the conventional methods of preparation. It is calculated by dividing the ASR (i.e., the electrode polarization resistance) or the overpotential, η of the nanostructured electrodes by the performance of the baseline electrodes fabricated by conventional mixing and sintering processes, measured under identical conditions.

f_p = ASR infiltrated electrode/ASR conventional electrode

or

f_p = η infiltrated electrode/η conventional electrode.

What emerges immediately is the outstanding performance of the infiltrated electrodes with respect to those conventionally prepared and the variability of the f_p factors.

Table 3 Performance and promotion factors for electrodes prepared by infiltration

Impregnated nanoparticle	Scaffold/skeleton	Performance	Promotion factor, fp	References
Cathode of SOFC				
GDC (5.8 mg cm^{-2})	LSM	$R_\varepsilon = 0.21$ Ω cm^2 @ 700 °C	56 for O_2 reduction	Laosiripojana and Assabumrungrat (2007)
Pd (1.8 mg cm^{-2})	LSM/YSF	$R_\varepsilon = 0.9$ Ω cm^2 @ 600 °C	78 for O_2 reduction	Marbán and Valdés-Solís (2007)
Pd (1.2 mg cm^{-2})	LSCF	$R_\varepsilon = 2.9$ Ω cm^2 @ 600 °C	1.9 for O_2 reduction	Mahato et al. (2015)
GDC (1.5 mg cm^{-2})	LSCF	$R_\varepsilon = 1.6$ Ω cm^2 @ 600 °C	3.4 for O_2 reduction	Mahato et al. (2015)
LSM (∼2 mg cm^{-2})	YSZ	$R_\varepsilon = 1.6$ Ω cm^2 @ 600 °C	44 for O_2 reduction[a]	Marbán and Valdés-Solís (2007)
LSCF (1.1 mg cm^{-2})	YSZ	$R_\varepsilon = 0.54$ Ω cm^2 @ 600 °C		Lou et al. (2010)
LSCF (12.5 vol.%)	GDC	$R_\varepsilon = 0.25$ Ω cm^2 @ 600 °C	14 for O_2 reduction	Michaels and Vayenas (1984)
$La_{0.6}Sr_{0.4}CoO_3$ (30 wt%)	YSZ	$P = 2.1$ W cm^{-2} @ 800 °C in H_2/air		Mogensen and Kammer (2003)
$La_{0.6}Sr_{0.4}CoO_3$ (55 vol.%)	SDC	$R_\varepsilon = 0.36$ Ω cm^2 @ 600 °C		McIntosh and Gorte (2004)
Ag	LSCF/GDC	$R_\varepsilon = 0.98$ W cm^{-2} @ 600 °C	3.3 for H_2/air	McNaught and Wilkinson (1997)
	YSZ			

(continued)

Table 3 (continued)

Impregnated nanoparticle	Scaffold/skeleton	Performance	Promotion factor, fp	References
Pd (1.4 mg cm^{-2})		$R_\varepsilon = 0.22$ Ω cm^2 @ 700 °C		Michaels and Vayenas (1984)
BSCF (1.8 mg cm^{-2})	LSM	$R_\varepsilon = 1.3$ Ω cm^2 @ 700 °C	12 for O_2 reduction	Lu et al. (2006)
$Sm_{0.6}Sr_{0.4}CoO_3$	LSM/YSZ	$R_\varepsilon = 8.5$ Ω cm^2 @ 600 °C	2.3 for O_2 reduction	Ho (2014)
$Y_{0.5}Bi_{1.5}O_3$ (50 wt%)	LSM	$R_\varepsilon = 0.14$ Ω cm^2 @ 700 °C		Mogensen and Skarup (1996), Muradov and Veriroğlu (2005)
Anode of SOFC				
$Sm_{0.2}Ce_{0.8}O_2$ (~4 mg cm^{-2})	Ni/YSZ	$R_\varepsilon = 0.24$ Ω cm^2 @ 800 °C	7.3 for H_2 oxidation	He et al. (2004)
GDC (4 mg cm^{-2})	$(La_{0.75}Sr_{0.25})(Cr_{0.5}Mn_{0.5})O_3$	$R_\varepsilon = 0.44$ Ω cm^2 @ 800 °C	26 for CH_4 oxidation	Lin et al. (2005)
GDC (4 mg cm^{-2})	$(La_{0.75}Sr_{0.25})(Cr_{0.5}Mn_{0.5})O_3$	$R_\varepsilon = 0.12$ Ω cm^2 @ 800 °C	20 for H_2 oxidation	Lin et al. (2005)
GDC (1.42 mg cm^{-2})	Ni	$R_\varepsilon = 1.29$ Ω cm^2 @ 800 °C	25 for CH_4 oxidation	Lin et al. (2010)
Pd (0.36–0.46 mg cm^{-2})	$(La_{0.75}Sr_{0.25})(Cr_{0.5}Mn_{0.5})O_3$/YSZ	$R_\varepsilon = 0.88$ Ω cm^2 @ 800 °C	1.5 for H_2 oxidation	Lorente et al. (2012)
Pd (0.36–0.46 mg cm^{-2})	$(La_{0.75}Sr_{0.25})(Cr_{0.5}Mn_{0.5})O_3$/YSZ	$R_\varepsilon = 2.0$ Ω cm^2 @ 800 °C	4.6 for CH_4 oxidation	Lorente et al. (2012), Liu et al. (2011)
Pd (0.36–0.46 mg cm^{-2})	$(La_{0.75}Sr_{0.25})(Cr_{0.5}Mn_{0.5})O_3$/YSZ	$P = 0.111$ W cm^2 @ 800 °C	8 for C_2H_5OH/air	Lorente et al. (2012)
Pd (0.06 mg cm^{-2})	$(La_{0.7}Sr_{0.25})(Cr_{0.5}Mn_{0.5})O_3$/GDC	$R_\varepsilon = 1.1$ Ω cm^2 @ 750 °C	6.5 for CH_4 oxidation	Liu et al. (2013)
Pd (0.11 mg cm^{-2})	Ni/GDC	$R_\varepsilon = 0.6$ Ω cm^2 @ 700 °C	~5 for H_2 oxidation	Mann et al. (2006)
Pd (5 wt%) CeO_2 (5 wt%)	$(La_{0.75}Sr_{0.25})(Cr_{0.5}Mn_{0.5})O_3$/YSZ	$P = 0.52$ W cm^2 @ 700 °C	~5 for H_2/air	Medford et al. (2015)

(continued)

Table 3 (continued)

Impregnated nanoparticle	Scaffold/skeleton	Performance	Promotion factor, fp	References
Oxygen electrode of SOEC				
GDC ($2\ mg\ cm^{-2}$)	LSM	$R_\varepsilon = 0.15$ $\Omega\ cm^2$ @ 800 °C[b]	148 for O_2 oxidation	Nagata et al. (2001)
LSM	YSZ	$R_\varepsilon = 0.48$ $\Omega\ cm^2$ @ 800 °C[c]	2.3 with 50 % AH[c]	Neagu et al. (2013)

Reprinted with permission from © Elsevier Ltd. 2012 (Jiang 2012)

[a] Calculated based on the performance for LSM/YSZ composite cathodes

[b] Measured after polarized at 1000 mA cm^{-2}, 800 °C for 22 h

[c] Measured on a SOEC with 50 % absolute humidity (AH) and based on the performance for LSM/YSF oxygen electrode

In order to explain the above results, we can consider that:

1. The impregnation method affects positively the performance of an electrode since it yields to the deposition of high surface area nanostructured catalysts, which enlarge TPBs for the electrode reactions. This is confirmed by several models (Zhu et al. 2008; Nicholas and Barnett 2009). For instance, assuming a model based on the random packing of sphere particles and electronic or ionic conducting scaffold coated with a continuous nanoparticle layer, it is predicted a substantial increase in the TPB length, e.g., 6.2 and 14.6 times higher as compared to conventional mixed composites when the infiltrated nanoparticle size is 50 and 20 nm, respectively, assuming the particle size of a conventional composite phase is 1 mm).
2. The variability in the promotion effects is due to the significant differences in the microstructure, electronic and ionic properties of the scaffold, the distribution, and loading and catalytic activities of the infiltrated nanoparticles.
3. The observed significant promotion effect of the nanostructured electrodes with discretely distributed nanoparticles indicates that there must be significant catalytic effects of the infiltrated nanoparticles on the electrochemical performance of the electrodes in addition to the enhanced TPB. However, the broad range of results suggests that much more systematic and extensive studies are needed to fundamentally understand the promotion mechanism of the infiltrated nanoparticles.

Taking into account the above observations, the strategy for the development of an electrode through infiltration should consider the conduction properties of the backbone (ionic, electronic mixed conductor) in order to select the appropriate material to be infiltrated. Once the choice is made, it is necessary to have a careful control of the microstructure of the infiltrated materials in order to gain a significant increment in performance. The final microstructure is dependent on a careful optimization of the wettability of the impregnation solution toward the support. That implies the use of specific solvents and a careful control of variables that concur to the impregnation process (Lou et al. 2010; Jiang 2012).

Despite the several advantages offered by the infiltration technique, this approach is time-consuming, and developing effective ways to accelerate the impregnation process with an exact control of catalyst's microstructure represents the main demand to an industrial scale-up of the method. The use of a layer-by-layer (LbL)-assisted approach seems highly promising thanks to the versatile and tunable properties of LbL assembly (see box below, Xiang et al. 2012).

Layer by layer polymer electrolytes (PE) assisted impregnation

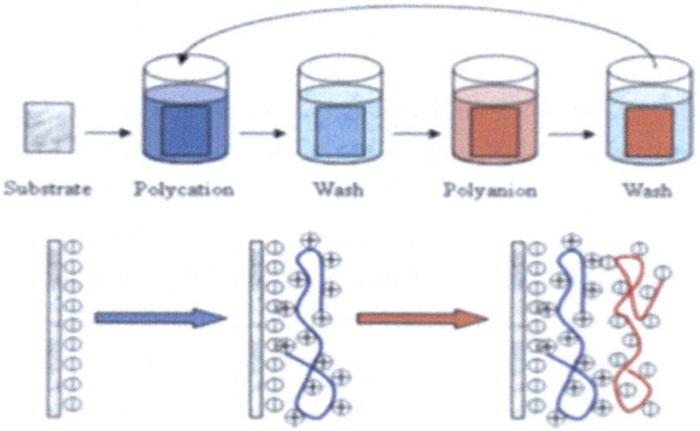

Two oppositely charged polymers or polyelectrolytes (PEs) dissolved in aqueous solution are alternatively deposited on a support surface by means of electrostatic attraction. After each dipping cycle, the surface charge is reversed, which enables deposition of a subsequent layer. The layer by layer (LbL) self-assembly deposition at a surface is largely controlled by a charge compensation mechanism and the final structure is primarily stabilized by strong electrostatic force and in certain cases by hydrogen bonding.

Chemical structures of polyelectrolytes commonly used to assemble multilayer films.

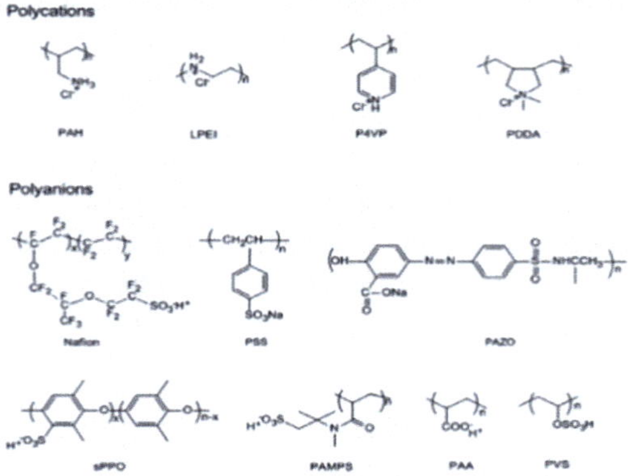

Polycations: *poly(allylamine hydrochloride) (PAH), linear poly(ethylene imine) (LPEI), poly(4-vinylpyridiine) (P4VP), poly(diallydimethylammonium chloride) (PDDA).*
Polyanions: *Nafion, poly(sodium styrene sulfonate) (PSS), poly(1-(4-(3-carboxy-hydroxyphenylazo)benzene sulfonaimdo)-1,2-ethanediyl, sodium salt) (PAZO), sulfonated poly(2,6-dimethyl-1,4-phenylene oxide) (sPPO), poly(2-acrylamido-2-methyl-1-propanesulfonic acid) (PAMPS), poly(acrylic acid) (PAA), poly(vinyl sulfate) (PVS).*

The method has been recently exploited to prepare $Pr_{0.7}Sr_{0.3}CoO_{3-\delta}$ (PSC) infiltrated cathodes, making possible a reduction of the fabrication time by a factor of 6–7 % (Choi et al. 2014).

The stability of infiltrated nanostructured catalysts in SOFC operating conditions is also an issue for a realistic scale-up of the method. The nanostructures often are prone to coarsening or to segregating on the electrode surface. Several solutions are proposed to avoid this phenomenon such as multicomponent infiltrations, selective alloying, or the engineering of specific nanostructures such as core–shell M@MOx catalysts (Bardini et al. 2016; Cargnello et al. 2010).

Cargnello et al. (2010) successfully prepared dispersible Pd-CeO_2 core–shell nanostructures by exploiting the self-assembly between functionalized Pd nanoparticles and cerium (IV) alkoxides. Such nanostructures when infiltrated in a YSZ scaffold showed an higher resistance to degradation with respect to a Pd-CeO_2 catalyst prepared through a conventional multi-steps infiltration. The core–shell nanostructure in fact inhibits the growth of Pd nanoparticles.

Promoting a strong interaction between infiltrated phase and the scaffold support has also shown beneficial effect (Serra and Buchkremer 2007).

To conclude among the methods of preparation of the SOFC electrodes, the impregnation technique has been growing in popularity since it offers several advantages:

It allows an extraordinary flexibility in the selection and combination of materials. Not only catalytic components but also ionic, electronic MIEC phases can be successively introduced to implement different functionalities and to meet stringent requirements of SOFC anodes and cathodes.

Low temperatures of calcination prevent catalysts from coarsening; thus, their specific surface area can be preserved, with beneficial effect on the size of TPB and on cell performance.

Impregnation relaxes thermal expansion coefficient (TEC) matching restrictions. A catalyst material with mismatched thermal expansion coefficient (TEC) as compared with the other cell components can be incorporated into the electrode, since the sintered backbone dominates the electrode's TEC.

Moreover, the method makes easier and feasible the decoupling of catalytic and charge transport functionalities, so it is a valid tool to investigate the complex interplay between the intrinsic catalytic activity of a material and its microstructural properties.

Developing effective methods to accelerate the impregnation process with an exact control of the catalyst microstructure represents the main challenge to make its scale-up possible.

3.2 *Preparation of Electrodes by in Situ Extrusion of Metallic Catalyst*

If infiltration is an effective way to prepare nanostructured catalyst, another recent approach with a great potential and further interesting development is the in situ growth of catalytic nanoparticles through a redox exsolution mechanism.

The infiltration method offers limited control over the size, distribution, and anchorage of the deposited species in the various steps of catalyst preparation. In addition, it is often time-consuming and costly.

In the latter approach, catalytic active transition metals are incorporated in a host oxide lattice to form solid solutions and released (exsolved) on the surface as metal particles, once the oxide lattice has been sufficiently reduced. The redox exsolution can be described by a decomposition process driven by reduction and controlled by the bulk and surface distribution of defects. Initially, the lattice is reduced, losing oxygen and gaining electrons until the point where metal nucleation becomes favorable. Nucleation occurs on surfaces rich of defects since the nucleation barrier is lower. The process drains exsolvable ions from the nearby lattice, so that additional ions diffuse to the surface to balance the compositional gradient leading to the growth of metal clusters. The process has been well demonstrated for perovskite oxides (ABO_3), in which the cations in position B can be partially exsolved to form catalytic nanoparticles. As compared to traditional deposition techniques, this process yields finer and better-distributed catalyst nanoparticles, and it is faster because it does not require multiple deposition steps. Typical systems demonstrated so far include metal particles predominantly consisting of Pd, Pt, Rh, Ru, Ni on ferrite, titanate, chromite, or chromite–manganite perovskite supports. It has been demonstrated for stoichiometric perovskite with A/B ratio = 1. In this case, exsolution occurs in the bulk; thus, it is not so effective to prepare catalysts.

Recently, the process was demonstrated also for A site-deficient perovskite (A/B < 1) (Sun et al. 2015; Neagu et al. 2013). In this case, the exsolution from B sites yields to nanoparticles on the surface, which are very stable and well distributed (Fig. 18). The mechanism of exsolution for this kind of materials was addressed by Neagu et al. (2013), which showed that in A site-deficient perovskites

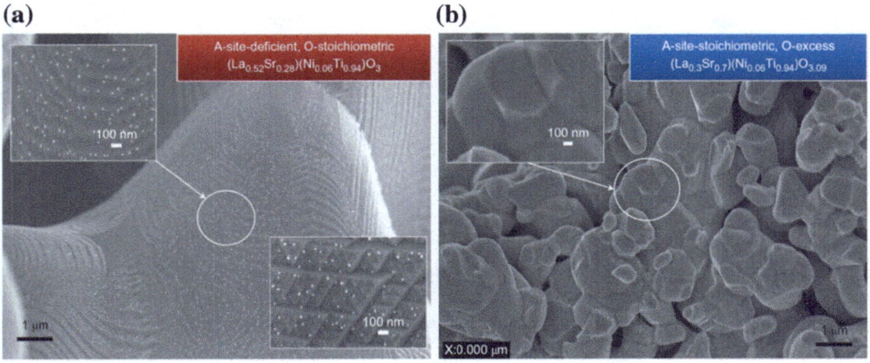

Fig. 18 The role of non-stoichiometry in the formation of exsolutions on stoichiometric and A site-deficient perovskites illustrated through SEM micrographs. **a**, Exsolutions from the initially A site-deficient, O-stoichiometric $La_{0.52}Sr_{0.28}Ni_{0.06}Ti_{0.94}O_3$ after reduction at 930 °C (20 h) in 5 % H_2/Ar. **b**, A site-stoichiometric, O-excess $La_{0.3}Sr_{0.7}Ni_{0.06}Ti_{0.94}O_{3.09}$ sample reduced at 930 ° C (20 h) in 5 % H_2/Ar indicates that no exsolution has occurred. Reprinted with permission from Macmillan Publishers Limited, © 2013 (Neagu et al. 2013)

exsolution acts to locally revert the perovskite toward a stable defect-free ABO_3 stoichiometry. At an atomic scale, the removal of oxygen from a perovskite cell with the A site vacant may be visualized as locally segregating the B site dopants from the main perovskite framework into an incipient exsolution (Fig. 19).

This mechanism implies that anchoring/coherence between the nucleating particles and the parent perovskite may easily occur because of reminiscent bonds between the two phases, which entails that particles can experience lattice strain. Electron microscopic analysis showed that particles are usually imbibed on the support for almost one-third (Fig. 20).

Such characteristics are expected to bring about peculiar chemical and physical properties to these catalysts. Not surprisingly, exsolved particles possess enhanced thermal stability and display low tendency to coalesce. Moreover, they can show a remarkable coking resistance while maintaining activity for desirable reactions. For instance, nickel-supported nanoparticles prepared by conventional methods underwent deactivation due to the formation and growth of carbon fibers at the nickel/support interface, that causes an uplifting of nanoparticles and the pulverization of catalyst. Conversely, when anchored because prepared by exsolution nickel nanoparticles show an outstanding resistance to coking deactivation (Negau et al. 2015).

(a)

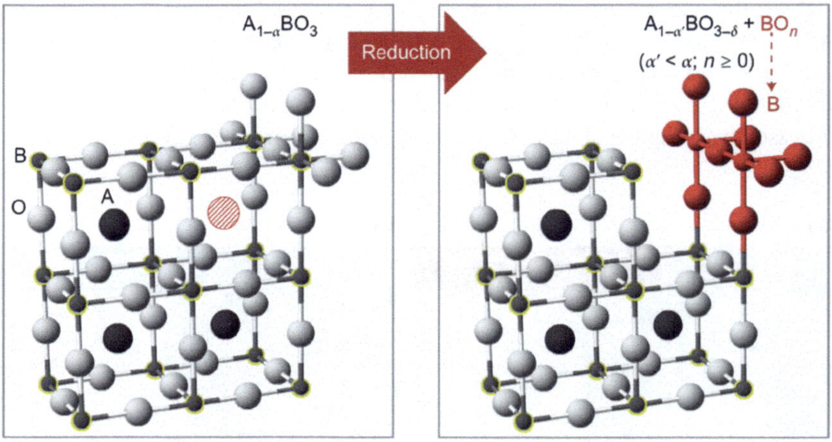

Fig. 19 Schematic representation of the exsolution of *B* site cations from *A* site-deficient perovskites (**a**). The *O* sites are depicted by *large silver spheres*, the *B* sites by small *gray spheres* highlighted in *yellow*, and the *A* sites by *large dark spheres*. A *large hashed red sphere*, shows an *A*-site vacancy. By removing oxygen from the *A* site-deficient unit cell through reduction, some *B* sites are locally isolated from the parent perovskite into an incipient BO_n exsolution (depicted by the *red* group of atoms in the *right panel*). Reprinted with permission from Macmillan Publishers Limited [Nature Chem.] (Neagu et al. 2013)

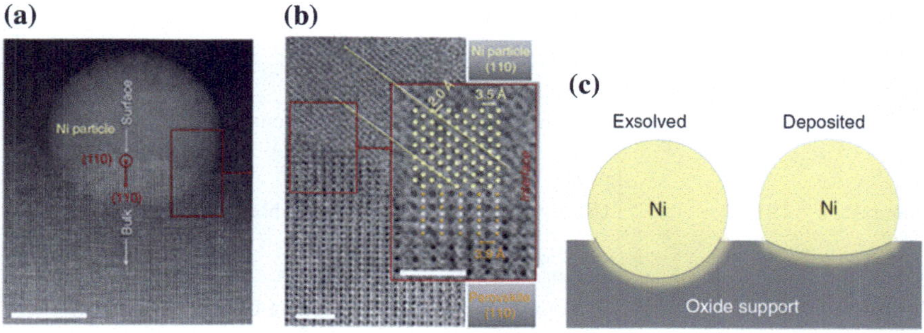

Fig. 20 Exsolved particle–substrate interface. **a** TEM micrograph (*dark field*) of a Ni particle exsolved on (110) native surface facet after aging (% H_2O/5 % H_2/Ar, 930 °C, 60 h); *scale bar* 10 nm. **b** TEM micrograph detail (*bright field*) of the metal–perovskite interface highlighting the corresponding atomic planes and orientations; *scale bar* 1 nm. **c** Schematic illustration of the particle–substrate interface for deposited and exsolved nickel particles. Reprinted from (Neagu et al. 2015), through Creative Commons CC-BY license

4 Use of Hydrocarbons and Biofuels in SOFC Anodes

4.1 Type of Fuels for SOFC Anodes

H_2 is the primary fuel for solid oxide fuel cell. The main advantage of its utilization is the absence of CO_2 emissions as well as of other pollutant emissions (NOx, etc.). The problem is that H_2 is very reactive and it does not exist chemically free in nature, it is instead mostly bound to either oxygen or carbon atoms, so H_2 can be obtained from natural compounds only with an energy expenditure. H_2 production is possible starting from the water molecule using renewable sources such as solar- or wind-related technologies or using thermochemical processes such as those related to nuclear plants (Bartels et al. 2010). However, at present, the most quantity of hydrogen is obtained by different processing technologies of hydrocarbon fuels. Besides the expensive production of hydrogen, there are difficulties related to its transportation and storage. This is because although hydrogen has the highest energy content per unit mass of any fuel, 140.4 MJ/Kg versus 48.6 MJ/Kg of gasoline, on a volume basis, the situation is reversed: 8.49 MJ/m^3 versus 31.15 MJ/m^3 of gasoline. Then, to improve penetration market of this fuel, it is desirable its in situ and on demand production.

Syngas, H_2 + CO gaseous mixture with a high content of H_2, can be successfully used instead of hydrogen in SOFC units. This is due to the high SOFC operating temperature at which CO can be directly oxidized according to Eq. (8).

Table 4 Types of fuel for SOFCs and related sources

Fuel	Chemical composition	Feedstocks
Hydrogen	H_2	Reforming processes of hydrocarbons and electrolysis of water with renewable technologies such as solar and wind or nuclear plants
Natural gas	CH_4	Underground reserves and renewable biogas
Coal		Geological
Propane	C_3H_8	A by-product of petroleum refining or natural gas processing
Biogas	CH_4–CO_2	Coming from the anaerobic fermentation of sludge and wastewater
Gasoline	C_4–C_{12}	Crude oil
Diesel	C_9–C_{25}	Fats and oils from sources such as soybeans, waste cooking oil, animal fats, and rapeseed
Methanol	CH_3OH	Natural gas, coal, or woody biomass
Ethanol	C_3H_7OH	Corn, grains, or agricultural waste (cellulose)
Glycerol	$C_3H_8O_3$	By-product of biodiesel

$$CO + O^{2-} \rightarrow CO_2 + 2e- \tag{8}$$

Therefore, CO is not a pollutant as in the case of low-temperature fuel cells such as PEMFCs (Steele and Heinzel 2001). This is a great advantage especially on lowering the costs related to H_2 purification required in the latter devices. It is commonly understood that in the reforming processes occuring in an anode, CO can be also consumed in the gas phase through the water gas shift reaction (WGSR) (9)

$$CO + H_2O \rightarrow CO_2 + H_2 \qquad \Delta H^\circ = -9,8\ \text{Kcal/mol} \tag{9}$$

and H_2 produced subsequently consumed electrochemically. This behavior depends on the anode composition and on its operating temperature, since WGSR is favored at low temperatures.

H_2 and syngas are generally produced mainly by the same reforming oxidative processes such as steam reforming (SR), partial oxidation (POx), autothermal reforming (ATR), dry reforming (DR), and coal gasification. We refer elsewhere for others non-oxidative processes such as thermal decomposition or refinery processes (Wang et al. 2005). Natural gas and fossil fuels are currently the two main feedstocks used for the production of these energetic molecules, and they satisfy 80 % of world energy demand. This scenario is expected to last up to 2050 as recent reports on fuel sources have emphasized (Muradov and Veriroğlu 2005; Marbán and Valdés-Solís 2007; Specchia 2014). This temporal window is considered as a period of transition that will lead to produce free carbon emission energy. Research projects are focused on maximizing the efficiencies of current technology in fuel processing and on developing technologies for the production of fuels from renewable resources and biomasses. In this perspective, nowadays more and more attention is given to the use of "biofuels" (i.e., fuels obtained from the processing of biologic matter) to power solid oxide fuel cells (Shiratori et al. 2010; Laosiripojana and Assabumrungrat 2007; Xuan et al. 2009). Biogas is one of the most used biofuels (Kowalik et al. 2015). It can be obtained via anaerobic digestion or fermentation of biodegradable materials such as municipal waste, green waste, and biomass, and via gasification of wood or other biomasses. Biogas is composed of CH_4 and CO_2 and other gases such as sulfur, halogenated and silica compounds which are considered as contaminants. CH_4 content is generally in the range of 50–65 vol.%.

Other biofuels used to feed directly/indirectly SOFC anodes are bioalcohols such as ethanol and methanol. These fuels have a high energy density and are readily produced from renewable resources; in addition, they have the advantage to be easily transported and stored as liquid fuels (Nigam and Singh 2011). Another molecule that is attracting interest as fuel for SOFCs is glycerol. Glycerol is obtained as a by-product of the transesterification reaction of vegetable oils to obtain biodiesel, and it is a molecule with high energy value (four times much powerful than methanol) (Pagliaro and Rossi 2010).

Another important power source is coal, which is generally used to produce heat and electricity by combustion. For SOFC applications, coke can be used directly as solid fuel or converted into syngas through a gasification process (Lorente et al. 2012). The use of solid carbon offers the advantage of extremely high electric efficiency (>65–70 %), high fuel utilization, and production of pure stream of CO_2 without the additional cost of capture and energy penalty (Giddey et al. 2012).

Some chemical–physical characteristics of the main fuels used in reforming processes for the production of syngas and H_2 are shown in Tables 4 and 5.

Table 5 Properties of main fuels for SOFC applications

Fuel	Autoignition temperature (°C)	ED[a] (MJ/m^3 of liquid)	GED[b] (MJ/Kg)	Carbon content wt %	Hydrogen content wt%
Hydrogen	585	8.491	140.4	0	100
Natural gas	540–630	20.92	43.6	75	25
Propane	450	23.488	28.3	82	18
Gasoline	260–460	31.15	48.6	85–88	12–15
Diesel	180–320	31.435	33.8	84–87	13–16
Methanol	460	15.8	20.1	39.5	12.6
Ethanol	365	22	26.4	52.2	13.1
Glycerol	260		16	39.1	8.6

[a]Energy density
[b]Gravimetric energy density

4.2 Hydrocarbon Reforming Processes in SOFC Anodes

Catalytic reforming of hydrocarbon fuels is widely recognized as a practical method to produce hydrogen-rich reformates to be fed in fuel cell systems, and it has the advantage of utilizing various fuel sources (e.g., natural gas, ethanol, propane, gasoline, diesel, biodiesel, and glycerol).

Reformer Dual-Chamber Cell Configurations A general scheme described as a SOFC can operate with different fuel processing methods is reported in Fig. 21.

There are three possible configurations. In case A, the fuel is processed in an external reactor providing a syngas mixture to feed the cell and produce energy through electrochemical oxidation of it; in case B, fuel cell is coupled with an internal reformer and the resulting reformate gas is used to feed the cell. This configuration is called indirect internal reforming SOFC (IIR-SOFC); finally, in case C, the cell is powered directly with the fuel (DIR-SOFC) and reforming and electrocatalytic reactions can occur simultaneously inside the anode (Dokmaingram, references therein 2014).

The choice of the right configuration to be used depends on many factors, such as the operating temperature, the type of fuel used, and the final desired power. The criteria of choice should be addressed to maximize the efficiency with a lower cost of operation. What finally emerges from many dedicated studies is that the choice of the fuel plays the main role in defining the best configuration. However, heat transfer management issues can also address the choice.

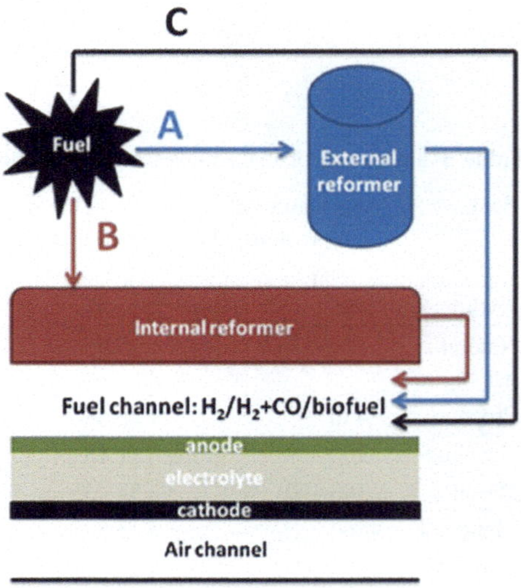

Fig. 21 Scheme of different methods of feeding carbon-based fuels to a dual-chamber SOFC

In the external reforming configuration (A), there is not a direct heat transfer between the reformer and the SOFC unit, but the heat developed during the electrochemical reaction can be recovered and utilized. Generally, only 50 % of the energy contained in the fuel is converted into electricity, and the rest is converted into heat that in turn may be used for the hydrocarbon reforming process or for other purposes such as gas turbine combined cycles (Choudhury et al. 2013).

In the IIR-SOFC configuration (B), the reforming reactions occur in a separate compartment but in close contact with the anode side. This approach is used to exploit the thermal exchange between the endothermic reforming reaction and the exothermic electro-oxidation reaction resulting in an autothermal operation; furthermore, material design can be optimized separately. However, also in this case, the temperature management is not easy due to the local mismatch between the rates of endothermic and exothermic reactions occurring in the two systems.

In the DIR-SOFC configuration (C), it is possible either to feed the SOFC anode with dry fuels that are electrochemically oxidized according to the general reaction (10):

$$C_nH_{2n+2} + (3n + 1)O^{2-} \rightarrow nCO_2 + (n + 1)H_2O + (6n + 2)e^- \quad (10)$$

or to premix hydrocarbons with small amount of steam or oxygen/air so that at the anode, the reforming reactions occur simultaneously with the electrochemical oxidation of CO and H_2. Both the approaches lead to a complex kinetics internally to the electrode, which is difficult to be rationalized. The issues related to the use of dry or humidified hydrocarbon fuels in SOFC anode have been critically reviewed by Mogensen and Kammer (2003) using methane as fuel. They noticed that with respect to a conventional external steam reforming process of methane, the direct use of steam/hydrocarbon premix in a SOFC presents further difficulties that go beyond the deactivation of the catalyst for the formation of carbon resulting from the side reactions that accompany the reforming process. They pointed out the disadvantages related to the use of an internal reforming process demonstrating that the water used in the process dilutes the fuel with a negative effect on the electromotive force of the cell. Moreover, the endothermicity of the process can cause large thermal gradients that can damage the cell. The management of temperature gradients in SOFC is currently an issue. The efficiency would be much higher with a direct conversion of dry hydrocarbons according to the general reaction (10). Reaction (10) is unlikely to occur in one single step. Hydrocarbon pyrolysis and oligomerization are not uncommon under typical SOFC operating temperature, and consequently, chemical conversions and electrochemical fuel oxidation pathways are very complex (see box below, Ge et al. 2012). Gas chromatographic analyses indicated that partial oxidation reactions are dominant for some oxide anodes (Shin et al. 2011).

Scheme of reactions of dry CH_4 inside a SOFC anode

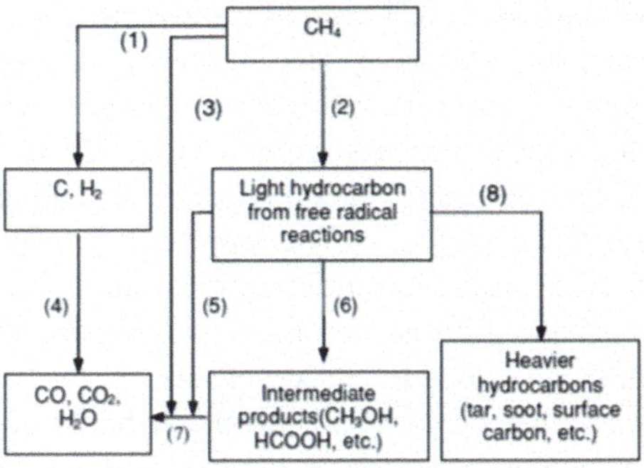

$CH_4 \rightarrow C + H_2$

$C + O^{2-} \rightarrow CO,\ CO_2$

$H_2 + O^{2-} \rightarrow H_2O$

$CH_4 + O^{2-} \rightarrow CH_3OH + 2e^-$

$CH_3OH + O^{2-} \rightarrow HCOOH + H_2O + 4e^-$

$HCOOH + O^{2-} \rightarrow CO_2 + H_2O + 2e^-$

Electrochemical hydrocarbon fuel utilization in SOFC may proceed as:

(i) oxidation of cracked carbon and hydrogen (Step 1 and 4)

(ii) oxidation of the intermediates resulted from free radical reactions (Step 5 and 7)

(iii) oxidation of oxygenated compounds (Step 6 and 7).

(iV) Carbon deposition (Step 8), becomes significant if SOFC is fed with dry CH_4, heavier hydrocarbon-laden fuels such as city coal gas and biomass-gasified syngas.

The exact reaction could be a combination of all the reactions mentioned above.

Shiratori et al. (2010) offered an example of internal reforming of biogas with and without O_2. In this case, the addition of the oxidant was to prevent carbon formation and to avoid deactivation of active sites of the catalyst. Using oxygen, they observed a power loss of the cell, due to the predominance of the reforming reactions rather than of the electrochemical processes. This would contribute to dilute the feeding.

The direct utilization of hydrocarbon fuels was extensively considered by Gorte et al. (2000) using copper as active metal and they emphasized that the penetration market of SOFC technology strongly depends on the development of DIR-SOFC.

Single-Chamber Cells An alternative configuration for SOFC is the use of one gas chamber where both the anode and the cathode are exposed to the same mixture of fuel and air (Fig. 22).

This configuration is referred to as a single-chamber SOFC (SC-SOFC). This design has the advantage of improving mechanical and thermal cell properties due to its simplified structure and of reducing carbon deposition, thanks to the level of oxygen in the mixture. Moreover, it overcomes the problem of sealing and allows the use of porous electrolyte with a significant simplification in the construction of stacks. However, there is the risk of forming explosive mixture between H_2 and O_2. The use of much diluted gases may mitigate the problem. Also, it would be safer working with mixtures of air and methane or air and propane (Riess 2015).

Figure 23 shows a schematic representation of the operating principle of SC-SOFCs fed with a mixture of methane and air. The principle relays on

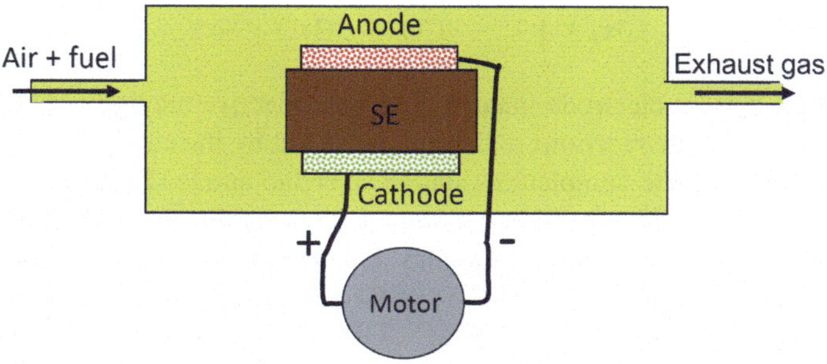

Fig. 22 Schematic of a single-chamber cell (SC–FC). Same gas mixture of fuel and air (*oxygen*) is supplied to both anode and cathode. Anode and cathode can be located in the same gas compartment. Reprinted with permission from © World Scientific Publishing Company (Riess 2015)

Fig. 23 Schematic representation of the single-chamber working principle. Reprinted from (Kuhn and Napporn 2010) open access licenses (http://creativecommons.org/licenses/by/3.0/)

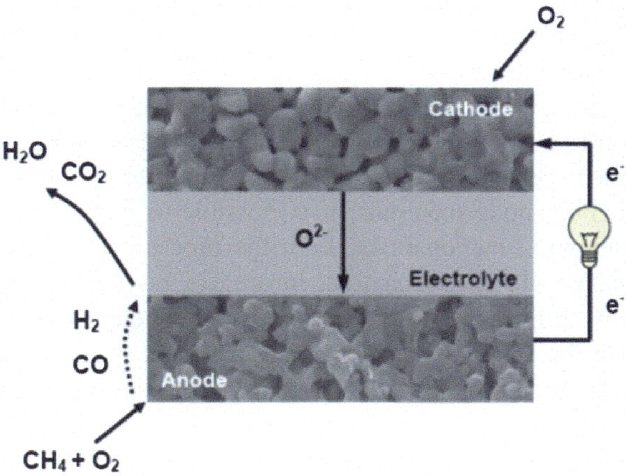

selectivity of the electrode catalysts toward the electrocatalytic processes. In this case, anode catalyst promotes the partial oxidation of methane to CO and H_2. The syngas formed then reacts electrochemically with oxygen ions at the anode. The cathode is inactive toward the fuel and promotes the electrochemical reduction of oxygen.

The partial fuel oxidation and depletion of incoming oxygen lead to a drop of the oxygen partial pressure at the anode, whereas a high oxygen partial pressure is registered at the cathode due to its inertness toward the fuel. This gradient in oxygen partial pressure is the driving force for the fuel cell operation and is at the origin of the OCV according to the Nernst equation. In closed-circuit conditions, the electrochemical reactions lead to a net flow of current through the electrolyte and in the external circuit. In case the anode is ideally selective, OCV would be the same of that measured in a dual-chamber cell and maximum current could be drawn at elevated efficiencies:

$$CH_4 + 4O_2^- \rightarrow CO_2 + 2H_2O + 8e^- \quad (11)$$

In practice, current electrode materials are not selective enough to catalyze only the respective electrode reactions and fuel can be lost by the catalytic activity of the cathode materials or the complete oxidation over the anode. Transport of reaction gases from one electrode to another can impede the establishment of an open-circuit voltage (OCV), and gas intermixing may enhance reactions that should not be catalyzed on the respective electrode. As consequence of these phenomena, the fuel utilization and cell efficiency of real SC-SOFCs are generally lower than 10 %. Selectivity can be controlled and improved not only through a proper choice of materials and cell design but also by playing with the working conditions. For instance, the lack of selectivity of one of the two electrodes can be managed at detriment of the fuel utilization. If the anode is not selective, then excess fuel in the gas mixture has to be used, the excess fuel being sacrificed in order to remove oxygen from the anode side. Conversely, if the anode is selective but not so the cathode, excess oxygen has to be introduced in order to remove (by reaction) all the fuel near the cathode. In both cases, this strategy demands wasting energy in terms of fuel utilization.

A tutorial review on how to select and to investigate materials suitable for this kind of cells is given by Riess (2015). The approach is interesting since it introduces general concepts that can be applied also to other type of cells.

For anode materials, it is possible to distinguish between electrocatalytic and catalytic functionalities. Both the processes are characterized by peculiar mechanisms: (1) Electrocatalytic processes involve incorporation/excorporation of the ions in/from the solid electrolyte, and in heterogeneous catalytic processes, this step can be possible using oxides with a good oxygen storage capability (i.e., ceria-based oxides); (2) chemisorption is an important step both in heterogeneous catalysis and in electrocatalysis; and (3) a unique property of the electrocatalytic process is the electrical current.

Therefore, in search for an electrocatalyst rather than a catalyst, it is important to characterize chemisorption, redox, and catalytic properties of the materials with ex situ characterization techniques for a preliminary screening, but it is also mandatory to test materials in operating conditions since the presence of an electric field can alter their surface properties and the type of adsorbed species.

4.3 *Fuel Reforming Processes*

The main fuel processing technologies used to produce H_2 or syngas to feed externally the fuel cell unit are briefly revised below.

Steam Reforming Processes From the decomposition of any hydrocarbon fuel, it is possible to obtain H_2 according to the following general equation:

$$C_nH_m \rightarrow nC + m/2H_2 \quad (12)$$

The drawback of this reaction is the formation of solid carbon, so water as steam is introduced in the process to have a different decomposition reaction and to avoid solid carbon formation, and so the general reaction (12) becomes

$$C_nH_m + nH_2O + nCO + (m/2 + n)H_2 \quad (13)$$

These reactions can occur with different hydrocarbon molecules; herein, we will consider methane (SMR) and ethanol (SER).

Steam Methane Reforming

In the case of methane, reactions (12) and (13) become, respectively

$$CH_4 \rightarrow C + 2H_2 \qquad \Delta H^\circ = 74,91\ \text{kJ/mol} \quad (14)$$

$$CH_4 + H_2O \rightarrow CO + 3H_2 \qquad \Delta H^\circ = 206\ \text{kJ/mol} \quad (15)$$

Another reaction that normally occurs is

$$CH_4 + 2H_2O \rightarrow CO_2 + 4H_2 \qquad \Delta H^\circ = 165\ \text{kJ/mol} \quad (16)$$

The process was also accompanied by the moderately exothermic water gas shift reaction and the Boudouard reaction:

$$CO + H_2O \leftrightarrow CO_2 + H_2 \qquad \Delta H^\circ = -41\ \text{kJ/mol.} \quad (17)$$

$$2CO \leftrightarrow C + CO_2 \qquad \Delta H^\circ = -172,54\ \text{kJ/mol.} \quad (18)$$

The latter is one of the main sources of coal deposition and deactivation of the catalysts used in the reforming process. The SMR is a highly endothermic process, and it is favored at high temperatures where reaction (15) becomes predominant and the water gas shift (17), which is favored at low temperature, becomes disadvantageous. The reactions involved in the process are never complete and tend to reach an equilibrium condition that depends on the temperature. To maximize the yield of reaction is possible to play with the steam-to-carbon ratio (generally indicated S/C) and with the pressure. The conversion of methane is complete at temperatures over 923 K, with a S/C ratio equal to 3 and 1 atm. Higher pressures over 1 MPa (10 bar) shift the temperature for the complete methane conversion above 1173 K. The operating parameters are the result of an optimization of the process to maximize methane conversion, selectivity, and durability and to minimize deactivation due, for example, to carbon formation, sulfur contamination, and catalyst degradation.

The main components of SMR plant are a furnace, where a reactor containing a nickel-based catalyst is placed and heated at 1123 K for the activation of reactions (14) and (15). Afterward, the reformate gas mixture is treated in a water gas shift reactor at low temperature where the mixture is enriched in H_2, and finally, H_2

is purified from CO_2 and unconverted CH_4 by CO_2 adsorbents and the use of a methanator, respectively, if ultrapure hydrogen is requested.

Steam Reforming of Ethanol

The overall steam reforming reaction of C_2H_5OH could be represented as follows:

$$C_2H_5OH + 3H_2O \rightarrow 2CO_2 + 6H_2 \qquad \Delta H^\circ = +347,4\ \mathrm{kJ/mol} \tag{19}$$

This strongly endothermic reaction is normally favored at high temperatures and low pressures; however, there are several reaction pathways that could occur during the reforming process, which generally take place at low temperature. Firstly, the dehydration of ethanol to ethylene, followed by its polymerization to form coke, can occur:

$$C_2H_5OH \rightarrow C_2H_4 + H_2O \tag{20}$$

$$C_2H_4 \rightarrow \text{coke} \tag{21}$$

Ethanol decomposes to methane which is simultaneously converted by steam.

$$C_2H_5OH \rightarrow CH_4 + CO + H_2 \tag{22}$$

$$CH_4 + 2H_2O \rightarrow 4H_2 + CO_2 \tag{23}$$

Ethanol can dehydrogenate to acetaldehyde that can then react with steam; furthermore, acetaldehyde can decompose to methane:

$$C_2H_5OH \rightarrow C_2H_4O + H_2 \tag{24}$$

$$C_2H_4O + H_2O \rightarrow 3H_2 + 2CO \tag{25}$$

$$C_2H_4O \rightarrow CH_4 + CO \tag{26}$$

Ethanol can decompose to acetone, and the latter reacts with water to give syngas:

$$2C_2H_5OH \rightarrow CH_3COCH_3 + CO + 3H_2 \tag{27}$$

$$CH_3COCH_3 + 2H_2O \rightarrow 5H_2 + 3CO \tag{28}$$

Finally, ethanol can react with water to give directly syngas, and the reaction products are involved in different reactions such as water gas shift, methanation, methane decomposition, and Boudouard reaction:

$$C_2H_5OH + H_2O \rightarrow 2CO + 4H_2 \tag{29}$$

$$CO + H_2O \rightarrow CO_2 + H_2 \tag{30}$$

$$CO + 3H_2 \rightarrow CH_4 + H_2O \tag{31}$$

$$CO_2 + 4H_2 \rightarrow CH_4 + 2H_2O \tag{32}$$

$$CH_4 \rightarrow 2H_2 + C \tag{33}$$

$$CO_2 \rightarrow O_2 + C \tag{34}$$

The overall process is also affected by the dissociative adsorption of water to form acetic acid

$$C_2H_5OH + H_2O \rightarrow CH_3COOH + 2H_2 \tag{35}$$

Therefore, the process is very complex and the final goal is to obtain an H_2-rich gas mixture, but the yield of the reaction strongly depends on the catalyst used. It is also important to avoid the formation of C_4 and C_2H_4 species (Contras et al. 2014) to hinder the formation of coke and therefore the deactivation of the catalyst.

Catalytic Partial Oxidation Partial oxidation technologies (POx) are used to form hydrogen, carbon monoxide, carbon dioxide, and water from liquids and highly viscous hydrocarbons. Partial oxidation consists of an oxidation reaction where the hydrocarbon fuel is mixed with just enough oxygen to convert the fuel to carbon monoxide according to the following general reaction:

$$C_nH_m + n + m/2O_2 \rightarrow nCO + m/2H_2O \tag{36}$$

When the reaction occurs with a catalyst (CPOx), it becomes

$$C_nH_m + n/2O_2 \rightarrow nCO + m/2H_2 \tag{37}$$

Fuel reacts with air over a catalyst, and the combustion is prevented from going to completion by controlling the amount of oxygen and residence time. Due to the fast reaction rates, the CPOX reformer has short response times; these reactors are very compact, and contact times are typically milliseconds. The CPOX system is comparatively more fuel flexible than SR or ATR and can tolerate higher levels of sulfur contaminants in the hydrocarbon fuels. Two disadvantages of this technology are that the reformate has low hydrogen content and the high operating temperatures could lead to catalyst degradation. Moreover, at high temperatures, the reforming reaction was followed by secondary reactions forming CO_2 as product:

$$C_nH_m + (n + m/4)O_2 \rightarrow nCO_2 + m/2H_2O \tag{38}$$

$$H_2 + 1/2\,O_2 \rightarrow H_2O \tag{39}$$

$$CO + 1/2\,O_2 \rightarrow CO_2 \tag{40}$$

Also, in this process, catalysts were deactivated by carbon deposition owing to the reduction of CO at high temperatures, and to the Boudouard reaction at low temperatures.

Autothermal Reforming The autothermal reforming (ATR) is a kind of combination of the both reforming methods explained above; in other words, in this process, the endothermic reaction of steam reforming takes advantage of the heat developed from the exothermic reaction of partial oxidation.

The reaction normally occurs inside a vessel containing a catalyst bed where the fuel, oxidant, and water are fed together (Fig. 24).

The process, which can be used to process both hydrocarbons and oxygenate fuels, is described by the following general reaction:

$$\begin{aligned} C_nH_mO_p + x(O_2 + 3.76N_2) + (2n{-}2x{-}p)H_2O \rightarrow \\ \rightarrow nCO_2 + (2n{-}2x{-}p + m/2)H_2 + 3.76x\ N_2 \end{aligned} \tag{41}$$

In this equation, as reported by Ahmed and Krumpelt (2001), x corresponds to the oxygen-to-fuel molar ratio, and this variable controls the quantity of water necessary to convert the fuel in CO_2, H_2, and heat. When x is equal to 0, the reaction corresponds to the steam reforming reaction.

Dry Reforming The dry reforming of methane (DR) is the largest and the most economical way to produce hydrogen, and the process produces an equimolar synthesis gas (H_2:CO = 1), from carbon dioxide and methane (CH_4), according to the following reaction:

$$CH_4 + CO_2 \rightarrow 2CO + 2H_2 \quad \Delta H^\circ_{1023} = 261\ \text{kJ mol}^{-1} \tag{42}$$

Thermodynamics for the DR reaction are not as favorable as the ATR or even the SMR reactions. However, the proportional consumption of one mole of carbon dioxide per mole of invested methane could reduce the carbon impact, which could overall lead to a "greener" consumption of natural gas. Thermodynamic aspects of this reaction were carefully analyzed by several scientists (Wang and Lu 1996; Pakhare and Spivery 2014), and the obtained results can be used to maximize the advantages coming from this kind of fuel processing approach.

These studies revealed that the reaction is not spontaneous below 640 °C, and the side reactions (43), (44), and (45), listed below, strongly affect the process yield.

$$CO_2 + H_2 \rightarrow CO + H_2O \quad (43)$$

$$CH_4 \rightarrow C + 2H_2 \quad (44)$$

$$2CO \rightarrow C + CO_2 \quad (45)$$

All reactions took place at a significant rate between 633 and 700 °C. Therefore, a higher temperature ($T \geq 700$ °C) was employed to minimize the effect of these side reactions. From the point of view of the final performance, the reforming of natural gas with CO_2 at atmospheric pressure is considered inefficient, because part of the hydrogen produced is consumed by the reverse water gas shift (Oyama et al. 2012). This yields to H_2:CO ratio lower than the unity and variable depending on the operating conditions and the catalyst used.

Coal Gasification Gasification may be one of the most flexible technologies to produce clean-burning hydrogen for tomorrow's automobiles and power-generating fuel cell systems (Higman et al. 2014). In the process, carbon-based feedstocks are converted into the gasifier in the presence of steam and oxygen at high temperatures and moderate pressures to synthesize gas containing a mixture of CO, H_2, CO_2, H_2O, CH_4, and also trace amounts of nearly all of the naturally occurring elements depending on the rank and geographic origin of the coal. The produced syngas is extremely dependent on the type of coal as well as on the gasification process used to produce it (Cayan et al. 2008). During coal gasification, the following main reactions may be considered:

$$C + CO_2 \rightarrow 2CO \quad (46)$$

$$C + H_2O \rightarrow CO + H_2 \quad (47)$$

$$C + O_2 \rightarrow CO_2 \quad (48)$$

$$CO + H_2O \leftrightarrow CO_2 + H_2 \quad (49)$$

$$C + 2H_2 \rightarrow CH_4 \quad (50)$$

The endothermic reactions (46) and (47) can be considered as the most important in a gasification process and they are promoted by the heat released from the exothermic reaction (48). During the process, if high steam concentrations are present, water gas shift reaction occurs (49). Finally, reaction (50) is more important at high pressures. In the work of Molina and Mondragón (1998), coal gasification process is well described, and particularly, they analyzed the parameters that influenced the process such as the coal rank, the thermal history of char, pores, and chemical structure of raw material. The syngas produced by the process was also used as a reagent for chemicals synthesis. Depending on the gasification technology employed, significant quantities of water (H_2O), carbon dioxide (CO_2) and methane (CH_4) can be present in the synthesis gas as well as several minor and trace components such as sulfur, ammonia, and a small amount of hydrogen cyanide, arsenic, and mercury. The obtained reformate is thus further purified with different technologies (e.g., Selexol™ and Rectisol™) depending on the purity requirements of downstream operations (Cayan et al. 2008).

Catalysts used in the Reforming Processes Nowadays, fuel processing technology is focused on the development of high-efficiency integrated systems with the objective to reduce costs, increase efficiency, and reduce operational temperatures.

Generally, catalysts based on nickel, due to its low cost and its high activity, have been extensively studied in all processes of reforming catalysts. Carbon deposition, fuel contaminants, and microstructure optimization are the issues addressed in many studies regarding nickel-based materials. Angeli et al. (2014) reported an overview on nickel, noble metals, and bimetallic-based catalysts for steam reforming at low temperature. They pointed out that although nickel was subjected to a strong carbon deposition, the incorporation of small amounts of other metals affected positively the catalytic activity.

Concerning the ethanol steam reforming, Haryanto et al. (2005) highlighted how crucial is the choice and the preparation method of the catalyst for the overall efficiency of the process. They concluded that the best catalyst for SER are ceria- and alumina-based catalysts loaded with Rh. Contreras et al. (2014) have given an exhaustive overview of catalysts tested in SER process.

Mechanisms involved in the use of noble metals as methane dry reforming catalysts were investigated by Pakhare and Spivery (2014). In their study, kinetics and thermodynamic issues were assessed and considered. They concluded that due to the high endothermicity of dry reforming reaction, it would be useful to support the process with another exothermic reaction such as partial oxidation to reach thermoneutral conditions. Catalysts for partial oxidation have been reviewed by Enger et al. (2008) with a special eye to reaction mechanisms and to the use of metal transition catalysts.

Concerning coal gasification, Irfan et al. (2011) investigated the utilization of CO_2 as gasification agent taking into account all parameters affecting the process such as the type of catalyst, pressure, and temperature.

4.4 Relationship Between Catalysis and Cell Performance

The results of all the studies cited above can serve as a reference and as a sound basis for the development of SOFC anodes operating with hydrocarbons or biofuels (see Sect. 2.2) since the catalysts and processes involved in a SOFC anode (configurations B and C) are similar to those examined here.

In this light, often a first screening of anode materials is obtained through an ex situ investigation of their catalytic properties toward some of the reforming processes described above, which depends on the type of fuel and on the final cell configuration. Steam reforming and autothermal reforming are often considered in the case of methane, light hydrocarbons, and alcohols, while dry reforming is taken into consideration using biogas. In fact, fuels are often fed to the anode with a certain degree of humidification, and the process of autothermal is considered the most suitable to simulate the concomitant occurrence of catalytic and electrochemical reactions at the anode (Boaro et al. 2010). In the ATR process, the partial oxidation would simulate the electrochemical oxidation in which O^{2-} anions are provided through the electrolyte.

Despite this approach is commonly adopted, the reader has to be aware that often ex situ catalytic studies can lead to misleading conclusions. Often, these studies do not take properly into account that electrode's microstructure plays a fundamental role in the electrochemical processes, since the catalytic tests are carried out using powders or configurations different from those present in the electrode.

In addition, gas-phase reactions, catalytic and electrochemical processes in an electrode can occur simultaneously with synergistic kinetics complicating the global reaction mechanism.

In relation to the interplay between catalysis and electrocatalysis in SOFC devices, we can also infer some useful conclusions from the pioneering studies of Vayenas on hydrocarbon-fed solid electrolytic cells focused on the electrocatalysis and chemical production rather than on the electric power generation (Stoukides and Vayenas 1984; Michaels and Vayenas 1984). In these studies, it was demonstrated that the application of a potential can increase the catalytic activity and selectivity of reactions taking place on electrodes composed of a metal supported on an oxygen ion-conducting electrolyte, with an enhancement that can, in some cases, be several orders of magnitude. The phenomenon is called NEMCA (**N**on-faradic, **E**lectrochemical **M**odification of **C**hemical **A**ctivity) and has found controversial explanations (Gorte and Vohs 2011). Its discussion goes beyond the scope of this work. The NEMCA effect has not been convincingly documented for working electrodes in SOFC devices. Despite that, it is worth pointing out how the overall kinetics of the electrode reactions can be different from that involved in a conventional catalytic process due to surface modifications related to the presence of an electrical field.

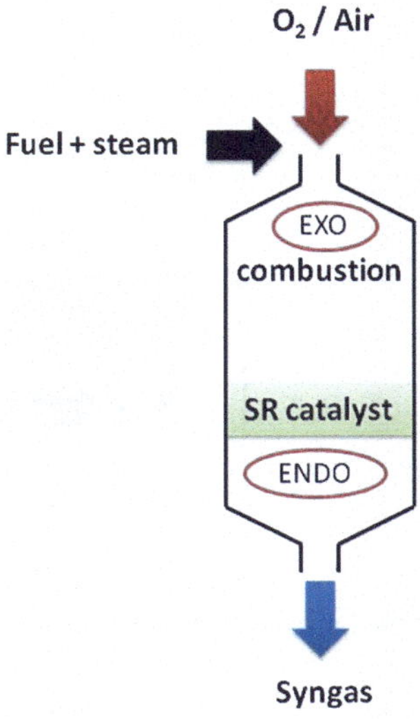

Fig. 24 Scheme of an autothermal reactor

All the above observations allow us to conclude that to distinguish and to understand catalytic processes inside the anode and their effect on the electrochemical process and on cell performance is very complicated. In situ studies and sophisticated techniques of analysis to individuate intermediates and define mechanisms would be necessary. In this regard, modeling could be a useful tool to predict and prove results not always rationalizable (see following chapters).

5 Introduction to Modeling Electrodic Processes in Fuel Cells

At first sight, a fuel cell is not very different from an electrochemical cell in solution. In both cases, one needs four basic elements to form a circuit: an anode and a cathode to perform oxidation and reduction reactions, respectively, an external circuit to conduct electrons, and an electrolyte to conduct ions between the electrodes (Fig. 25).

For both solid-state and solution-based systems, in order for a reaction to occur, redox active species will have to reach the electrodes: This means that in both cases, the influence of mass transport to the electrodes might contribute significantly to the apparent reaction rate; the fact that reactants diffuse through a liquid in one case and from the gas phase in the other does not change the fact that diffusion will ultimately determine the availability of the reactant at the reaction site, i.e., in both situations, if the redox reaction happens faster than reactants can diffuse at the electrode, the process will be diffusion-limited. Similarly, ions will diffuse in order to maintain electroneutrality and, for solid-state systems, react themselves at the interface; the electrolyte being a solid in one case and a liquid in the other means

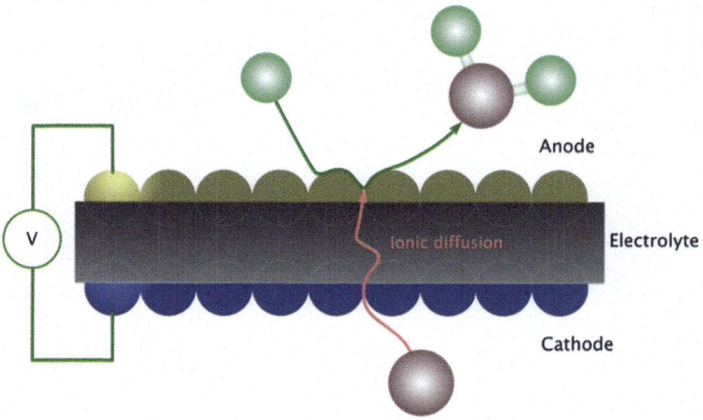

Fig. 25 Simplified scheme of a solid oxide fuel cell

that ionic conduction will involve different kinetics and activation energies, but does not change the topological similarity of the systems.

Imperfections in one or more of the above processes will decrease the V_{theo}, that is, the theoretical voltage developed by an ideal device, equivalent to the oxygen reduction potential in a H_2–O_2 fuel cell, thereby acting as bottlenecks that do not allow the fuel cell to reach its theoretical limits. This is usually expressed using the following expression:

$$V_{cell} = V_{theo} - \eta_{act,c} - \eta_{act,a} - \eta_{ohm} - \eta_{dif,c} - \eta_{dif,c} \tag{51}$$

where contributions termed a and c are relative to the anode and cathode, respectively; η_{act} defines activation losses due to slow reaction kinetics, η_{ohm} defines ohmic losses due to slow ionic transport, and η_{dif} defines concentration losses due to insufficient mass transport of fuel molecules to reaction sites. One of the main issues in achieving high enough accuracy when modeling a SOFC (or a fuel cell in general) is that none of these contributions can be ignored (Ni et al. 2007; Gawel et al. 2015; Chan and Xia 2002): Each of them can, in fact, significantly affect performance, particularly as a consequence of material degradation during the operation of the device. Moreover, the contributions are highly correlated. As shown in Fig. 26, the diffusion of gases, which depends on their partial pressure P and electrode geometry, determines the concentration of reactants at active sites (c at the anode for oxygen ions and P_f, P_O for the partial pressure of the fuel in the anode and oxygen in the cathode, respectively), which in turn affects electrode kinetics.

Electrode kinetics then determines the cell voltage which drives the diffusion of oxygen ions through the electrolyte. All these factors contribute to the cell voltage and output current. Last but not least, cell temperature also influences all these variables at the same time.

Given the high number of processes that must be simulated concurrently, in order to build computational models able to replicate the functioning of a device in a reasonable amount of time, approximations have to be introduced. However, an oversimplification can lead to the exclusion of important factors that affect fuel cell

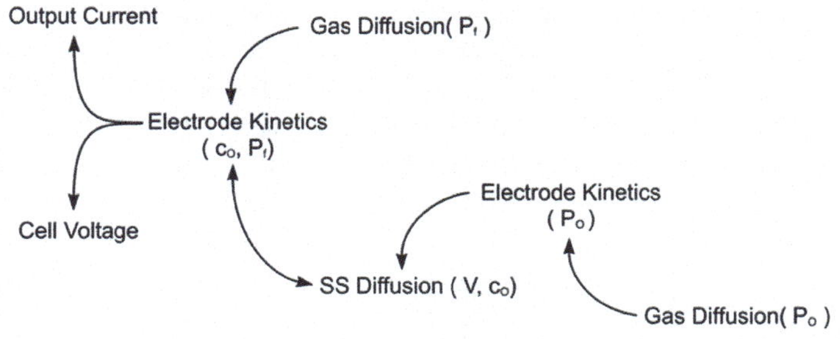

Fig. 26 Simplified scheme of the processes driving a fuel cell which highlights their interdependence

performance, thereby providing skewed results. The aim is then to achieve a compromise between accuracy and efficiency in order to isolate the most important bottlenecks that ultimately determine the performance in a device. In the following sections, the most common methods to achieve this compromise will be briefly introduced and their limitations will be exposed.

5.1 Non-electrochemical Contributions to Fuel Cell Performance

Diffusion The current that can be generated by a fuel cell strongly depends on the availability of the reactants (see the next section). In turn, the concentration of the reactants depends on how quickly they can reach the interface at which they can react: This contribution to fuel cell performance is termed concentration polarization and corresponds to the term η_{dif} in Eq. 5.1. Fick's laws are widely employed for the theoretical description of mass transport phenomena in many electrochemical techniques in solution as well as in the modeling of gas flows in solid oxide fuel cell systems (Izzo et al. 2008). Fick's laws define the relation of the molecular flux to the concentration gradient of the species of interest ϕ (Eq. 52) and of the time derivative of the concentration ϕ to the Laplacian of the concentration ϕ (Eq. 53) through the diffusion coefficient.

The strength of this approach lies in being able to employ a relatively simple model which can be expanded to take into account the effect of other phenomena (i.e., porosity and convection) or simplified to one dimension as needed, thereby minimizing the computational effort required to describe the system at hand by avoiding unnecessary complications.

$$J = -D\nabla\phi \tag{52}$$

$$\frac{\delta\varphi}{\delta t} = D^2\nabla\phi \tag{53}$$

For example, Fick's equations are commonly modified to include the effect of the porosity of SOFC anodes and cathodes on diffusion by substituting the diffusion coefficient in Eq. 52 with an effective diffusion coefficient (Eq. (54)), which is a combination of a coefficient related to molecular diffusion (D_m) and a coefficient related to diffusion in pores, called Knudsen diffusion (D_k); depending on the system at hand, different methods can be used to calculate these parameters (Yakabe et al. 2000).

$$D^{\text{eff}} = \left(\frac{1}{D_m} + \frac{1}{D_k}\right)^{-1} \tag{54}$$

Using the same approach, the model can be further modified for binary and ternary systems (i.e., systems in which two or three different diffusing species are present); in this case, however, the limits of Fick's model become apparent when the condition of equimolar counter diffusion does not hold anymore. In a binary mixture of two components A and B, equimolar counterdiffusion means that the diffusivity of A in B is similar to the diffusivity of B in A. This condition is true until the molecular weights of A and B are relatively similar. In situations commonly encountered in fuel cells, such as a gas feed composed of H_2 and CO/CO_2, this is not the case. In such cases, a model based on the more general Maxwell–Stefan equation can be employed.

$$-\Delta\varphi = \sum_{j=1,j\neq i}^{n} \frac{\varphi_j N_i - \varphi_i N_j}{CD_{i,j}} \tag{55}$$

The summation in Eq. 55 goes over all species i and j and uses a binary diffusion coefficient for each couple: This means that the model allows the diffusivity of species i in species j to be different from the diffusivity of species j in species i. This characteristic makes the model suitable for situations in which the molecular weights of the diffusing species are significantly different, such as in the case of humidified hydrogen (H_2–H_2O mixtures). The Maxwell–Stefan equation, however, does not include the Knudsen diffusivity, meaning that it ignores pore size and its effect on the flux.

If the contribution of pore size cannot be excluded from the treatment, as in the case of small pores, a modified Maxwell–Stefan equation is often used: the dusty gas model (DGM).

$$-\Delta\varphi_i = \frac{N_i}{D_{i,k}} + \sum_{j=1,j\neq i}^{n} \frac{\varphi_j N_i - \varphi_i N_j}{CD_{i,j}} \tag{56}$$

In Eq. 56, the basic formulation of the dusty gas model is presented, with φ_i indicating the concentration of species i, $D_{i,k}$ the Knudsen coefficient for species i, and $D_{i,j}$ the binary diffusivity of species i in species j. The effect of pores and counterdiffusion are both accounted for; this comes at the cost of more computational overhead and a more complex model which requires a numerical solution (as opposed to analytical solutions available for the Fick and Stefan–Maxwell models).

Even though the model itself is arguably more complete, the DGM approach is not always recommended. The choice of the model largely depends on operating parameters such as the size of pores in the anodic scaffold, the fuel, and the current density at which the fuel cell is simulated; i.e., if the current density is low and the molecular weights of the species involved are not very different (i.e., if equimolar counter diffusion holds), simpler models can provide equally accurate results with less computational overhead (Suwanwarangkul et al. 2003; Bertei and Nicolella 2015). Similarly, if pores are sufficiently large, the Stefan–Maxwell formulation can give accurate results.

Geometry Geometry includes two broad areas, namely the macroscopic cell geometry and the microscopic geometry of the electrodes. Concerning the macroscopic geometry of the fuel cell, two configurations are commonly encountered: planar and tubular. From a computational perspective, the macroscopic geometry is often overlooked in favor of a more thorough description of the microstructural and electrochemical aspects of the device. This being said, studies that analyze the impact of different fuel cell geometries on current output through modeling have been conducted and successfully managed to highlight the different parameters that can be used to optimize performance (Stiller et al. 2005).

The microstructural aspect, on the other hand, is more complex. Efforts toward the formulation of realistic 3D models of the electrodes based on experimental data are ongoing and are providing interesting insight into the relative importance of processes in fuel cell electrodes at a microscopic scale (Al-Masri et al. 2014; Cai et al. 2011, Claire et al. 2011). A 3D description of the fuel cell is also indispensable to correctly account for heat dissipation in the device (Ho 2014) which, particularly in fuel cell stacks, can pose a serious threat to the lifetime of the system. The experimental and computational effort required to realistically reproduce the electrode structure, however, makes this approach difficult to use routinely, making it more suitable for an a priori analysis to rationalize the engineering of the cells before their construction; following this line of thought, specialized models to optimize the electrode microstructure by predicting the optimal values of parameters such as the anode-sintering temperature have also been proposed (Suzue et al. 2007), as well as studies investigating the relationship between grain size and the area of catalytically active regions (Deng and Petric 2005).

Even though reducing the number of dimensions can affect the ability of the model to precisely quantify the concentration overpotential (Tseronis et al. 2008), 2D and 1D models are the most widely employed due to the reduced computational resources they require (Campanari and Iora 2004). The use of low-dimensional models, 1D in particular, usually means that the geometry of the electrodes is disregarded and mass transport through the porous electrode is computed solely on the basis of the gas diffusivities. The microstructure of electrodes, however, including thickness, porosity, and pore size, can significantly hamper the ability to generate large currents even in the presence of good catalysts (Noh et al. 2011). The ability to include these parameters in the description of a fuel cell is then desirable. An example of a model that includes both the description of gas flows and that of the solid state is the cross-linked pore model (Haugaard and Livbjerg 1998; Feng and Stewart 1973; Johnson and Stewart 1965). Equation 57 presents the model in the case of binary diffusion.

$$N_A = \frac{c\rho_P}{\tau}\left[\int \left[\frac{1-\chi_A}{D_{AB}} + \frac{1}{D_k}\right]^{-1} v' dr\right]\frac{d\chi_A}{d\xi} \tag{57}$$

The core of the equation is similar to the dusty gas model, featuring the flux as the inverse of the sum of the reciprocal binary and Knudsen diffusivities D_{AB} and

D_k; the flux, however, is integrated over the pore volume distribution ν'(r) and weighted by the apparent catalyst particle density ρ_p and a tortuosity factor τ, thereby coupling it to the effective geometry of the porous scaffold; these parameters allow one to gauge the influence of the electrodic structure on mass transport and make the cross-linked model particularly versatile (Haugaard and Livbjerg 1998).

Monte Carlo methods have also been successfully employed to investigate the dynamics of various aspects of the fuel cell, such as the microstructural evolution of anodes due to thermal aging (Zhang et al. 2015) and the correlation between the variation in different parameters (i.e., temperature, activation energy of the anodic and cathodic reactions, and electrolyte thickness) and the average current density (He et al. 2014, 2016). In all cases, simulations were able to provide insights which could be employed to predict the properties of real-life device and hence optimize their synthesis.

Ionic Transport Ionic transport in solid oxide fuel cells is a complex process. Oxygen atoms are generated at the cathode where molecular gaseous oxygen is reduced and incorporated into a metal oxide. Driven by the potential difference present between anode and cathode, the oxygen atoms then diffuse through the cathode, the solid electrolyte itself, and the electrolyte–anode interface before they can reach the active sites where the fuel is oxidized. If the diffusion process is too slow, it can act as a bottleneck which will increase the internal resistance of the fuel cell. Many different parameters can affect the mobility of ions through the fuel cell. For example, specialized models have been developed to investigate the interplay between ionic conductivity and factors such as particle size and lattice distortion on the ionic conduction of the anodic and cathodic scaffolds (Costamagna et al. 1998); similarly, the transition of ions between the electrode/electrolyte interfaces can create significant resistive and capacitive contributions; models have been developed to target the behavior of ions and the potential distribution at the electrode/electrolyte specifically (Hong et al. 2015). While these are important details which can hamper the performance of devices, incorporating them into a model of a fuel cell is generally not computationally viable.

Accurate descriptions of the movement of oxygen atoms have been carried out as atomistic simulations, which try to map the energy landscape diffusing ions face in order to discern the effect of the structure and chemistry of materials on the diffusivity (Dholabhai and Adams 2012; Yuan et al. 2015). In practice, the effect of slow ionic transport in solid oxide fuel cells is usually described by using Ohm's law, which shows a linear decrease in the current that can be extracted from the fuel cell at any given cell voltage.

$$I = \frac{V}{R} \tag{58}$$

The resistance R is then usually associated with the resistance posed by the electrolyte against diffusion of oxygen ions. The resistance also includes current leakage through the electrolyte, that is, a flow of electrons from the anode toward the cathode. This phenomenon is particularly relevant when reducible materials are used in the electrolyte, such as ceria-based oxides for applications in intermediate-temperature fuel cells. Examples are also present in the literature in which the two effects are quantified separately (Donazzi et al. 2016).

Internal Reforming When alkanes are used as fuels, the gases can be transformed into more reactive species in a pre-reforming stage before the fuel enters the fuel cell (Zhu et al. 2008; Finnerty et al. 2000) or inside the fuel cell in a process called internal reforming (Ge et al. 2012). The simplest example of the latter case is the reforming of methane by water or carbon dioxide to yield carbon monoxide and hydrogen. As shown in Eq. 59, reforming will contribute to determine the concentration and the nature of the species that will react in the anode, with the effect that the SOFC will burn more reactive gases, thereby allowing for an increased current output.

$$CH_4 + H_2O \rightarrow CO + 3H_2 \tag{59}$$

At intermediate temperature, the homogeneous reforming of methane is usually slow enough that it can be disregarded and only the heterogeneous reaction can be included in a model (Hecht et al. 2005); when more reactive alkanes, such as propane and butane, are employed, gas-phase reactions can become important and they also need to be accounted for. Since internal reforming occurs inside the anodic compartment, besides increasing power output (Zhu et al. 2015), the process also has a significant influence on the thermal management of the fuel cell, because it absorbs a considerable amount of the heat generated in the anode (Nagata et al. 2001).

The most straightforward approach is to assume a steady state in order to avoid cumbersome calculations of gas-phase kinetics (Aguiar et al. 2002). This approach, however, is a severe approximation since the gas phase is not near thermodynamic equilibrium (Girona et al. 2015). Moreover, by not describing gas-phase reactions, it is not possible to take into account the possibility of carbon deposition, one of the major limiting factors to the lifetime of a SOFC operating on organic fuels.

$$CH_4 \rightarrow C + 2H_2 \tag{60}$$

$$2CO \rightarrow C + CO_2 \tag{61}$$

Equations 60 and 61, methane cracking and the Boudouard reaction, respectively, show the most important reactions which lead to carbon deposition: By correctly modeling gas-phase kinetics, important information regarding these processes can be obtained, which can aid in the prediction of cell performance over time (Klein et al. 2007; Girona et al. 2015).

5.2 Electrochemical Contributions to Fuel Cell Performance

The Butler–Volmer Equation Following the similarities between solution-based and solid-state electrochemical systems outlined at the beginning of Sect. 5, analytical current–potential relationships developed for the description of electrodes immersed in a liquid are often employed to describe the electrochemical processes which take place at a solid/gas interface. Among these, the most well known and widespread is the Butler–Volmer equation. The Butler–Volmer formalism derives from the assumption that a reactant A transforms into a product B by moving along a one-dimensional, continuous reaction path with the transfer of one electron via an outer sphere mechanism (Fig. 3), that is, one where no significant interaction occurs between reactants in the transition state (McNaught and Wilkinson 1997). The Butler–Volmer formalism, in fact, fails if applied to reactions which follow an inner sphere mechanism, that is reactions where there are significant rearrangements in the bond configuration of the molecules involved in the redox reaction. If sufficient energy is provided, the reactant reaches a saddle point termed the transition state; the energy required to reach the transition state is defined with an exponential. Two such exponentials are employed: one for the forward reaction A → B and one for the backward reaction B → A. This description translates mathematically in Eq. 62:

$$j = j_0 \exp\left[-\frac{\alpha n F v}{RT}\right] - \exp\left[\frac{(1-\alpha) n F v}{RT}\right] \tag{62}$$

where j is the current density, F is the Faraday constant, R is the ideal gas constant, and j_0 is the exchange current density. The energetics for the two reactions are defined by two parameters, v and α.

The overpotential $v = E - E_{eq}$ quantifies the driving force or how much the applied potential E differs from the equilibrium potential E_{eq} that is from the potential at which no current flows across the cell. Multiplying the overpotential by the Faraday constant gives the difference in Gibbs' free energy $\Delta G°$ between the reactant and the product at that overpotential (see Fig. 27).

The parameter α is the charge transfer coefficient, which varies from 0 to 1 and stems from the assumption that a linear transformation along a single reaction coordinate connects the reactant to the transition state and then to the product. Depending on the molecule and the type of reaction, the position of the transition state might be closer to the reactant or to the product, where being closer means being more similar in structure. α quantifies this similarity: $\alpha < 0.5$ indicates that the transition state is structurally similar to the reactant, while $\alpha > 0.5$ indicates that the transition state is structurally similar to the product. It should be noted that since the structural reorganization of a molecule requires energy, being similar in structure implies being similar in energy; therefore, if α is close to 1, a large amount of energy will be required for the reaction to happen (a large overpotential in electrochemical terms), that is, in order to change the reactant enough to bring it to

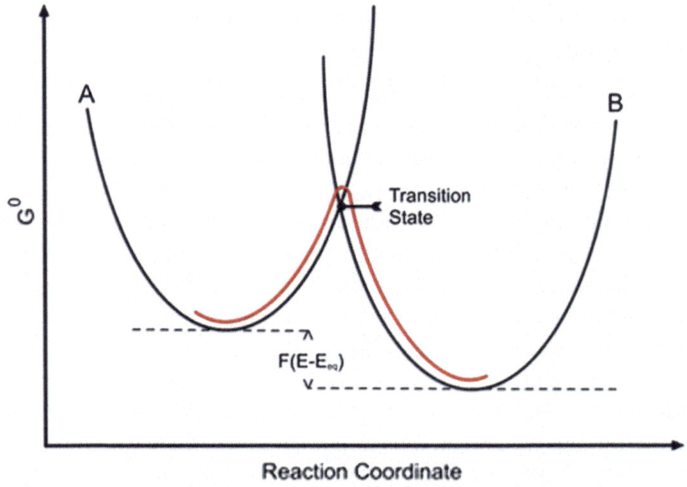

Fig. 27 Simplified scheme of a reaction involving a reactant *A*, a product *B*, and a transition state at the intersection of the two *parabolae*. The reaction path along the reaction coordinate is highlighted in *red*

the transition state. In most cases, α is more or less arbitrarily set equal to 0.5, which implies that the transition state is structurally (and thus energetically) equidistant from reactants and products.

In specific situations, the Butler–Volmer equation can also be simplified, thereby decreasing the use of computational resources. Two of the most common variants of the formula are the high and low polarization approximations: The former can be applied when one exponential in the B–V equation is significantly larger than the other and the latter when the exponents of the exponentials are sufficiently small, thereby allowing one to use the approximation ex = 1 + x (Eq. 63).

$$\ln\frac{j}{j_0} = \frac{\mathrm{RT}}{\alpha\mathrm{nF}\nu} \tag{63}$$

A more in-depth analysis of when it is appropriate to use each approximation can be found in the literature (Noren and Hoffman 2005).

Expanding the Butler–Volmer Formalism The Butler–Volmer equation has been employed extensively to describe electrochemical processes in fuel cells because it is assumed that if one concentrates on the rate-determining step, interfacial redox reactions occur in the same way, regardless of the medium. The differences that do arise because of the different diffusion dynamics are accounted for in the description of mass transport, while other elementary steps that constitute the overall redox reaction are compressed in a single A → B event described by a single apparent rate constant. This practice, albeit practical, is misleading if one wants to understand and correctly describe the inner workings of fuel cell anodes and cathodes. In

Fig. 28, a schematic reaction scheme for the hydrogen oxidation reaction in a SOFC anode is presented, depicting the main steps that lead to the formation of water. A hydrogen molecule (orange) diffuses from the gas phase and chemisorbs to the catalyst, where it decomposes. Hydrogen atoms subsequently diffuse on the surface of the catalyst to meet the oxygen atoms diffusing from the cathode at the triple-phase boundary (TPB), that is, the boundary between the gas phase, the catalyst, and the support. At the TPB, water can form and then desorb. The first problem one encounters in applying the BV equation to this scheme is that of actually identifying the reactant and the product in the A $\rightarrow$ B scheme. Taking the hydrogen oxidation reaction as example, defining the anodic process as H_2 + 1/2 $O_2 \rightarrow 2H_2O$ would not be, in fact, correct. As will be discussed in the next section, in order for the product to form, gas reactants must first adsorb on the solid catalyst and dissociate.

This means that the assumption that the reaction follows an outer sphere mechanism is also invalid: In solid/gas systems, in fact, the adsorption and decomposition of the fuel are an essential part of the catalytic process, which implies a strong interaction between the reactants at the transition state. Even if one tries to take this into account by defining the reactants as dissociated hydrogen and oxygen $2H^{\cdot} + O^{2-} \rightarrow H_2O$, one is still overlooking what often are the rate-determining steps of these reactions, that is, the adsorption, the dissociation, and sometimes the desorption of the reaction products (Mogensen and Skarup 1996), or the formation of particularly stable reaction intermediates (Nørskov et al. 2004; Gorski et al. 2011).

On top of this, the solid catalyst is treated as an inert substrate, when it is clear that it actively participates in the reactions by undergoing redox and structural transitions itself. This complexity is compressed in the BV equation into a single rate constant, which should represent the slower and thus rate-determining step.

Multistep Reactions In order to try to overcome these limits, variations to the Butler–Volmer formalism have been developed that allow to take the factors discussed above into account in some measure. Gas–solid reactions are complex. They invariably feature at least three separate stages: adsorption, decomposition of reactants, and recombination of reactants' fragments into the products. Even for relatively small molecules such as methane, the decomposition and recombination stages can be complex multistep processes. This means that describing electrode kinetics with only one rate constant effectively masks the complexity of the electrochemical processes occurring in the anode. In order to describe the redox kinetics more realistically, the Butler–Volmer equation can be modified to accommodate an arbitrary number of reaction steps and many electron transfer processes by acting on the definition of the charge transfer coefficient (Mann et al. 2006). This approach involves defining the charge transfer coefficient as shown in Eq. 64

$$\alpha = \frac{\beta(n - n_b - n_a)}{v} \tag{64}$$

where β is the symmetry factor, n the number of electrons transferred in the overall reaction, n_b and n_a the number of electrons transferred before and after the rate-determining step, respectively, and ν the stoichiometric coefficient defined as the number of times the rate-determining step must take place in order for the overall reaction to occur. A detailed description of this approach can be found in Bockris et al. (1970).

Adsorption and Chemisorption The first step in the reaction of gas molecules is the adsorption of the gas onto a solid catalyst. In fuel cells, this adsorption is often a chemisorption, that is, a chemical reaction between the surface and the fuel molecule (Nørskov 1990). In Fig. 28, the chemisorption of hydrogen is shown as an example: In the reaction, the hydrogen molecule is broken into individual hydrogen atoms, each bound to the active site (shown in red). In the case of larger molecules such as methane, the decomposition is a multistep process which involves the formation of various intermediates. The fuel and oxygen atoms must then meet and recombine to form the product(s). Once formed, the reaction product is still attached to the surface and must desorb in order for the reaction to be complete and the active site to become available for the adsorption of another fuel molecule. In order to account for the limited number of reaction sites available, the Butler–Volmer equation can be modified to incorporate a function describing the surface coverage of catalyst particles.

$$j = j_0 f_1(\theta, \theta_0) \exp\left[-\frac{\alpha n F v}{RT}\right] - f_2(\theta, \theta_0) \exp\left[\frac{(1-\alpha) n F v}{RT}\right] \tag{65}$$

$f(\theta, \theta_0)$ depends on the number of surface sites available and on the number of fuel molecules that can adsorb on those sites at any given time. The function $f(\theta, \theta_0)$ can be derived from classical adsorption theories, such as those of Langmuir and Frumkin. It should be noted that the choice of the function describing adsorption

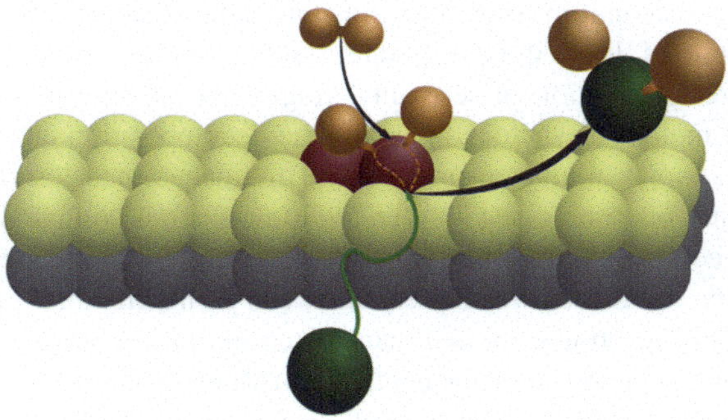

Fig. 28 Schematic depiction of the reaction of a hydrogen molecule (shown in *orange*) in solid oxide fuel cell anode, highlighting the steps of chemisorption on a catalytically active substrate (*red spheres*); scission into two hydrogen atoms, each bound to the active site; diffusion to the triple-phase boundary and reaction with an oxygen atom (*green sphere*) to form water. The oxygen atom diffuses through the electrolyte (*gray spheres*) and the anodic scaffold (*yellowish spheres*)

has strong implications on the reaction mechanism proposed, e.g., describing adsorption through the Langmuir model means that the assumptions of the Langmuir model, that is, adsorbed molecules do not interact with one another and all adsorption sites are equivalent, will be applied to the processes occurring in the anode as well.

It should be kept in mind that no matter what the sophistication of these extensions, the Butler–Volmer formalism models a macroscopic quantity, that is, the output current and the cell voltage. What the Butler–Volmer equation allows one to do is to discriminate between the macroscopic factors that influence cell performance as described in Fig. 26, but cannot provide unambiguous information on the reaction mechanism or the reaction intermediates.

To be able to describe the basic processes that drive the fuel cell at the level of detail shown in Fig. 28, the time and length scales investigated must be changed, and atomistic simulations based on ab initio methods, mostly based on density functional theory (DFT) and molecular dynamics, are required. An in-depth analysis of the elementary steps that take place inside a fuel cell electrode can provide complementary information that can help interpret experimental data and clarify reaction mechanisms (Peng et al. 2016).

Poisoning In the previous section, it was said that in order to react, gas molecules must first chemisorb to the surface. If the bond between a reactant or a product and the surface is too strong, however, the reaction will proceed slowly, since a larger energy will be needed to desorb the reaction product from the catalyst. Eventually, if the intermediates accumulate on the surface, the catalyst will be rendered inactive not because of a missing interaction between catalyst and fuel, but as a consequence of the formation of a bond which requires too much energy to be broken: The active site will then be permanently occupied and will not be able to react with other molecules. This observation was formalized in the Sabatier principle, which states that the interaction between substrate and catalyst should be neither too weak nor too strong, but "just right" (Roduner 2014; Medford et al. 2015). Experimentally, the Sabatier principle is observed through the formation of nonvolatile side products (i.e., carbon deposition) which can physically impede the contact between catalyst and reactants, as well as the chemical modification of the surface composition of the solid catalyst. The effect of having too strong a catalyst–substrate interaction can be investigated experimentally by comparing the activity of different catalysts with the heat of formation of the bond between the catalyst and a target reactant: Such experiments can cause the appearance of the so-called volcano plots: These plots show how catalytic activity is highest at intermediate catalyst–substrate bonding energies and decreases when the bonding energy becomes too strong. These experiments provide a useful guide for the screening of new catalyst formulations (Toulhoat and Raybaud 2003; Jiang et al. 2003). Computationally, information on the bonding energy between catalyst and reactants or intermediates can only be accessed through atomistic simulations, which can allow us to predict whether a catalyst will be poisoned by calculating a priori the enthalpy of bond formation between the active site and reaction intermediates (Medford et al. 2015).

Understanding the reaction at the molecular and atomic levels allows one to rationalize the activity of catalysts by revealing the rate-determining step(s) of the reactions and the intermediates involved (Rossmeisl and Bessler 2005). The atomistic and multiphysics modeling of cells acts as complementary tools: the former to optimize existing catalysts as well to discover new ones, the latter to monitor the functioning of the entire fuel cell device and make sure that other factors, e.g., mass transport, are sufficiently optimized.

Acknowledgments The authors would like to thank the Italian Ministry of University and Research (MIUR) for funding BIO-ITSOFC project (2010KHLKFC 003, PRIN 2010–2011).

References

Adijanto, L., Kungas, R., Bidrawn, F., Gorte, R. J., & Vohs, J. M. (2001). Stability and performance of infiltrated $La_{0.8}Sr_{0.2}Co_xFe_{1-x}O_3$ electrodes with and without $Sm_{0.2}Ce_{0.8}O_{1.9}$ interlayer. *Journal of Power Sources, 156*, 5797–5802.

Adler, S. B. (2004). Factors governing oxygen reduction in solid oxide fuel cell cathodes. *Chemical Reviews, 104*, 4791–4843.

Aguiar, P., Chadwick, D., & Kershenbaum, L. (2002). Modelling of an indirect internal reforming solid oxide fuel cell. *Chemical Engineering Science, 57*, 1665–1677.

Ahmed, S., & Krumpelt, M. (2001). Hydrogen from hydrocarbon fuels for fuel cells. *International Journal of Hydrogen Energy, 26*, 291–301.

Al-Masri, A., Peksen, M., Blum, L., & Stolten, D. (2014). A 3D CFD model for predicting the temperature distribution in a full scale APU SOFC short stack under transient operating conditions. *Applied Energy, 135*, 539–547.

An, W., Gatewood, D., Dunlap, B., & Turner, C. H. (2011). Catalytic activity of bimetallic nickel alloys for solid-oxide fuel cell anode reactions from density-functional theory. *Journal of Power Sources, 196*, 4724–4728.

Angeli, D. S., Monteleone, G., Giaconia, A., & Leminidou, A. A. (2014). State of the art catalysts for CH_4 steam reforming at low temperature. *International Journal of Hydrogen Energy, 39*, 1979–1997.

Atkinson, A., Barnett, S., Gorte, R. J., Irvine, J. T., Mcevoy, A. J., Mogensen, M., et al. (2004). Advanced anodes for high-temperature fuel cells. *Nature Materials, 3*, 17–23.

Bardini, L., Pappacena, A., Dominguez-Escalante, M., Llorca, J., Boaro, M., & Trovarelli, A. (2016). Structural and electrocatalytic properties of molten core Sn@SnOx nanoparticles on ceria. Applied Catalysis B: Environmental, *197*, 254–261.

Bartels, J. R., Pate, M. B., & Olson, N. K. (2010). An economic survey of hydrogen production from conventional and alternative energy sources. *International Journal of Hydrogen Energy, 35*, 8371–8384.

Bastidas, D. M., Tao, S. W., & Irvine, J. T. S. (2006). A symmetrical solid oxide fuel cell demonstrating redox stable perovskite electrodes. *Journal of Materials Chemistry, 16*, 1603–1605.

Bertei, A., & Nicolella, C. (2015). Common inconsistencies in modeling gas transport in porous electrodes: The dusty-gas model and the Fick law. *Journal of Power Sources, 279*, 133–137.

Boaro, M. (2014). Unpublished communication.

Boaro, M., Modafferi, M., Pappacena, A., Llorca, J., Baglio, V., Frusteri, F., et al. (2010). Comparison between Ni–Rh/gadolinia doped ceria catalysts in reforming of propane for anode implementations in intermediate solid oxide fuel cells. *Journal of Power Sources, 195*, 649–661.

Boaro, M., Pappacena, A., Abate, C., Ferluga, M., Llorca, J., & Trovarelli, A. (2014). Effect of redox treatments on $Ce_{0.50}Zr_{0.50}O_2$ based solid oxide fuel cell anodes. *Journal of Power Sources, 270*, 79–91.

Bockris, J. O. M., & Reddy, A. K. N. (1970). *Modern electrochemistry*, Vol. 1. Mac-Donald/Plenum Press.

Bøgild Hansen, J., & Rostrup-Nielsen, J. (2010). Sulfur poisoning on Ni catalyst and anodes. In W. Vielstich & A. Hubert (Eds.), *Handbook of Fuel Cells-Fundamentals, Technology and Applications*, New York: John Wiley & Sons.

Brett, D. J. L., Atkinson, A., Brandon, N. P., & Skinner, S. J. (2008). Intermediate temperature solid oxide fuel cells. *Chemical Society Reviews, 37*, 1568–1578.

Burnat, D., Heel, A., Holzer, L., Kata, D., Lis, J., & Graule, T. (2012). Synthesis and performance of A-site deficient lanthanum-doped strontium titanate by nanoparticle based spray pyrolysis. *Journal of Power Sources, 201*, 26–36.

Busawon, A. N., Sarantaridis, D., & Atkinson, A. (2008). Ni infiltration as a possible solution to the redox problem of SOFC anodes. *Electrochemical and Solid-State Letters, 11*, B186–B189.

Cai, Q., Adjiman, C. S., & Brandon, N. P. (2011). Investigation of the active thickness of solid oxide fuel cell electrodes using a 3D microstructure model. *Electrochimica Acta, 56*(28), 10809–10819.

Campanari, S., & Iora, P. (2004). Definition and sensitivity analysis of a finite volume SOFC model for a tubular cell geometry. *Journal of Power Sources, 132*, 113–126.

Canales-Vázquez, J., Ruiz-Morales, J. C., Marrero-López, D., Pe˜na-Martínez, J., Nú˜nez, P., & Gómez-Romero, P. (2007). Fe-substituted (La, Sr) TiO_3 as potential electrodes for symmetrical fuel cells (SFCs). *Journal of Power Sources, 171*, 552–557.

Cargnello, M., Wieder, N. L., Montini, T., Gorte, R. J., & Fornasiero, P. (2010). Synthesis of dispersible Pd@CeO2 nanostructures by self-assembly. *Journal of the American Chemical Society, 132*, 1402–1409.

Cayan, F. N., Zhi, M., Pakalapati, R., Celik, I., Wu, N., & Gemmen, R. (2008). Effects of coal impurities on anode of solide oxide fuel cells. *Journal of Power Sources, 185*, 595–602.

Chan, S. H., & Xia, Z. T. (2002). Polarization effects in electrolyte/electrode-supported solid oxide fuel cells. *Journal of Appled Electrochemistry, 32*, 339–347.

Choi, Y., Choi, S., Jeong, H. Y., Liu, M., Kim, B.-S., & Kim, G. (2014). Highly efficient layer-by-layer-assisted infiltration for high-performance and cost-effective fabrication of nano-electrodes. *ACS Applied Materials & Interfaces, 6*, 17352–17357.

Choudhury, A., Chandra, H., & Arora, A. (2013). Application of solid oxide fuell cell technology for power generation—A Review. *Renewable and Sustainable Energy Reviews, 20*, 430–442.

Claire, Q. C., Adjiman, S., & Brandon, N. P. (2011). Modelling the 3D micro-structure and performance of solid oxide fuel cell electrodes: Computational parameters. *Electrochimica Acta, 56*, 5804–5814.

Contreras, J. L., Salmones, J., Colin-Luna, J. A., Nuño, L., Quintana, B., Cordova, I., et al. (2014). Catalysts for H_2 production using the ethanol steam reforming (a review). *International Journal of Hydrogen Energy, 39*, 18835–18853.

Costamagna, P., Costa, P., & Antonucci, V. (1998). Micro-modelling of solid oxide fuel cell electrodes. *Electrochimica Acta, 43*, 375–394.

Da Silva, A. L., & Heck, N. C. (2015). Thermodynamics of sulfur poisoning in solid oxide fuel cells revisited: The effect of H_2S concentration, temperature, current density and fuel utilization. *Journal of Power Sources, 296*, 92–101.

Deng, X., & Petric, A. (2005). Geometrical modeling of the triple-phase-boundary in solid oxide fuel cells. *Journal of Power Sources, 140*, 297–303.

Dholabhai, P., & Adams, J. (2012). A blend of first-principles and kinetic lattice Monte Carlo computation to optimize sarium-doped ceria. *Journal of Materials Science, 47*, 7530–7541.

Ding, D., Li, X., Lai, S. Y., Gerdes, K., & Liu, M. (2014). Enhancing SOFC cathode performance by surface modification through infiltration. *Energy & Environmental Science, 7*, 552–575.

Dokmaingram, P. (2014). Configuration development of Autothermal solid oxide fuel cell: A Review. *Engineering Journal, 19*, 1–13.

Donazzi, A., Rahmanipour, M., Maestri, M., Groppi, G., Bardini, L., Pappacena, A., et al. (2016). Experimental and model analysis of the co-oxidative behavior of syngas feed in an intermediate temperature solid oxide fuel cell. *Journal of Power Sources, 306*, 467–480.

Enger, B. C., Lødeng, R., & Holmen, A. (2008). A review of catalytic partial oxidation of methane to synthesis gas with emphasis on reaction mechanism over metal catalysts. *Applied Catalysis A: General, 346*, 1–27.

Feng, C., & Stewart, W. E. (1973). Practical models for isothermal diffusion and flow of gases in porous solids. *Industrial & Engineering Chemistry Fundamentals, 12*, 143–147.

Finnerty, C., Tompsett, G. A., Kendall, K., & Ormerod, R. M. (2000). SOFC system with integrated catalytic fuel processing. *Journal of Power Sources, 86*, 459–463.

Gawel, D. A. W., Pharoah, J. G., & Beale, S. B. (2015). Development of a SOFC performance model to analyze the powder to power performance of electrode microstructures. *ECS Transactions, 68*, 1979–1987.

Ge, X.-M., Chan, S.-H., Liu, Q.-L., & Sun, Q. (2012). Solid oxide fuel cell anode materials for direct hydrocarbon utilization. *Advanced Energy Materials, 2*, 1156–1181.

Ge, X. M., Zhang, L., Fang, Y. N., Zeng, J., & Chan, S. H. (2011). Robust solid oxide cells for alternate power generation and carbon conversion. *RSC Advances, 1*, 715–724.

Giddey, S., Badwal, S. P. S., Kulkarni, A., & Munnings, C. (2012). A comprensive review of direct carbon fuel cell technology. *Progress in Energy and Combustion Science, 38*, 360–399.

Girona, K., Sailler, S., Gelin, P., Bailly, N., Georges, S., & Bultel, Y. (2015). Modelling of gradual internal reforming process over Ni-YSZ SOFC anode with a catalytic layer. *The Canadian Journal of Chemical Engineering, 93*, 285–296.

Gong, M. Y., Liu, X. B., Trembly, J., & Johnson, C. (2007). Sulfur-tolerant anode materials for solid oxide fuel cell application. *Journal of Power Sources, 185*, 1086–1093.

Goodenough, J. B. (2004). Electronic and ionic transport properties and other physical aspects of perovskites. *Reports on Progress in Physics, 67*, 1915–1993.

Gorski, A., Yurkiv, V., Starukhin, D., & Volpp, H.-R. (2011). H_2O chemisorption and H_2 oxidation on yttria-stabilized zirconia: Density functional theory and tempera-ture-programmed desorption studies. *Journal of Power Sources, 196*, 7188–7194.

Gorte, R. J., & Vohs, J. M. (2009). Nanostructured anodes for solid oxide fuel cells. *Current Opinion in Colloid & Interface Science, 14*, 236–244.

Gorte, R. J., & Vohs, J. M. (2011). Catalysis in solid oxide fuel cells. *Annual Review of Chemical and Biomolecular Engineering, 2*, 9–30.

Gorte, R. J., Park, S., Vhos, J. M., & Wang, C. (2000). Anodes for direct oxidation of dry hydrocarbons in a solid-oxide fuel cell. *Advanced Materials, 12*, 1465–1469.

Gross, M. D., Vohs, J. M., & Gorte, R. J. (2007). A Strategy for Achieving High Performance with SOFC Ceramic Anodes. *Electrochemical and Solid-State Letters, 10*, B65–B69.

Haryanto, A., Fernando, S., Murali, N., & Adhikari, S. (2005). Current Status of hydrgen production techniques by steam reforming of ethanol: A review. *Energy & Fuels, 19*, 2098–2106.

Haugaard, J., & Livbjerg, H. (1998). Models of pore diffusion in porous catalysts. *Chemical Engineering Science, 53*, 2941–2948.

He, H. P., Huang, Y. Y., Regal, J., Boaro, M., Vohs, J. M., & Gorte, R. J. (2004). Low temperature fabrication of oxide composites for solid-oxide fuel cells. *Journal of the American Ceramic Society, 87*, 331–336.

He, Z., Li, H., & Birgersson, E. (2014). Correlating variability of modeling parameters with non-isothermal stack performance: Monte Carlo simulation of a portable 3d planar solid oxide fuel cell stack. *Applied Energy, 136*, 560–575.

He, Z., Li, H., & Birgersson, E. (2016). Correlating variability of modeling parameters with cell performance: Monte Carlo simulation of a quasi-3D planar solid oxide fuel cell. *Renewable Energy, 85*, 1301–1315.

Hecht, E. S., Gupta, G. K., Zhu, H., Dean, A. M., Kee, R. J., Maier, L., et al. (2005). Methane reforming kinetics within a Ni–YSZ SOFC anode support. *Applied Catalysis A: General, 295*, 40–51.

Higman, C., & Tam, S. (2014). Advances in coal gasification, hydrogenation, and gas treating for the production of chemicals and fuels. *Chemical Reviews, 114*, 1673–1708.

Ho, T. X. (2014). A three-dimensional model for transient performance of a solid oxide fuel cell. *International Journal of Hydrogen Energy, 39*, 6680–6688.

Hong, L., Hu, J.-M., Gerdes, K., & Chen, L.-Q. (2015). Oxygen vacancy diffusion across cathode/electrolyte interface in solid oxide fuel cells: An electro-chemical phase-field model. *Journal of Power Sources, 287*, 396–400.

Irfan, M. F., Usman, M. R., & Kusakabe, K. (2011). Coal gasification in CO_2 atmosphere and its kinetics since 1948: A brief review. *Energy, 1*, 12–40.

Irvine, J. T. S., & Connor, P. (2013). Alternative materials for SOFCs, opportunities and limitations in SOFC. In J. T. Irvine & P. Connor (Eds.), *Solid Oxide Fuel Cells: Facts and Figures* (pp. 163–180). London: Springer-Verlag: Green Energy and Technology series.

Irvine, J. T. S., Neagu, D., Verbraeken, M. C., Chatzichristodoulou, C., Graves, C., & Mogensen, M. B. (2016). Evolution of the electrochemical interface in high-temperature fuel cell and elctrolyser. *Nature Energy, 1*, 1–13.

Izzo, J. R., Peracchio, A. A., & Chiu, W. K. S. (2008). Modeling of gas transport through a tubular solid oxide fuel cell and the porous anode layer. *Journal of Power Sources, 176*, 200–206.

Jiang, S. P. (2008). Development of lanthanum strontium manganite perovskite cathode materials of solid oxide fuel cells: A review. *Journal Materials Science, 43*, 6799–6833.

Jiang, S. P. (2012). Nanoscale and nano-structured electrodes of solid oxide fuel cells by infiltration: Advances and challenges. *International Journal of Hydrogen Energy, 37*, 449–470.

Jiang, S. P., Callus, P. J., & Badwal, S. P. S. (2000). Fabrication on performance of Ni/3 mol% Y_2O_3–ZrO_2 cermet anodes for Solid Oxide fuel Cell. *Solid State Ionics, 132*, 1–14.

Jiang, S. P., Duan, Y. Y., & Love, J. G. (2002). Fabrication of high-performance NiO-Y_2O_3-ZrO_2 cermet anodes of solid oxide fuel cells by ion impregnation. *Journal of the Electrochemical Society, 149*, A1175–A1183.

Jiang, L., Wang, G.-C., Guan, N.-J., Wu, Cai Z.-S., Pan, Y.-M., Zhao, X.-Z., et al. (2003). DFT studies of coadsorption and activation on some transition metal surfaces. *Acta Physicochimica Sinica, 19*, 393–397.

Jiang, Z., Xia, C., & Fanglin, Fanglin Chen. (2010). Nano-structured composite cathodes for intermediate-temperature solid oxide fuel cells via an infiltration/impregnation technique. *Electrochimica Acta, 55*, 3595–3605.

Johnson, M. F. L., & Stewart, W. E. (1965). Pore structure and gaseous diffusion in solid catalysts. *Journal of Catalysis, 4*, 248–252.

Jun, A., Kim, J., Shin, J., & Kim, G. (2016). Perovskite as a cathode material: A review of its role in solid-oxide fuel cell technology. *ChemElectroChem, 3*, 511–530.

Jung, S., Lu, C., He, H., Ahn, K., Gorte, R. J., & Vohs, J. M. (2006). Influence of composition and Cu impregnation method on the performance of Cu/CeO2/YSZ SOFC anodes. *Journal of Power Sources, 154*, 42–50.

Kim, S.-D., Lee, J.-J., Moon, S.-H., Hyun, S.-H., Moona, J., Kimb, J., et al. (2007). Effects of anode and electrolyte microstructures on performance of solid oxide fuel cells. *Journal of Power Sources, 169*, 265–270.

Kim, G., Lee, S., Shin, J. Y., Corre, G., Irvine, J. T. S., Vohs, J. M., et al. (2009). Investigation of the structural and catalytic requirements for high-performance SOFC anodes formed by infiltration of LSCM. *Electrochemical and Solid-State Letters, 12*, B48–B52.

Klein, J.-M., Bultel, Y., Georges, S., & Pons, M. (2007). Modeling of a SOFC fuelled by methane: From direct internal reforming to gradual internal reforming. *Chemical Engineering Science, 62*, 1636–1649.

Kowalik, P., Antoniak-Juraka, K., Błesznowskib, M., Herrerac, M. C., Larrubiac, M. A., Alemanyc, L. J., et al. (2015). Biofuel steam reforming catalyst for fuel cell application. *Catalysis Today, 254*, 129–134.

Kuhn, M., & Napporn, T. W. (2010). Single-chamber solid oxide fuel cell technology—From its origins to today's state of the art. *Energies, 3*, 57–134.

Laosiripojana, N., & Assabumrungrat, S. (2007). Catalytic steam reforming of methane, methanol and ethanol over Ni/YSZ: The possible use of these fuels in internal reforming SOFC. *Journal of Power Sources, 163*, 943–951.

Lee, J.-H., Moon, H., Lee, H.-W., Kim, J., Kim, J.-D., & Yoon, K. H. (2002). Quantitative Analysis of microstructure and its related electrical properties of SOFC anode. *Solid State Ionic, 148*, 15–26.

Lin, B., Wang, S., Liu, X., & Meng, G. (2010). Simple solid oxide fuel cells. *Journal of Alloys and Compounds, 490*, 214–222.

Lin, Y., Zhan, Z., Liu, & Barnett, S. A. (2005). Direct operation of solid oxide fuel cells with methane fuel. *Solid State Ionics, 176*, 1827–1835.

Liu, Z., Beibei, Liu, Ding, D., Mingfei, Liu, Chen, F., & Xia, C. (2013). Fabrication and modification of solid oxide fuel cell anodes via wet impregnation/infiltration technique. *Journal of Power Sources, 237*, 243–259.

Liu, M., Lynch, M. E., Blinn, K., Alamgir, F. M., & Choi, Y. (2011). Rational SOFC materials design: New advances and tools. *MaterialsToday, 14*, 534–546.

Lorente, E., Millan, M., & Brandon, N. P. (2012). Use of gasification syngas in SOFC: Impact of real tar on anode materials. *International Journal of Hydrogen Energy, 37*, 7271–7278.

Lou, X., Liu, Z., Wang, S., Xiu, Y., Wong, C. P., & Liu, M. (2010). Controlling the morphology and uniformity of a catalyst-infiltrated cathode for solid oxide fuel cells by tuning wetting property. *Journal of Power Sources, 195*, 419–424.

Lu, C., Sholklapper, T. Z., Jacobson, C. P., Visco, S. J., & De Jonghe, L. C. (2006). LSM-YSZ cathodes with reaction-infiltrated nanoparticles. *Journal of the Electrochemical Society, 153*, A1115–A1119.

Mahato, N., Banerjee, A., Gupta, A., Omar, S., & Balani, K. (2015). Progress in material selection for solid oxide fuel cell technology: A review. *Progress in Materials Science, 72*, 141–337.

Mann, R. F., Amphlett, J. C., Peppley, B. A., & Thurgood, C. P. (2006). Application of Butler-Volmer equations in the modelling of activation polarization for PEM fuel cells. *Journal of Power Sources, 161*, 775–781.

Marbán, G., & Valdés-Solís, T. (2007). Towards the hydrogen economy*?*. *International Journal of Hydrogen Energy, 32*, 1625–1637.

McIntosh S., & Gorte, R. J. (2004). Direct hydrocarbon solid oxide fuel cells. *Chemical Reviews, 104*, 4845–4865.

McNaught, A. D., & Wilkinson, A. (1997). *IUPAC compendium of chemical terminology*, 2nd edn. Oxford: © Blackwell Scientific Publications.

Medford, A. J., Vojvodic, A., Hummelshøj, J. S., Voss, J., Abild-Pedersen, F., Studt, F., et al. (2015). From the Sabatier principle to a predictive theory of transition-metal heterogeneous catalysis. *Journal of Catalysis, 328*, 36–42.

Michaels, J. N., & Vayenas, C. G. (1984). Styrene production from ethylbenzene on platinum in a zirconia electrochemical reactor. *Journal of the Electrochemical Society, 131*, 2544–2550.

Mogensen, M., & Kammer, K. (2003). Conversion of hydrocarbons in solid oxide fuel cells. *Annual Review of Materials Research, 33*, 321–331.

Mogensen, M., & Skarup, S. (1996). Kinetic and geometric aspects of solid oxide fuel cell electrodes. *Solid State Ionics, 86*, 1151–1160.

Molina, A., & Mondragón, F. (1998). Reactivity of coal gasification with steam and CO_2. *Fuel 77* (15), 1831–1839.

Muradov, N. Z., & Veriroğlu, T. N. (2005). From hydrocarbon to hydrogen-carbon to hydrogen economy. *International Journal of Hydrogen Energy, 30*, 225–237.

Nagata, S., Momma, A., Kato, T., & Kasuga, Y. (2001). Numerical analysis of output characteristics of tubular SOFC with internal reformer. *Journal of Power Sources, 101*, 60–71.

Neagu, D., Oh, T.-S., Miller, D. N., Ménard, H., Bukhari, S. M., Gamble, S. R., et al. (2015). Nano-socketed nickel particles with enhanced coking resistance grown in situ by redox exsolution. *Nature Communications, 6*(1–8), 2015.

Neagu, D., Tsekouras, G., Miller, D. N., Ménard, H., & Irvine, J. T. (2013). In situ growth of nanoparticles through control of non-stoichiometry. *Nature Chemistry, 5*, 916–923.

Ni, M., Leung, M. K. H., & Leung, D. Y. C. (2007). Parametric study of solid oxide fuel cell performance. *Energy Conversion and Management, 48*, 1525–1535.

Nicholas, J. D., & Barnett, S. A. (2009). Finite-Element modelling of Idealized infiltrated composite solid oxide fuel cell cathodes. *Journal of the Electrochemical Society, 156*, B458–B464.

Nigam, P. S., & Singh, A. (2011). Production of liquid biofuels from renewable resources. *Progress in Energy and Combustion Science, 37*, 52–58.

Nikolla, E., Schwank, J., & Linic, S. (2007). Promotion of the long-term stability of reforming Ni catalysts by surface alloying. *Journal of Catalysis, 250*, 85–93.

Noh, H.-S., Lee, H., Kim, B.-K., Lee, H.-W., Lee, J.-H., & Son, J.-W. (2011). Microstructural factors of electrodes affecting the performance of anode-supported thin film yttria-stabilized zirconia electrolyte (~ 1 μm) solid oxide fuel cells. *Journal of Power Sources, 196*, 7169–7174.

Noren, D. A., & Hoffman, M. A. (2005). Clarifying the Butler-Volmer equation and related approximations for calculating activation losses in solid oxide fuel cell models. *Journal of Power Sources, 152*, 175–181.

Norskov, J. K. (1990). Chemisorption on metal surfaces. *Reports on Progress in Physics, 53*, 1253.

Nørskov, J. K., Rossmeisl, J., Logadottir, A., Lindqvist, L., Kitchin, J. R, Bligaard, T., & Jonsson, H. (2004). Origin of the overpotential for oxygen reduction at a fuel-cell cathode. *The Journal of Physical Chemistry B, 108*, 17886–17892.

Oyama, S. T., Hacarlioglu, P., Gu, Y., & Lee, D. (2012). Dry reforming of methane has no future for hydrogen production: Comparison with steam refroming at high pressure in standard amd membrane reactors. *International Journal of Hydrogen Energy, 37*, 10444–10450.

Pagliaro, M., & Rossi, M. (2010). *The future of glycerol: New usages for a versatile raw material.* Copyright: 2010 RCD Publisher.

Pakhare, D., & Spivery, J. (2014). A review of dry (CO_2) reforming of methane over noble metal catalysts. *Chemical Society Reviews, 43*, 7813–7837.

Piao, J., Sun, K., Zhang, N., Chen, X., Xu, S., & Zhou, D. (2007). Preparation and characterization of $Pr_{1-x}Sr_xFeO_3$ cathode materials for intermediate temperature solid oxide fuel cells. *Journal of Power Sources, 172*, 633–640.

Peng, Y., Si, W., Li, X., Luo, J., Li, J., Crittenden, J., et al. (2016). Comparison of MoO_3 and WO_3 on arsenic poisoning V_2O_5/TiO_2 catalyst: DRIFTS and DFT study. *Applied Catalysis B: Environmental, 181*, 692–698.

Primdahl, S., & Mogensen, M. (1997). Oxidationof hydrogen on Ni/YSZ cermet anodes. *Journal of the Electrochemical Society, 144*, 3409–3419.

Qin, H., Tan, X., Huang, W., Jiangn, J., & Jiangn, H. (2015). Application of urea precipitation method in preparation of advanced ceramic powders. *Ceramic International, 41*, 11598–11604.

Riess, I. (2015). Selectivity and mixed reactant fuel cells. *Functional Materials Letters, 8*, 1540010 (8 p).

Roduner, E. (2014). Understanding catalysis. *Chemical Society Reviews, 43*, 8226–8239.

Rossmeisl, J., & Bessler, W. G. (2005). In catalytic activity for SOFC anode materials. *Solid State Ionics, 178*, 1694–1700.

Ruiz-Morales, J. C., Canales-Vázquez, J., Ballesteros, B., Peña-Martínez, J., Marrero-López, D., Irvine, J. T. S., & Núñez, P. (2007). LSCM–(YSZ–CGO) composites as improved symmetrical electrodes for solid oxide fuel cells. *Journal of the European Ceramic Society, 27*, 4223–4227.

Ruiz-Morales, J. C., Canales-Vázquez, J., Marrero-López, D., Pérez-Coll, D., Peña-Martínez, J., & Núñez, P. (2008). An all-in-one flourite-based symmetrical solid oxide fuel cell. *Journal of Power Sources, 177*, 154–160.

Ruiz-Morales, J. C., Canales-Vázquez, J., Peña-Martínez, J., Marrero-López, D., & Núñez, P. (2006). On the simultaneous use of $La_{0.75}Sr_{0.25}\,Cr_{0.5}Mn_{0.5}O_{3-\delta}$ as both anode and cathode material with improved microstructure in solid oxide fuel cells. *Electrochimica Acta, 52*, 278–284.

Ruiz Morales, J. C., Canales-Vazquez, J., Savaniu, C., Marrero-Lopez, D., Zhou, W. Z., & Irvine, J. T. S. (2006). Disruption of extended defects in solid oxide fuel cell anodes for methane oxidation. *Nature, 439*, 568–571.

Ruiz-Morales, J. C., Marrero-López, D., Gálvez-Sánchez, M., Canales-Vázquez, J., Cristian, Savaniu C., & Savvin, S. N. (2010). Engineering of materials for solid oxide fuel cells and other energy and environmental applications. *Energy & Environmental Science, 3*, 1670–1681.

Sengodan, S., Choi, S., Jun, A., Shin, T. H., Ju, Y.-W., Jeong, H. Y., et al. (2015). Layered oxygen-deficient double perovskite as an efficient and stable anode for direct hydrocarbon solid oxide fuel cells. *Nature Materials, 14*, 205–209.

Serra, J. M., & Buchkremer, H. P. (2007). On the nanostructuring and catalytic promotion of intermediate temperature solid oxide fuel cell (IT-SOFC) cathodes. *Journal of Power Sources, 172*, 768–774.

Shao, Z. P., & Haile, S. M. (2004). A high-performance cathode for the next generation of solid-oxide fuel cells. *Nature, 431*, 170–173.

Shin, T. H., Ida, S., & Ishihara, T. (2011). Doped CeO_2-$LaFeO_3$ composite oxide as an active anode for direct hydrocarbon-type solid oxide fuel cells. *Journal of the American Chemical Society, 133*(48), 19399–19407.

Shiratori, Y., Ijichi, T., Oshima, T., & Sasaki, K. (2010). Internal reforming SOFC running on biogas. *International Journal of Hydrogen Energy, 32*, 7905–7912.

Sholklapper, T. Z., Jacobson, C. P., Visco, S. J., & De Jonghe, L. C. (2008). Synthesis of dispersed and contiguous nanoparticles in solid oxide fuel cell electrodes. *Fuel Cells, 5*, 303–312.

Skinner, S. J., Cook, S., & Kilner, J. A. (2013). Materials for next generation SOFC. In J. T. Irvine & P. Connor (Eds.), *Solid Oxide Fuel Cells: Facts and Figures*. London: Springer-Verlag: Green Energy and Technology series, reference therein.

Specchia, S. (2014). Fuel processing activities at European level: A panoramic overview. *International Journal of Hydrogen Energy, 39*, 17953–17968.

Steele, B. C. H., & Heinzel, A. (2001). Matrials for fuel cell technologies. *Nature, 414*, 345–351.

Stiller, C., Thorud, B., Seljeboe, S., Mathisen, O., Karoliussen, H., & Bolland, O. (2005). Finite-volume modeling and hybrid-cycle performance of planar and tubular solid oxide fuel cells. *Journal of Power Sources, 141*, 227–240.

Stoukides, M., & Vayenas, C. G. (1984). Electrocatalytic rate enhancement of propylene epoxidation on porous silver electrodes using a zirconia oxygen pump. *Journal of the Electrochemical Society, 131*, 839–884.

Sun, C., Hui, R., & Roller, J. (2010). Cathode materials for solid oxide fuel cells: A review. *Journal of Solid State Electrochemistry, 14*, 1125–1144.

Sun, Y., Li, J., Zeng, Y., Amirkhiz, B. S., Wang, M., Behnamian, Y., et al. (2015). A-site deficient perovskite: The parent for *in situ* exsolution of highly active, regenerable nano-particles as SOFC anodes. *Journal of Materials Chemistry A, 3*, 11048–11056.

Suwanwarangkul, R., Croiset, E., Fowler, M. W., Douglas, P. L., Entchev, E., & Douglas, M. A. (2003). Performance comparison of fick's, dusty-gas and stefan-maxwell models to predict the concentration overpotential of a SOFC anode. *Journal of Power Sources, 122*, 9–18.

Suzue, Y., Shikazono, N., & Kasagi, N. (2007). Modeling of sofc anodes based on the stochastic reconstruction scheme. *ECS Transactions, 7*, 2049–2055.

Takahashi, S., Nishimoto, S., Matsuda, M., & Miyake, M. (2010). Electrode properties of Ruddlesden-Popper series $Ln_{n+1}Ni_nO_{3n+1}$ (n = 1,2 and 3) as intermediate temperature solid oxide fuel cells. *Journal of the American Ceramic Society, 93*, 2329–2333.

Tao, S., & Irvine, J. T. S. (2004). Discovery and Characterization of Novel Oxide Anodes for Solid Oxide Fuel Cells. *Chemical Record, 4*, 83–95.

Toulhoat, H., & Raybaud, P. (2003). Kinetic interpretation of catalytic activity patterns based on theoretical chemical descriptors. *Journal of Catalysis, 216*, 63–72.

Tseronis, K., Kookos, I. K., & Theodoropoulos, C. (2008). Modelling mass transport in solid oxide fuel cell anodes: A case for a multidimensional dusty gas-based model. *Chemical Engineering Science, 65*, 626–5638.

Tsipis, E. V., & Kharton, V. V. (2008). Electrode Materials and Reaction mechanism in solid oxide mechanism: A brief review. *Journal of Solid State Electrochemistry, 12*, 1367–1391.
Vohs, J. M., & Gorte, R. J. (2009). High performance SOFC cathodes prepared by infiltration. *Advanced Materials, 21*, 943–956.
Wang, S., & Lu, G. M. Q. (1996). Carbon dioxide reforming of methane to produce synthesis gas over metal-supported catalysts: State of the art. *Energy & Fuels, 1996*(10), 896–904.
Wang, Y., Shah, N., & Huffman, G. P. (2005). Simultaneous production of hydrogen and carbon nanostructures by decomposition of propane and cyclohexane over alumina supported binary catalysts. *Catalysis Today, 99*, 359–364.
Wei, B., Lv, Z., Huang, X. Q., Miao, J. P., Sha, X. Q., Xin, X. S., et al. (2006). Crystal structure, thermal expansion and electrical conductivity of perovskite oxides $BaxSr_{1-x}Co_{0.8}Fe_{0.2}O_{3-\delta}$ (0.3<x<0.7). *Journal of the European Ceramic Society, 26*(13), 2827–2832.
Xiang, Y., Lu, S., & Jiang, S. P. (2012). Layer-by-layer self-assembly in the development of electrochemical energy conversion and storage devices from fuel cells to super capacitors. *Chemical Society Reviews, 41*, 7291–7321.
Xuan, J., Leung, M. K. H., Leung, D. Y. C., & Ni, M. (2009). A review of biomass-derived fuel processors for fuel cell systems. *Renewable and Sustainable Energy Reviews, 13*, 1301–1313.
Yakabe, H., Hishinuma, M., Uratani, M., Matsuzaki, Y., & Yasuda, I. (2000). Evaluation and modelling of performance of anode-supported solid oxide fuel cell. *Journal of Power Sources, 86*, 423–431.
Yan, A. Y., Cheng, M. J., Dong, Y. L., Yang, W. S., Maragou, V., Song, S. Q., et al. (2006). Investigation of a $Ba_{0.5}$ $Sr_{0.5}Co_{0.8}Fe_{0.2}O_{3-\delta}$ based cathode IT-SOFC: I. The effect of CO_2 on the cell performance. *Applied Catalysis, B: Environmental, 66*, 64–71.
Yang, C., Yang, Z., Jin, C., Xiao, G., Chen, F., & Han, M. (2012). Sulfur-tolerant redox-reversible anode material for direct hydrocarbon solid oxide fuel cells. *Advanced Materials, 24*, 1439–1443.
Yoo, S., Choi, S., Shin, J., Liu, M., & Kim, G. (2012). Electrical properties, thermodynamic behavior, and defect analysis of $La_{n+1}Ni_nO3_{n+1+d}$ infiltrated into YSZ scaffolds as cathodes for intermediate temperature SOFCs. *RSC Advances, 2*, 4648–4655.
Yuan, F., Zhang, Y., & Weber, W. J. (2015). Vacancy-vacancy interaction induced oxygen diffusivity enhancement in undoped non-stoichiometric ceria. *Journal of Physical Chemistry C, 119*, 13153–13159.
Zhan, Z., Lin, Y., Pillai, M., Kim, I., & Barnett, S. A. (2006). High-rate electrochemical partial oxidation of methane in solid oxide fuel cells. *Journal of Power Sources, 161*, 460–465.
Zhao, Y., Xia, C., Jia, L., Wang, Z., Li, H., Yu, J., et al. (2013). Recent progress on solid oxide fuel cell: Lowering temperature and utilizing non-hydrogen fuels. *International Journal of Hydrogen Energy, 38*, 16498–16517.
Zhang, Y., Ni, M., Yan, M., & Chen, F. (2015). Thermal aging stability of infiltrated solid oxide fuel cell electrode microstructures: A three-dimensional kinetic monte carlo simulation. *Journal of Power Sources, 299*, 578–586.
Zhu, W., Ding, D., & Xia, C. (2008). Enhancement in three-phase boundary of SOFC electrodes by an ion impregnation method: A modelling comparison. *Electrochemical and Solid-State Letters, 11*, B83–B86.
Zhu, T., Yang, Z., & Han, M. (2015). Performance evaluation of solid oxide fuel cell with in-situ methane reforming. *Fuel, 161*, 168–173.

Energy System Analysis of SOFC Systems

Andrea Lanzini, Domenico Ferrero and Massimo Santarelli

Abstract A solid oxide fuel cell (SOFC) system involves phenomena that take place at different scales. The electrochemical active cell—a multicomposite structure that consists of different microscopic layers made of different materials—is thus at the center of a broader and complex energy system, whose efficient and durable performance much depends on several other integrated subsystems and auxiliaries which both provide reactants to the fuel cell and extract exhausts from the same. This is the reason energy analysis focuses on the overall plant behavior and not only on cell and stack performance. This chapter will review modeling tools and approaches for system performance evaluation.

Abbreviations

List of Symbols

ASR	Area specific resistance (Ωcm^2)
F	Faraday's constant (=96485) (C mol^{-1})
FU	Fuel utilization
I	Current (A)
R	Ideal gas constant (=8.314) (J mol^{-1} K^{-1})
T	Temperature (K)
V	Voltage (V)
W	Electric power (W)
$\Delta\bar{g}$	Molar Gibbs free energy change (J mol^{-1})
ΔH	Total enthalpy change (W)
ΔS	Entropy rate change (W K^{-1})
j	Current density (A cm^{-2})
$\bar{n}$, $\dot{n}$	Molar flow rate (mol s^{-1})
n_{el}	Number of electrons

A. Lanzini (✉) · D. Ferrero · M. Santarelli (✉)
Energy Department, Polytechnic of Turin, Turin, Italy
e-mail: andrea.lanzini@polito.it

M. Santarelli
e-mail: massimo.santarelli@polito.it

M. Boaro and A.S. Aricò (eds.), *Advances in Medium and High Temperature Solid Oxide Fuel Cell Technology*, CISM International Centre for Mechanical Sciences 574,
DOI 10.1007/978-3-319-46146-5_6

$\bar{h}$ Molar specific enthalpy (J mol^{-1})
$\bar{s}$ Molar specific entropy (J mol^{-1} K^{-1})
p (partial) pressure, bar
x Cathodic recirculation ratio
y Anodic recirculation ratio

Greek Symbols

Σ_{irr} Rate of irreversibilities production (W)
Φ Thermal power (W)
λ Air excess ratio

List of Subscripts and Superscripts

an	Anode
aux	Auxiliaries
cat	Cathode
f	Fuel
i	i-component of a mixture
in	Inlet
mix	Mixture
N	Total number of components (chemical species) in a mixture
N, Nernst	Nernst potential
op	Operating
out	Outlet
r	Reaction
tot	Total
0	Reference thermodynamic condition (25 °C, 1 bar)

List of Acronyms

A-SOFC	Atmospheric SOFC
BoP	Balance of Plant
DIR-SOFC	Direct internal reforming SOFC
ER-SOFC	External reformer SOFC
GT	Gas turbine
IGFC	Integrated gasifier fuel cell
IIR-SOFC	Indirect internal reformer SOFC
LHV	Lower heating value
NETL	National Energy Technology Lab
NG	Natural gas
OCV	Open-circuit voltage

P-SOFC	Pressurized SOFC
S(M)R	Steam (methane) reforming
SOFC	Solid oxide fuel cell
WGS	Water gas shift
WWTP	Wastewater treatment plant

1 Energy System Analysis of SOFC Systems

The scope of this chapter is to introduce the reader to the tools and techniques that are widely used to model the energy and electrochemical performance of solid oxide fuel cell (SOFC) systems. A brief description of the intrinsic SOFC economics in order to assess the plant profitability is also included.

The perspective offered is that of the professional engineer or scientist who is interested in modeling the performance of the SOFC system and compares it to that of other conventional power technologies. In addition, tools for the preliminary design, design improvement, and system optimization of SOFC plants are also presented.

The energy and techno-economic performance of SOFC systems can be assessed through mass and energy balances of control volumes that include the SOFC power unit and the required Balance of Plant (BoP). The control volume of an SOFC generator and its auxiliaries can be easily modeled through process engineering software (e.g., Aspen Plus®, Aspen Hysys®, and Chemcad®). Underlying assumptions and equations required for this type of energy modeling will be covered in this section.

The first section of this chapter provides a description of the main BoP components of SOFC plants. Different SOFC plant configurations and design features are introduce in the following of this chapter. Finally, the energy model of the SOFC is described in detail with related equations for the 0-D modeling of the fuel cell plant.

2 SOFC Stack and Balance of Plant (BoP) Components

The SOFC system mainly includes the following components:

- Gas cleanup sections;
- Chemical reactors (e.g., catalytic reformer and anode off-gas burner).

- SOFC module(s): Each module consists of several stack towers either series- or parallel-connected;
- Heat exchangers;
- Turbomachinery equipment (fuel/air blowers or compressors, gas turbines for hybrid cycles, etc.);

Instrumentation, control and automation devices, piping, valves, and electrical equipment complete the list presented above. The system layout for the SOFC fueled by compressed natural gas (CNG) is shown in Fig. 1. The main input streams are cathode air (dried and filtered), fuel gas, and demineralized water. The requirement of external water depends on whether anodic recirculation is used or not. In case of hot anodic recirculation (or alternatively water condensation from the exhaust gas), the need for external water is eliminated.

Gas cleanup system. A high purity in the fuel stream is required for the SOFC anode and reformer catalyst. The amount and typology of fuel contaminants vary widely depending on the fuel type. For gaseous feeds, sulfur compounds are generally the most abundant. In particular, a high H_2S concentration is found in coal syngas and in digester or landfill gas (see Table 1). Concerning fossil gas, most of the sulfur in European natural gas is removed at the well. NG from Russia and the North Sea contains some H_2S and COS, while most overseas supply of NG arrives as liquefied NG and thus contains no sulfur. For instance, in Italy, the gas pipeline

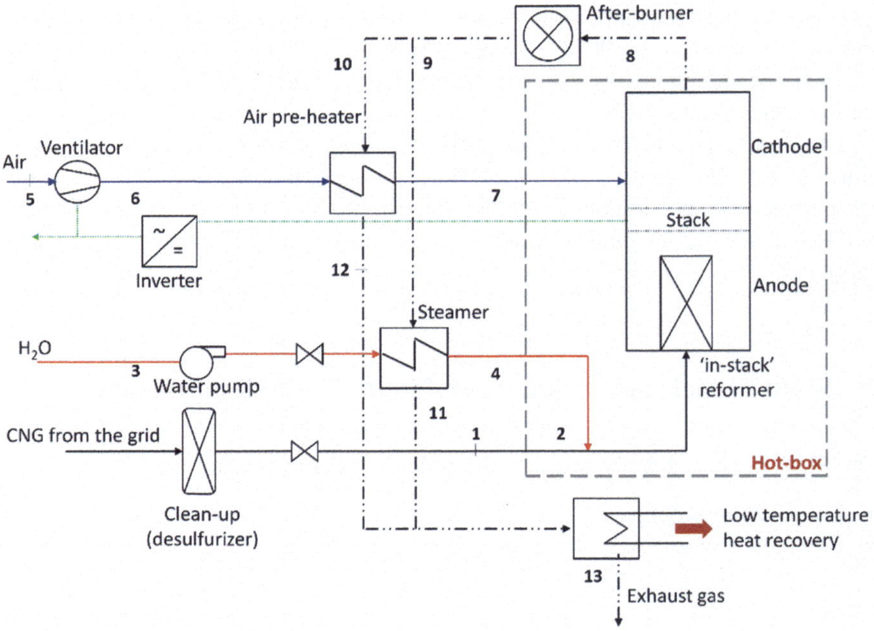

Fig. 1 SOFC system layout

Table 1 Sulfur type and amount in gaseous fuels

Fuel type	Sulfur contaminant(s)	Typical sulfur concentration
Natural gas	H_2S, COS, odorants	2–12 ppm(v)
LPG	Odorants	20–100 ppm(v)
Digester gas (manure)	H_2S, COS, mercaptans	500–2000 ppm(v)
Sewage biogas	H_2S, mercaptans	<100 ppm(v) with in situ abatement in the digester
Landfill gas	H_2S, COS, mercaptans	>100 ppm(v)
Coal syngas	H_2S	>100 ppm(v)

network operator specifies for H_2S an upper limit of 6.6 mg/m^3 that corresponds to less than 5 ppm (v).

In natural gas (NG), sulfur odorants are added in the gas distribution lines for safety reasons. Tetrahydrothiophene (THT) is a cyclic sulfide compound with the chemical formula (CH2)4S that is often used for the odorization of domestic gas. The minimum range is 32 mg/m3 according to UNI EN ISO 19739. Tert-butyl mercaptan (TBM)—(CH3)3CSH—is the most used component of gas odorants. TBM has to be blended with other types of odorant due to its rather high melting point of 0.5 °C. It is thus always utilized as a blend of other compounds, typically dimethyl sulfide (DMS), methyl ethyl sulfide, tetrahydrothiophene, or other mercaptans such as isopropyl mercaptan, sec-butyl mercaptan, and/or n-butyl mercaptan. For a blend with 75–80 % TBM, the minimum TMB concentration only to provide sufficient odorization is 9 mg/m^3.

Organosulfur compounds will decompose to H_2S in the anode or reformer H_2-rich environment, thus causing sulfur passivation of the catalyst. For this reason, the natural gas is passed through fixed bed reactors filled with adequate adsorbent materials to remove the sulfur odorants. Common adsorbent materials for fossil gas desulfurization are activated carbons, activated alumina, and zeolites. Higher sulfur capacity is obtained with zinc oxide as sorbent material. The involved chemisorption reaction is the following:

$$H_2S + ZnO \rightarrow H_2O + ZnS \tag{1}$$

The ZnO polisher generally is operated at 200–350 °C.

Currently, there are non-sulfur gas odorants being tested in Europe to avoid the burden of gas desulfurization. This would be highly beneficial for fuel cell systems.

For biogas fuels (e.g., digester gas from wastewater treatment plant or landfill gas), both capital and operating costs of the cleanup system become significant as several contaminants are found. Sulfur (H_2S and other organic sulfur compounds), siloxanes, and halogenated materials are the impurities that have the greatest impact on gas cleanup system reliability and capital and operating costs. In Tables 2, 3 and 4, the average and maximum concentrations of sulfur and halogenated and organic silicon compounds are given for biogas that is produced from sludge digestion in a wastewater treatment plant (WWTP).

Table 2 Sulfur contaminants: average and maximum concentrations in typical WWTP biogas

Formula	Compound	Average (ppm)	Max. (ppm)
H_2S	Hydrogen sulfide	70	200
COS	Carbonyl sulfide	n.a.	n.a.
CS_2	Carbon disulfide	n.a.	n.a.
C_2H_6S	Dimethyl sulfide (DMS)	n.a.	n.a.
$C_2H_6S_2$	Dimethyl disulfide (DMDS)	n.a.	n.a.
CH_4S	Methanethiol (methyl mercaptan)	1.70	7.90
C_2H_6S	Ethanethiol (ethyl mercaptan)	0.77	1.30
C_3H_8S	Propanethiol (propyl mercaptan)	0.33	3.55
$C_4H_{10}S$	Tert-butyl mercaptan (TBM)	0.94	8.38

Table 3 Siloxane contaminants: average and maximum concentrations in typical WWTP biogas

Formula	Compound	Average (ppm)	Max. (ppm)
$C_6H_{18}O_3Si_3$	(D3) Hexamethylcyclotrisiloxane	0.101	0.200
$C_8H_{24}O_4Si_4$	(D4) Octamethylcyclotetrasiloxane	0.095	0.268
$C_{10}H_{30}O_5Si_5$	(D5) Decamethylcyclopentasiloxane	0.346	1.221
$C_{12}H_{36}O_6Si_6$	(D6) Dodecamethylcyclohexasiloxane	n.a.	n.a.
$C_6H_{18}OSi_2$	(L2) Hexamethyldisiloxane	0.032	0.060
$C_8H_{24}O_2Si_3$	(L3) Octamethyltrisiloxane	0.079	0.192
$C_{10}H_{30}O_3Si_4$	(L4) Decamethyltetrasiloxane	0.283	0.980
$C_{12}H_{36}O_4Si_5$	(L5) Dodecamethylpentasiloxane	0.023	0.041
$C_4H_{12}Si$	Tetramethylsilane	n.a.	n.a.
$C_3H_{10}OSi$	Trimethylsilanol (TMS)	0.037	0.149

Table 4 Halogen contaminants: average and maximum concentrations in typical WWTP biogas

Formula	Compound	Average (ppm)	Max. (ppm)
HCl	Hydrogen chloride	0.24	0.59
C_2Cl_4	Tetrachloroethylene	0.029	0.081
$C_2H_4Cl_2$	1,2-dichloroethane	0.088	0.109

COS, C2S, DMS, and DMDS are also expected to be present in the raw biogas; however, they were excluded in the analyses carried out in the WWTP.

Biogas from manure or the organic fraction of municipal solid waste (OFMSW) is generally higher in H_2S and other sulfur contaminants. Even more critical is the situation for landfill gas where higher loads of contaminants are present also in term of halocarbons.

Lifetime and durability issues connected with SOFC technology are strongly related to the amount of contaminants that reach the stack. H_2S underdoes a dissociative chemisorption process on the Ni surface that is well described by a Temkin-like isotherm (see Hansen 2008). Atomic sulfur adsorbs on the Ni surface in the anode electrode, thus reducing the three-phase boundary (TPB). A sudden drop of cell voltage occurs as the Ni surface gets saturated with sulfur. Anode deactivation is generally reversible if not protracted over the time. Sluggish degradation in cell voltage is instead observed with HCl (see Trembly et al. 2007). HCl is a rather stable compound in the anode environment that has a limited interaction with Ni. Siloxanes are instead extremely detrimental for the SOFC performance. High degradation rates are observed already at ppb(v) level of contaminant in the fuel stream. Posttest analysis revealed that Si (as silica) is mostly deposited at the inlet of the fuel channel on both the interconnect and the anode side of the cell suggesting a relatively fast condensation-type process (see Madi et al. 2015).

In Fig. 2, a possible configuration for the cleanup of biogas is shown. First, a gravel filter is placed to remove particulates and liquid droplets from the as-received biogas in order to reach a water-saturated stream. An additional scrubber unit can be inserted upstream the chiller to wash out NH_3. In case of high loads of H_2S, a chemical wash would be required to reduce the H_2S content below 50–100 ppm (v). The chiller serves to reduce the water content of the biogas. A dew point of 3–5 °C is achieved in the biogas stream at the outlet of the chiller. Finally, cleanup vessels with activated carbons as adsorbent materials are used to remove sulfur compounds and siloxanes. A lead-and-lag configuration is often preferred for the vessels which are responsible of the S/Si removal. The lead reactor is responsible for removing most of the contaminant. The lag reactor, instead, acts as a guard bed in case of temporary higher loads of contaminants reaching the cleanup section. Once the breakthrough concentration is measured, the current lag reactor becomes the new lead vessel. Saturated catalyst is replaced with a fresh amount in the previous lead reactor that will be eventually put online again as the new lag one.

Chemical reactors. Chemical reactors in the SOFC plant include among others the catalytic reformer and the anode off-gas afterburner. Gas cleanup reactors are the other relevant chemical reactors that are usually found in the plant and that were already described in the previous section. In addition, cartridges with ion exchange resins are used for the water demineralization.

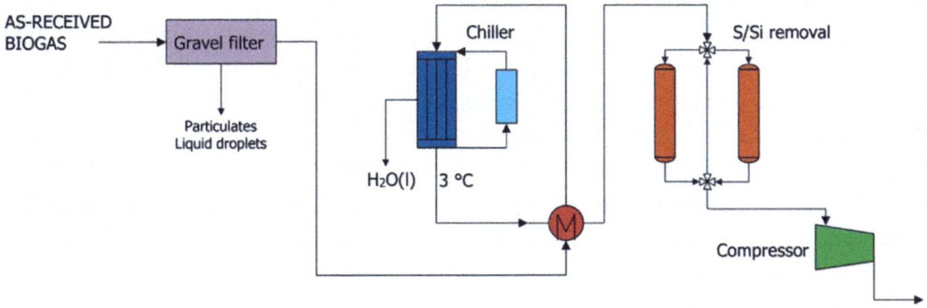

Fig. 2 Example of biogas cleanup unit for fuel cell applications

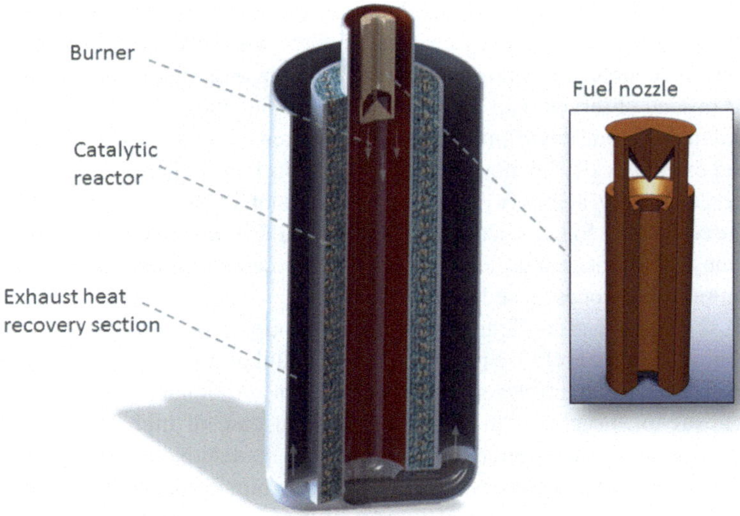

Fig. 3 Schematic a fuel processor reactor integrated with a fuel burner and heat recovery section for the hot exhaust gas

In Fig. 3, the schematic of a catalytic steam reformer is shown. Ni-based pellets are contained in the annular region. The inner part of the reactor contains the combustion chamber where the gas fuel is burnt with air. The hot exhaust generated from fuel combustion flows through the external annulus to provide heat to the endothermic reformer volume. The fuel feed at the catalytic annulus is generally a mixture of natural gas (or biogas) and steam.

Otherwise, the reformer is integrated internally to the stack so as to achieve the 'in-stack reformer' configuration. According to this arrangement, the reformer is heated directly by the SOFC solid structure through combined radiative and convection heat transfer phenomena. For instance, the 'in-stack' reformer configuration was used in the Siemens Westinghouse 5 kW unit (see Fig. 4).

SOFC modules. Single cells are piled together and series connected through metallic interconnects in order to form larger units called stacks. Stacks can be further bundled together (either in series or in parallel connection) to form the power core unit of the SOFC module. The bundled stacks are then insulated and enclosed in metallic vessels to form the full SOFC module. In Fig. 5, the bundling of stack units together to build large power modules is shown. A thick insulation is eventually required to reduce heat losses toward the environment. Ultra-low conductivity materials are used for the SOFC insulation with values in the range 0.2–0.4 $Wm^{-1}\,K^{-1}$ at 600 °C.

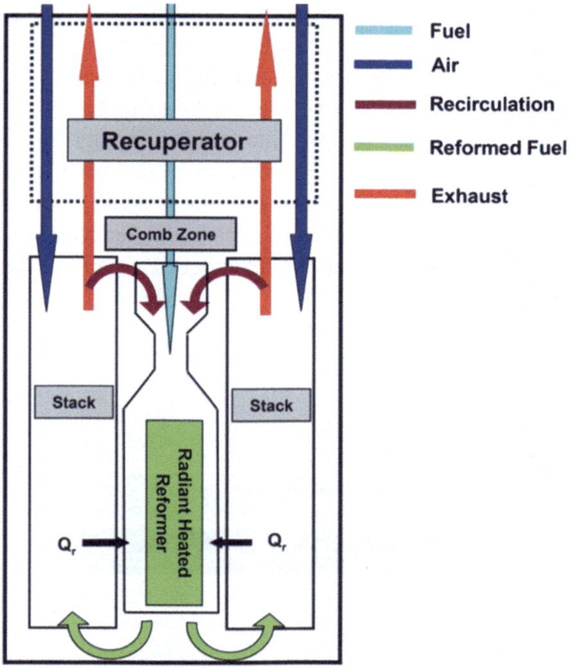

Fig. 4 Layout of the 5 kWe Siemens Westinghouse SOFC generator

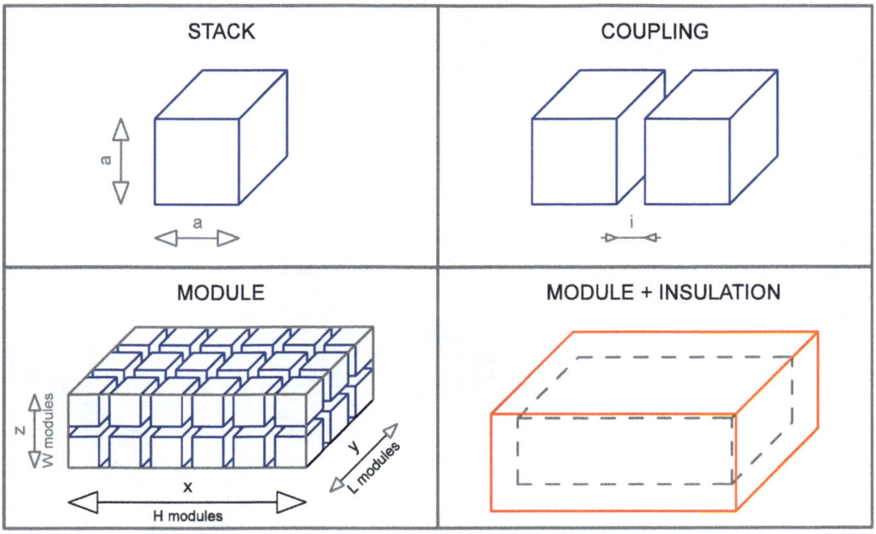

Fig. 5 Bundling of SOFC stacks in power core modules

In Fig. 6, the layers that constitute the stack repeat unit (SRU) are shown. On the anode side, a contact layer (e.g., Ni mesh or Ni foam) is used to achieve a good contact between the anode and the interconnect. At the cathode side, the oxidant atmosphere is more severe in terms of metal corrosion. The cathode electrode is generally connected to the interconnect through a contact layer (e.g., contact paste). Also, the interconnect is protected by a special electric conducting coating (e.g., Co–Mn spinel) to reduce corrosion while preventing/reducing Cr evaporation which irreversibly poisons the cathode electrode.

In Fig. 7, a planar anode-supported cell is shown together with the metal interconnect plate. In Fig. 8, a naked stack that consists of several planar cells connected together is shown.

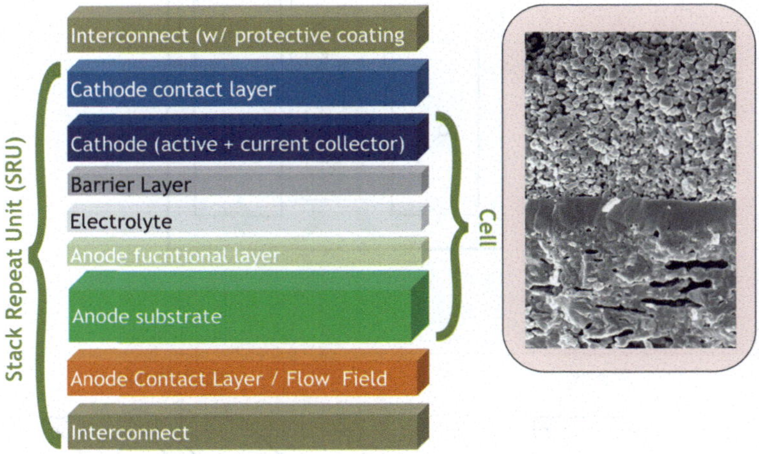

Fig. 6 Layers in the stack repeat unit (SRU)

Fig. 7 Planar anode-supported cell with interconnector (courtesy of SOLIDpower S.p.a.)

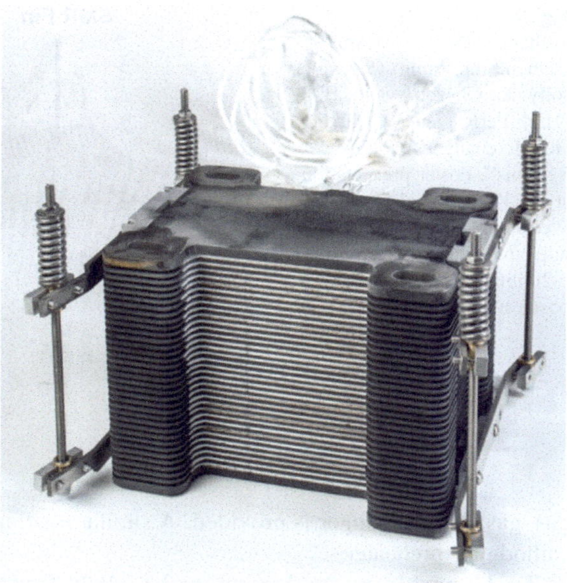

Fig. 8 Naked SOFC stack (courtesy of SOLIDpower S.p.a.)

Heat exchangers. High-temperature heat exchangers are required for SOFC applications to preheat the feed gases. The air preheater is often one the largest devices by volume, and for this reason, one of the most expensive items of the overall BoP. In order to reduce the required heat exchange area, the temperature difference of the hot/cold streams should be the largest possible in this component. This is the reason why the air preheater is often placed downstream from the afterburner; the hot afterburner off-gas enters the air preheater and flows through countercurrent with respect to the fresh cathode air.

Key material selection criteria for the air preheater are listed below:

- oxidation resistance;
- joining via conventional methods;
- resistance to chromium evaporation;
- good creep strength;
- availability; and
- cost (e.g., $/kg).

Ni alloys are often employed due to the good mechanical properties even at temperature above 650 °C. The reformer also requires heat exchange surface to recover sensible heat of the hot exhaust gas from the afterburner and to sustain the endothermic reforming reactions. In Fig. 9, the photograph of a high-surface area

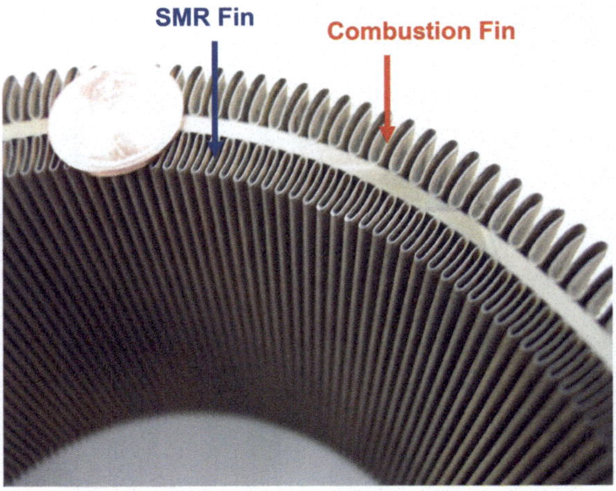

Fig. 9 Catalytic heat exchanger (SMR = steam methane reforming) (*Source* Jolly, S., Ghezel-Ayagh, H., "High Performance Catalytic Heat Exchanger for SOFC Systems", Poster presentation at the 12th Annual SECA Workshop. July 26–28, 2011. https://www.netl.doe.gov/file%20library/events/2011/seca/poster-abstracts/Jolly.pdf)

gas–gas heat exchanger is provided. A similar configuration is also feasible for the cathode air preheater.

Turbomachinery equipment and fuel/air feeding systems. The typical turbomachinery equipment for a fuel cell plant consists in blowers or compressors for the fuel and air feeding to the SOFC. For atmospheric systems, the typical pressure drop is limited to few hundreds of mbar in the overall system. For this reason, low-Δp components are selected to avoid excessive pressurization of the SOFC internal volume against atmospheric pressure. SOFC planar stacks in fact are generally incapable to withstand a differential pressure higher than 50–100 mbar across the fuel and air volumes. In particular, the portions of the cell jointed with glass-ceramic sealant to the interconnects are among the weakest points from a mechanical point of view. SOFC stacks are generally air-cooled. This means that excess waste heat generated within the stack is removed by cathode air that enters the stack volume at a cooler temperature than the stack inner temperature. The air blower/compressor thus must be capable to rapidly adjust the airflow rate in order to maintain constant stack core temperature. For the fuel feeding, NG is generally available at 5 bar or more at distribution lines for industrial users. If this is the case for the involved SOFC plant installation, then a mass flow controller device is able to control the fuel rate of the fuel cell. An ejector that takes the pressurized fuel as primary fluid can be used then to entrain a portion of the anode exhaust to provide steam to the reformer. For residential applications, the NG is instead available at only few mbars. Demineralized and pressurized water is generally used in a venturi device to mix with the NG feed, thus assuring enough pressure in the anode side to feed the mixed and then evaporated fuel/steam mixture. The anode exhaust is often cooled upstream of the afterburner to recover condensed water. The drained water is filtered, demineralized, and finally recirculated to a venturi device so as to avoid the need for external water source. In Fig. 10, the main connections and components of an SOFC stack power core are shown.

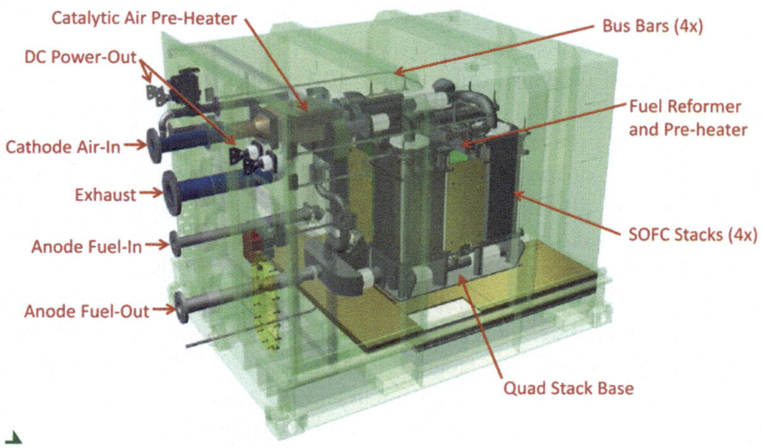

Fig. 10 50 kW proof-of-concept module design from FCE (*Source* Jolly, S., Ghezel-Ayagh, H., "High Performance Catalytic Heat Exchanger for SOFC Systems", Poster presentation at the 12th Annual SECA Workshop. July 26–28, 2011. https://www.netl.doe.gov/file%20library/events/2011/seca/poster-abstracts/Jolly.pdf)

3 SOFC Plant Design Features

Different SOFC plant configurations are mainly based on one of the following distinguishing design features listed below:

- Cell type (tubular, planar, flat tubular, delta design).
- Fuel processing arrangement; direct internal (DIR-SOFC) or indirect internal reformer (IIR-SOFC) and external reformer (ER-SOFC).
- Anodic and/or cathodic recirculation.
- Atmospheric versus pressurized operation (A-SOFC vs. P-SOFC).
- Hybridization with a bottoming cycle to enhance electrical efficiency.
- CO_2 capture from the CO_2-rich anode exhaust.

Cell types. Siemens Westinghouse—one of the first company that attempted the commercialization of SOFC systems—has gradually evolved their cell design to enhance the cell power density (see Vora 2005). The original tubular design was gradually replaced by a delta design that was able to minimize the electronic path of electric charges, thus increasing the performance of the SOFC.

By increasing the power density, the SOFC-specific cost (that is expressed in $ per m^2 of active cell/stack area) is proportionally decreased. Also, having more compact systems is beneficial to minimize heat losses toward the external environment and to reduce the volume size of other equipment.

The planar cell design has gradually demonstrated to be the easiest one to mass-manufacture with high power density performance. The success of the planar design is attested by the many cell and stack manufacturers presently relying on this design. In particular, the Ni-YSZ anode-supported planar cell is now the dominant design, even though metal-supported cells and different anode compositions are available. The electrolyte-supported cell is a robust alternative to the anode-supported cell type also often employed.

Fuel processor. The most efficient design for an SOFC plant is to have the fuel processor internally arranged with respect to the SOFC hot volume. In this way, a considerable amount of the SOFC waste heat—that comes from the heat of electrochemical reactions plus the internal irreversibilities given by the cell polarization—is internally recycled to the fuel processor. In fact, endothermic steam-reforming reactions require a high-temperature source that is provided by the SOFC hot volume. The fuel processor can be either direct or indirect depending on the role of the anode electrode. In the direct internal reforming (DIR) configuration, the SOFC Ni anode itself is the catalyst that promotes steam-reforming reactions. The twofold function of Ni is to provide active sites for both chemical and electrochemical reactions. Ideally, the DIR-SOFC configuration is the most appealing one since a dedicated fuel processor unit is avoided at all. Presently, this is a widely explored and partially proven solution with CH_4-based fuels (i.e., NG and biogas). In fact, the kinetics of CH_4 steam-reforming reaction on Ni is very fast so that the risk of a local cold spot at the fuel inlet region is observed. To overcome this issue, either the fuel is partially prereformed before reaching the anode (partial DIR-SOFC) or a staged fuel feed across the fuel channel is adopted. The degree of internal reforming is often limited to 50 % to avoid excessive cooling of the stack at the fuel inlet region.

For the IIR-SOFC configuration, the reformer section is thermally integrated with the SOFC stack; however, a different catalyst is used that is physically located in a volume other than that of the anode electrode. In this way, the kinetics of steam-reforming reactions is controlled independently from the SOFC anode microstructure, which is optimized to provide the best electrochemical performance. A typical arrangement of the IIR-SOFC is having the reforming section in-between two stack towers. An additional configuration is the in-stack reforming option, i.e., the reforming section is placed within the stack tower and it is repeated after each block, or section, of *n* identical cells. The external reformer configuration is employed instead for simpler SOFC systems for which a less thermally-integrated 'hot-box' arrangement is sought. An external reformer (i.e., a catalytic fuel processor) can also be more practical for dealing with unconventional fuels that might require reforming options different from steam-reforming. For instance, the catalytic partial oxidation (POX) fuel processor is often employed for fuels such as kerosene or diesel. The advantage of the C-POX reformer is that no external heat is required for the reforming reactions.

Atmospheric versus pressurized operation. SOFC modules can be operated at relatively high pressures in order to reap the thermodynamic benefits of close integration with large gas turbines. From a practical standpoint, pressurized operation might also be required in order to reduce the considerable footprint of a GW^e-scale SOFC power island (see Traverso et al. 2010).

Operating planar SOFCs at the relatively high pressures ($\sim$20 bar) are deemed to be feasible as long as the thin ceramic electrolyte membranes are not exposed to significant pressure differentials (see Seidler et al. 2011). SOFC manufacturers—

e.g., Rolls-Royce Fuel Cell Division (see Trasino et al. 2011) and the German Aerospace Center DLR (Seidler et al. 2011)—are currently operating single cells and planar stacks at pressures of 3–7 bara. When operating a pressurized SOFC/GT hybrid (i.e., with a gas turbine providing both cathode pressurization and power recovery from the hot cathode exhaust), transient fluctuations in pressure are perhaps the most critical concern, as even modest pressure differences between the anode and cathode chambers can damage or entirely destroy the fuel cell. Higher SOFC operating pressures require more accurate and costly control systems and are likely to engender higher O&M costs and reduced system availability. Furthermore, as a result of elevated partial pressures of the reactant and product gases, high-pressure operation may also accelerate common chemical reaction-based SOFC degradation mechanisms, such as interconnector oxidation and electrode phase instability. Such considerations require careful investigation before high-pressure SOFC/GT operation can be considered a realistic option.

The plant depicted in Fig. 11 is an atmospheric plant with anodic recirculation through a hot recirculator. A fraction of the anode exhaust is indeed required to achieve a safe steam-to-carbon (S/C) ratio at the inlet of the reformer. The remaining exhaust gas is burnt with the cathode exhaust in the afterburner. The hot flue gas is thus used to preheat feeding fuel and cathode air. The reformer is thermally sustained by the SOFC waste heat based on an internal indirect reforming configuration.

In Fig. 12, the pressurized SOFC-GT hybrid cycle is shown. Both the anode and cathode feeding streams are pressurized. The burner off-gas is thus expanded in a turbine to recover additional electric power. The SOFC fuel utilization is generally lower in atmospheric plant in order to have sufficient waste heat in the expanded

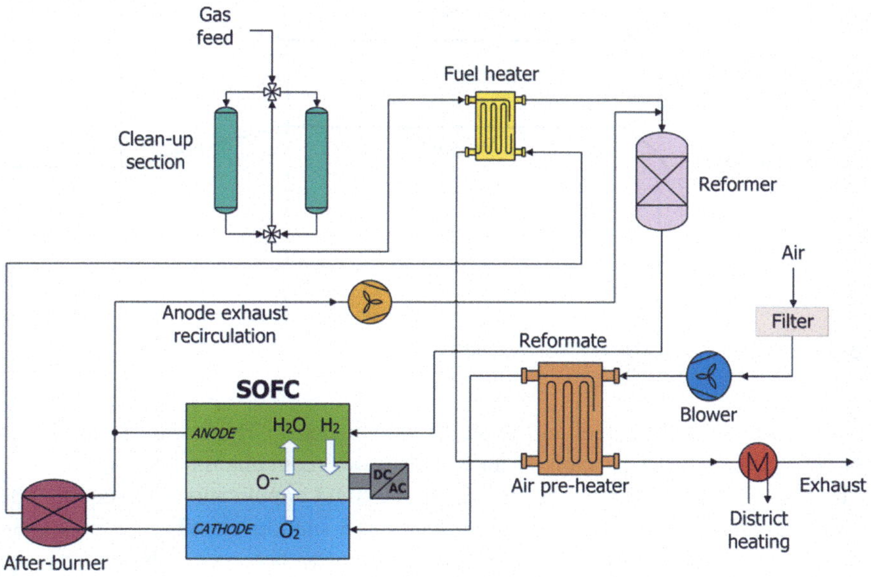

Fig. 11 Atmospheric SOFC plant layout with anodic recirculation

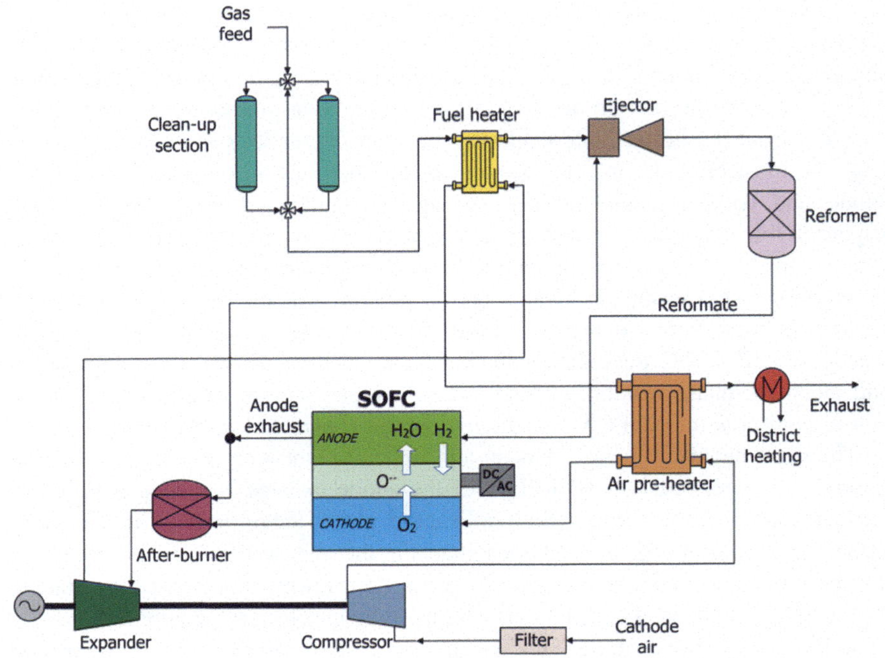

Fig. 12 Pressurized hybrid SOFC-GT plant layout with ejector-based anodic recirculation

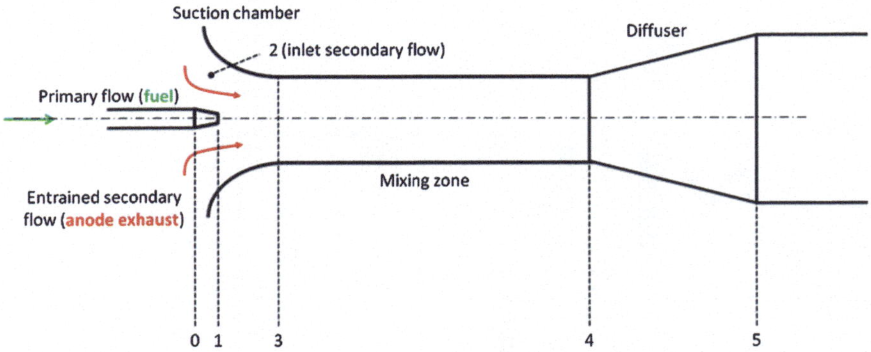

Fig. 13 Ejector-based anode exhaust recirculation

exhaust to preheat fresh fuel and fresh cathode air feeds. The anodic recirculation can be accomplished by means of an ejector-based recirculation system.

In Fig. 13, the schematic of an ejector for SOFC applications is shown. The primary flow is the pressurized anode feed. The suction chamber is the region where the secondary flow (i.e., the anode exhaust) becomes entrained by the high-pressure fuel stream that is accelerated by passing it through the throat (or nozzle) of the ejector. The primary fuel, once injected through the nozzle into the suction chamber, will expand and mix with the entrained anode exhaust stream. The resulting pressure of the mixed fluid will much reduced compared to the pressure of the anode feed.

CO_2capture. CO_2 capture from large emitters such as power plants is an important task to be accomplished over the next few decades in order to curb greenhouse gas emissions in the electricity sector (see IEA, 2012). In light of this goal, novel energy systems that promise high efficiency and low carbon emissions, such as fuel cell hybrid cycles with CCS (Lanzini 2011; Discepoli et al. 2012; Campanari et al. 2010), merit research, development, and detailed analysis of their prospective economics. Of particular interest are SOFC-based power cycles, which have the potential to significantly increase the fuel-to-electricity conversion efficiency of next-generation power plants. Moreover, capturing CO_2 from SOFC-based power exacts a notably smaller efficiency penalty than that associated with conventional power cycles + CCS.

More than a decade ago, Singhal (1999) and Hassmann (2001) found that NG-fueled Siemens Westinghouse hybrid gas turbine–fuel cell cycles (at scales >2–3 MW) without CO_2 capture could achieve ideal conversion efficiencies as high as 70 % LHV; the system investigated by Hassmann (2001) employed a 'sophisticated' turbine cycle (rather than, e.g., a microturbine). More recently, a 2011 DOE-NETL report (NETL 2011b) on NG SOFC plants with CO_2 capture calculated a LHV efficiency of 70.3 % for a pressurized plant with external reforming of methane. According to a NETL review of conventional fossil energy power plants (NETL 2011a), the LHV efficiency penalty due to CCS is almost 8 percentage point (p.p.) in NG-fed combined cycles and ~8–11 p.p. for IGCC plants (depending on the specific gasifier employed). For NG pressurized hybrid SOFC cycles, Franzoni et al. (2008) calculated a CCS efficiency penalty of less than 3 p.p. using post-SOFC CO_2 capture. Similarly, Park et al. (2011) calculated a LHV efficiency of 67.4 % for a NG hybrid SOFC plant venting CO_2 and 65.0 % when CO_2 was captured after the SOFC. In post-SOFC capture, the anode exhaust is oxy-combusted, yielding a stream containing mostly steam and CO_2, from which the latter is separated after steam condensation and water knockout.

Regarding coal IGFC plants, Grol (2009) investigated a pressurized SOFC integrated with a low-temperature catalytic gasifier (producing a syngas with a relatively high methane content, ~18 %), with/without post-SOFC CCS; LHV efficiencies of 62.4 and 59.2 % were obtained in the vented and CCS cases, respectively. In another recent report, DOE-NETL (NETL 2011c) calculated a LHV efficiency of 51.6 % for a pressurized IGFC-CCS plant that employs an enhanced gasifier able to produce a syngas with up to 11 % volume of methane.

Spallina et al. (2011a) designed and analyzed several IGFC plant configurations, all based on a shell gasifier and a pressurized SOFC, with different gas turbine inlet temperatures (TITs); LHV efficiencies were 52–54 %. Spallina et al. (2011b) extended this work by adding post-SOFC CO_2 capture via oxy-combustion of the anode exhaust and calculated a 6 % drop in efficiency. Romano et al. (2011) studied a novel IGFC plant configuration that captures CO_2 before the SOFC through physical absorption in a Selexol-based AGR unit; the plant design also features a methanation process upstream of the SOFC (to increase the methane content of the syngas to ~26 % vol.), a complete recycle of the shifted anode exhaust to the AGR unit for CO_2 capture, and a state-of-the-art cooled gas turbine capable of exploiting both anode and cathode hot exhaust gases with high efficiency. With this configuration, a LHV efficiency of 51.7 % was achieved, 4.5 percentage points higher than that of the plant proposed by Spallina et al. (2011b).

Rao (1991) was the first to propose the use of a methanator to convert part of the chemical energy of the synthesis gas into heat and then to recover power from it efficiently through an expander. More recently, Li et al. (2010) investigated the role of a methanator and syngas expander topping cycle placed upstream an IGFC plant featuring a low-temperature catalytic gasifier, e.g., see Grol (2009); however, no significant improvement in efficiency was observed (with respect to the case without methanation) because the methane content of the syngas coming from the catalytic hydrogasifier was already so high (36 % vol.) to obscure the potential benefit of methanation upstream of the SOFC.

Carbon capture is a design option for SOFC systems that is relatively more straightforward to integrate then in conventional fossil power plants. A schematic of the SOFC-GT cycle with carbon capture through oxy-combustion is shown in Fig. 14.

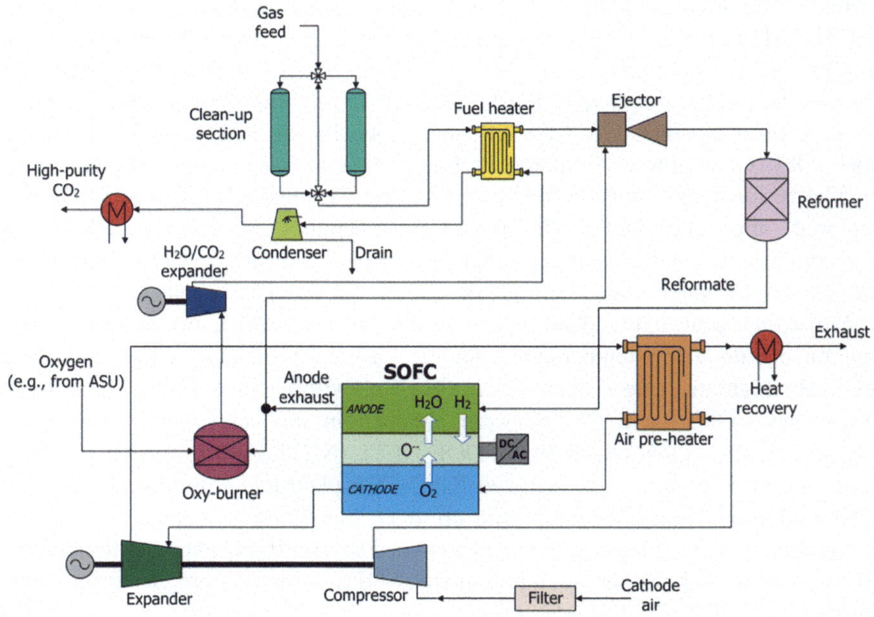

Fig. 14 Pressurized hybrid SOFC-GT plant layout with carbon capture through oxy-combustion of the anode exhaust

4 Energy System Analysis of SOFC Systems

In this section, all the equations that are required to build a lumped volume model of the SOFC are introduced.

Reversible potential of a fuel cell. The fundamental driving force of an electrochemical cell is the gradient of chemical potential that is established across the two electrodes (anode and cathode) which are separated by an ion-conducting electrolyte layer. Compared to a battery, the fuel cell is an open system where reactants (fuel and oxidant) are continuously supplied to sustain electrochemical reactions. A solid oxide cell operates by conducting oxygen ions (O^{--}) through the electrolyte layer that generally consists of zirconia (ZrO_2) doped with yttria (Y_2O_3)[1]. Referring to the fuel cell operation mode (SOFC), oxygen ions are the electric charges which are transported through the electrolyte layer from cathode to anode. In the electrolysis mode (SOEC), the oxygen flux through the electrolyte is reversed.

For SOFC operation, the involved reactions at the fuel electrode are listed below:

$$\text{(Fuel electrode reaction)} \quad H_2 \rightarrow 2H^+ + 2e^- \tag{2}$$

$$\text{(Fuel electrode reaction)} \quad 2H^+ + O^{--} \rightarrow H_2O \tag{3}$$

$$\text{(Oxygen electrode reaction)} \quad \frac{1}{2}O_2 + 2e^- \rightarrow O^{--} \tag{4}$$

For SOEC operation, the above-listed reactions are reversed:

$$\text{(Fuel electrode reaction)} \quad H_2O \rightarrow 2H^+ + O^{--} \tag{5}$$

$$\text{(Fuel electrode reaction)} \quad 2H^+ + 2e^- \rightarrow H_2 \tag{6}$$

$$\text{(Oxygen electrode reaction)} \quad O^{--} \rightarrow \frac{1}{2}O_2 + 2e^- \tag{7}$$

Clearly, the overall oxidation reaction is as follows:

$$H_2 + \frac{1}{2}O_2 \rightleftharpoons H_2O \tag{8}$$

The Nernst voltage expresses the ideal reversible potential available across a fuel cell membrane when no internal current and/or fuel leaks occur. It is expressed as a variation of the Gibbs free energy involved with the complete redox reactions of the reactants that feed the fuel cell, divided by the electric charges ($n^{el} \cdot F$) exchanged in it:

[1]The dopant is added to increase the ionic conductivity of zirconia.

$$V_{\text{Nernst}} = \frac{-\Delta^r \bar{g}(T,p)}{n^{\text{el}} \cdot F} = \frac{-\Delta^r \bar{g}(T)}{n^{\text{el}} \cdot F} - \frac{\text{RT}}{n^{\text{el}} \cdot F}\ln(\text{K}) \tag{9}$$

The sign '−' (minus) before the Gibbs free energy term is valid for the fuel cell mode. For electrolysis operation, the sign changes to a '+' (plus).

The term inside the natural logarithm establishes the dependency of the Gibbs free energy variation from the partial pressure of the reactant gases, and it is expressed as following:

$$K = \prod_j \left(\frac{p_j}{p_0}\right)^{v_j} \tag{10}$$

where v_j are the stoichiometric coefficients of the involved reaction.

Hence, a dependency of the Nernst voltage on both temperature and fuel composition is thus found.

Remark 1 Let us consider a H_2/H_2O binary mixture feeding the anode of the SOFC. The overall electrochemical redox reaction occurring is given in Eq. (8).

The Nernst voltage potential is thus calculated as follows:

$$\begin{aligned} V_{\text{Nernst}} &= \frac{-\Delta^r \bar{g}(T,p)}{2F} = \frac{-\Delta^r \bar{g}(T)}{2F} + \frac{\text{RT}}{2F}\ln\left(\frac{p_{H_2} p_{O_2}^{\frac{1}{2}}}{p_{H_2O}} \frac{1}{p_0^{\frac{1}{2}}}\right) \\ &= \frac{-\Delta^r \bar{g}(T)}{2F} + \frac{\text{RT}}{2F} ln\left[\left(\frac{y_{H_2} y_{O_2}^{\frac{1}{2}}}{y_{H_2O}}\right)\left(\frac{p}{p_0}\right)^{\frac{1}{2}}\right] \end{aligned} \tag{11}$$

where the Gibbs free term is written as follows:

$$\Delta^r \bar{g}(T) = \bar{g}_{H_2O}(T) - \bar{g}_{H_2}(T) - \frac{1}{2}\bar{g}_{O_2}(T) \tag{12}$$

Equation (11) can be rewritten as follows:

$$\begin{aligned} V_{\text{Nernst}} &= \frac{-\Delta^r \bar{g}(T)}{2F} - \frac{\text{RT}}{2F}\ln\left(\frac{\frac{p_{H_2O}}{p_0}}{\frac{p_{H_2}}{p_0} \cdot \left(\frac{p_{O_2}^{\text{cat}}}{p_0}\right)^{1/2}}\right) \\ &= -\frac{1}{2}\left(\frac{\Delta^r g(T)}{F} + \frac{\text{RT}}{F} ln\left(\frac{\frac{p_{H_2O}}{p_0}}{\frac{p_{H_2}}{p_0}}\right) - \frac{\text{RT}}{2F}\ln\left(\frac{p_{O_2}^{\text{cat}}}{p_0}\right)\right) \end{aligned} \tag{13}$$

Looking at the potential of the anodic electrode only, the partial pressure of oxygen at the anode side can be calculated by writing the equilibrium constant[2] of Eq. (8) as the following:

[2]The equilibrium condition is assumed for OCV condition.

$$-\frac{-\Delta^r \bar{g}(T)}{\mathrm{RT}} = \ln(\mathrm{K}) \tag{14}$$

Equation (13) can be then rewritten as follows:

$$0 = \frac{\Delta^r \bar{g}(T)}{\mathrm{RT}} + \ln\left(\frac{\frac{p_{H_2O}}{p_0}}{\frac{p_{H_2}}{p_0} \cdot \left(\frac{p_{O_2}^{an}}{p_0}\right)^{1/2}}\right) \tag{15}$$

From the previous equation, the following expression for $p_{O_2}^{an}$ is obtained as follows:

$$\ln\left(\frac{p_{O_2}^{an}}{p_0}\right) = 2\frac{\Delta^r \bar{g}(T)}{\mathrm{RT}} + 2\ln\left(\frac{\frac{p_{H_2O}}{p_0}}{\frac{p_{H_2}}{p_0}}\right) \tag{16}$$

Equation (16) can be rewritten as follows:

$$\frac{\mathrm{RT}}{2F} ln\left(\frac{p_{O_2}^{an}}{p_0}\right) = \frac{\Delta^r \bar{g}(T)}{F} + \frac{\mathrm{RT}}{F}\ln\left(\frac{\frac{p_{H_2O}}{p_0}}{\frac{p_{H_2}}{p_0}}\right) \tag{17}$$

By inserting Eq. (17) into Eq. (11), we obtain an expression of the Nernst voltage only in terms of the oxygen chemical potential across the SOFC electrolyte:

$$V_{\text{Nernst}} = -\frac{1}{2}\left[\frac{\mathrm{RT}}{2F}\ln\left(\frac{p_{O_2}^{an}}{p_0}\right) - \frac{\mathrm{RT}}{2F}\ln\left(\frac{p_{O_2}^{cat}}{p_0}\right)\right] = \frac{\mathrm{RT}}{4F}\ln\left(\frac{p_{O_2}^{cat}}{p_{O_2}^{an}}\right) \tag{18}$$

The expression of the Nernst voltage as depending on the ratio of oxygen partial pressure across the electrolyte is useful to evaluate the ideal reversible potential of multifuel mixtures—this is under the assumption that the gas reaches equilibrium inside the anode reactor. The partial pressure of oxygen at the anode chamber, i.e., the numerator of the natural logarithm of the last term of Eq. (18), is directly evaluated by calculating the equilibrium gas composition of the inlet mixture that reacts within the fuel cell.

Remark 2 The following section provides the mathematical derivation which shows how under equilibrium condition the same result for the Nernst potential is obtained either using the H_2 or the CO oxidation reactions.

The Nernst potential for H_2 and CO oxidation is given in Eqs. (19) and (20), respectively:

$$V_N^{H_2} = \frac{-\Delta\bar{g}(T)_{H_2}}{2F} + \frac{RT}{2F}\ln\left(\frac{p_{H_2}p_{O_2}^{\frac{1}{2}}}{p_{H_2O}}\frac{1}{p_0^{\frac{1}{2}}}\right) \tag{19}$$

$$V_N^{CO} = \frac{-\Delta\bar{g}(T)_{CO}}{2F} + \frac{RT}{2F}\ln\left(\frac{p_{CO}p_{O_2}^{\frac{1}{2}}}{p_{CO_2}}\frac{1}{p_0^{\frac{1}{2}}}\right) \tag{20}$$

The condition which must be satisfied is that $V_N^{H_2} = V_N^{CO}$, otherwise stated as follows:

$$V_N^{H_2} - V_N^{CO} = 0 \tag{21}$$

Equations (18) and (19) are thus combined together as the following:

$$V_N^{H_2} - V_N^{CO} = -\frac{\Delta\bar{g}_{H_2}}{2F} + \frac{\Delta\bar{g}_{CO}}{2F} - \frac{RT}{2F}\ln\left(\frac{p_{H_2O}p_{CO}}{p_{H_2}p_{CO_2}}\right) \tag{22}$$

$$\begin{aligned} V_N^{H_2} - V_N^{CO} = &-\frac{1}{2F}\left(\bar{g}_{H_2O} - \bar{g}_{H_2} - \frac{1}{2}\bar{g}_{O_2} - \bar{g}_{CO_2} + \bar{g}_{CO} + \frac{1}{2}\bar{g}_{O_2}\right) \\ &- \frac{RT}{2F}\ln\left(\frac{p_{H_2O}p_{CO}}{p_{H_2}p_{CO_2}}\right) \end{aligned} \tag{23}$$

Since the fuel gas is assumed to be under equilibrium, the relation expressed by the equilibrium constant of the water–gas shift reaction ($CO + H_2O \rightleftharpoons H_2 + CO_2$) applies:

$$\ln K_{WGS}^e = -\frac{-\Delta^{WGS}\bar{g}(T,p)}{RT} = \ln\left(\frac{p_{H_2}p_{CO_2}}{p_{H_2O}p_{CO}}\right) \tag{24}$$

$$K_{WGS}^e = \exp\left(-\frac{\Delta\bar{g}_{WGS}}{RT}\right) = \frac{p_{H_2}p_{CO_2}}{p_{H_2O}p_{CO}} \tag{25}$$

$$-\frac{\Delta\bar{g}_{WGS}}{2F} = \frac{RT}{2F}\ln\left(\frac{p_{H_2}p_{CO_2}}{p_{H_2O}p_{CO}}\right) \tag{26}$$

By substituting Eq. (26) in Eq. (23), the following expression yields:

$$V_N^{H_2} - V_N^{CO} = -\frac{1}{2F}(-\Delta\bar{g}_{WGS}) - \frac{1}{2F}\Delta\bar{g}_{WGS} = 0 \tag{27}$$

Fuel processor Chemical Model

A common and effective approach for the calculation of the fuel processor performance is to apply the chemical equilibrium condition to the reformer unit. This method is useful and computationally fast to evaluate the thermal load and outlet chemical composition of the reactor. The main drawback is that no information for the sizing of the reactor volume is gained. A multidimensional kinetic approach is required instead to design the reactor. For instance, a 1D plug-flow reactor model based on global reaction rates of the main reactions involved is generally sufficient to capture the gas composition change along the reactor's bed and thus to calculate the required length (and volume) of the reactor in order to approach the equilibrium composition.

Before starting to introduce the equations, the reactions for steam reforming of methane are presented:

$$\mathrm{CH_4 + H_2O \leftrightarrow 3H_2 + CO} \quad \Delta \mathrm{H}^0_{298} = 206\ \frac{\mathrm{kJ}}{\mathrm{mol}} \tag{28}$$

$$\mathrm{CO + H_2O \leftrightarrow H_2 + CO_2} \quad \Delta H^0_{298} = -41\ \frac{\mathrm{kJ}}{\mathrm{mol}} \tag{29}$$

$$\mathrm{CH_4 + 2H_2O} \leftrightarrow 4H_2 + CO_2 \quad \varDelta H^0_{298} = 165\ \frac{\mathrm{kJ}}{\mathrm{mol}} \tag{30}$$

Equilibrium Modeling Approach

The calculation of chemical equilibrium at fixed temperature and pressure for a multicomponent and multiphase system corresponds to the minimization of the Gibbs free energy of the mixture. For a system with N_P phases and N_C components, the total Gibbs free energy is as follows:

$$G_{\mathrm{mix}} = \sum_{i}^{\mathrm{NP}} \sum_{j}^{\mathrm{NC}} n_i^k \mu_i^k \tag{31}$$

where n_i^k is the number of moles of component i in k phase, and μ_i^k is the chemical potential of component i in k phase, in which the chemical potential is a function of k phase composition, temperature, and pressure (P).

By introducing the fugacity terms, the previous equation can be rewritten as follows:

$$G_{\mathrm{mix}} = \sum_{i}^{\mathrm{NP}} \sum_{j}^{\mathrm{NC}} n_i^k \left[\mu_i^0 + \overline{R}T\ \ln \frac{\hat{f}_i^k}{f_i^0} \right] \tag{32}$$

where μ_i^0 is the reference chemical potential of pure substance i, and f_i^0 is the fugacity of pure component i at standard state.[3]

For an ideal gas mixture, Eq. (32) becomes:

[3]The standard state for vapour phase is taken as an ideal gas at system temperature and pressure of 1 bar, for the liquid phase is taken as the fugacity of pure component i in liquid phase at system temperature and for solid phase the reference state is the solid phase at 298.15 K and 1 bar.

$$G_{\text{mix}} = \sum_{i}^{\text{NP}} \sum_{j}^{\text{NC}} n_i^k \left[\bar{g}_i^0(T,p) + \bar{\text{R}}\text{T} \ln \frac{y_i p}{p_{\text{ref}}} \right] \tag{33}$$

The equilibrium approach through the minimization of the Gibbs free energy of the reformer mixture is also able to predict the formation of solid carbon (Fig. 15).

Kinetic modeling approach. The macroscopic kinetic approach is used to predict the gas outlet composition, given a certain volume of the reactor and under known properties of the catalyst (e.g., Ni surface area and bed porosity). This approach generally aims to provide the intrinsic reaction rate of global reactions. For steam methane reforming (SMR) of methane onto Ni catalyst, Xu and Froment (1989) performed a series of experiments on a catalyst containing 15.2 % Ni, having a surface area of 9.3 m^2/g_{cat} and a void fraction of 0.528. The catalyst was crushed into small particles (0.18–0.25 mm) to avoid diffusion limitations and obtain pure intrinsic reaction rates.

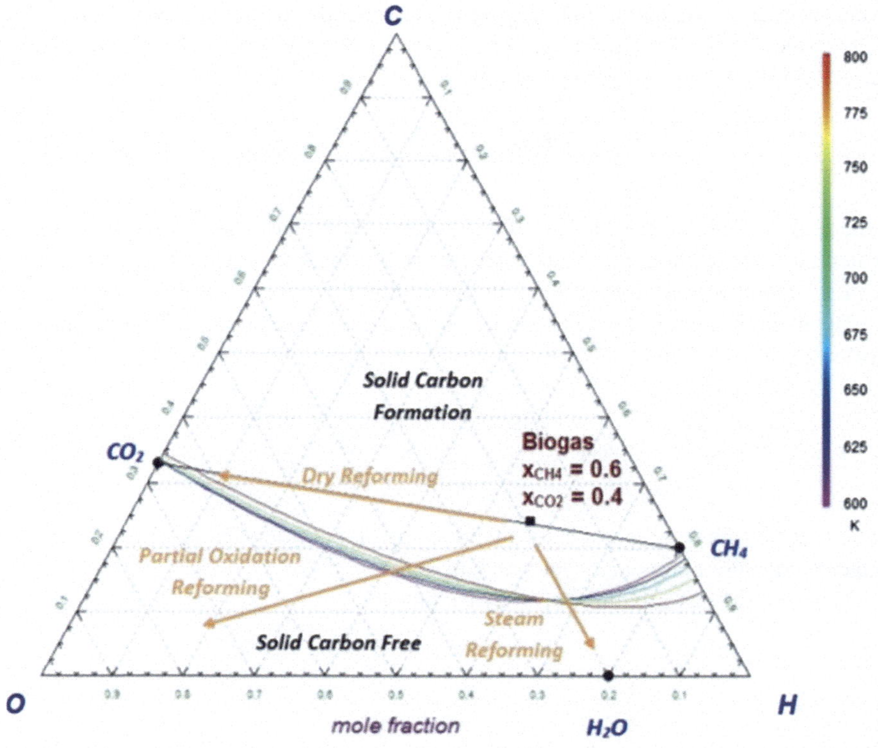

Fig. 15 Carbon formation boundaries for biogas fuel at different temperature and for steam-reforming, dry-reforming, and POX conditions

Xu and Froment developed the SMR reaction based on the following assumptions:

- H_2O reacts with surface nickel atoms to form adsorbed atomic oxygen and gaseous hydrogen;
- Methane is adsorbed onto surface nickel atoms. The adsorbed methane either reacts with the adsorbed oxygen or dissociates to form chemisorbed radicals CH_x with $x = 0–3$.
- The concentration of the carbon-containing radicals is much lower than the total available concentration of active sites.
- The adsorbed oxygen and the carbon-containing radicals react to form chemisorbed CH_2O, CHO, CO, or CO_2.
- The H_2 formed is directly released into the gas phase, and/or the gaseous H_2 is in equilibrium with adsorbed H and H_2.

Based on the above-mentioned assumptions, a Langmuir–Hinshelwood reaction mechanism is established which includes as many as 13 reaction steps:

*s*1:	$CH_4 + (S) = CH_4(S)$	
*s*2:	$H_2O + (S) = O(S) + H_2$	
*s*3:	$CO(S) = CO + (S)$	
*s*4:	$CO_2(S) = CO_2 + (S)$	
*s*5:	$H(S) + H(S) = H_2(S) + (S)$	
*s*6:	$H_2(S) = H_2 + (S)$	
*s*7:	$CH_4(S) + (S) = CH_3(S) + H(S)$	
*s*8:	$CH_3(S) + (S) = CH_2(S) + H(S)$	
*s*9:	$CH_2(S) + O(S) = CH_2O(S) + (S)$	
*s*10:	$CH_2O(S) + (S) = CHO(S) + H(S)$	
*s*11:	$CHO(S) + (S) = CO(S) + H(S)$	(rate determining step)
*s*12:	$CHO(S) + O(S) = CO_2(S) + H(S)$	(rate determining step)
*s*13:	$CO(S) + O(S) = CO_2(S) + (S)$	(rate determining step)

where (S) represents a site on the catalytic surface.

Using the rate-determining step assumption, the rate expression for the reactions in Eqs. (28), (29), and (30) is the following:

$$r_1 = \frac{\frac{k_1}{p_{H_2}^{2.5}}\left[p_{CH_4}p_{H_2O} - \frac{p_{H_2}^3 p_{CO}}{K_1}\right]}{\mathrm{DEN}^2} \tag{34}$$

$$r_2 = \frac{\frac{k_2}{p_{H_2}}\left[p_{CO}p_{H_2O} - \frac{p_{H_2} p_{CO_2}}{K_2}\right]}{\mathrm{DEN}^2} \tag{35}$$

$$r_3 = \frac{\frac{k_3}{p_{H_2}^{3.5}}\left[p_{CH_4}p_{H_2O}^2 - \frac{p_{H_2}^4 p_{CO_2}}{K_3}\right]}{DEN^2} \quad (36)$$

$$\mathrm{DEN} = 1 + K_{CH_4}p_{CH_4} + K_{CO}p_{CO} + K_{H_2}p_{H_2} + \frac{K_{H_2O}p_{H_2O}}{p_{H_2}} \quad (37)$$

The reaction rates are expressed in [kmol kg^{-1} (of catalyst)], while the partial pressures of components are expressed in [bar].

The kinetic constants $k_{i(i=1,2,3)}$ are calculated according to the Arrhenius form (pre-exponential terms and activation energies are given in Table 5):

$$k_i = A_i \exp\left(-\frac{E_{a,i}}{\mathrm{RT}}\right) \quad (38)$$

The equilibrium constants of the three reactions are given by the following expressions:

$$K_1 = \exp(30.42 - 27106/T) \quad [\mathrm{bar}^2]$$
$$K_2 = \exp(-3.798 + 4160/T) \quad [-]$$
$$K_3 = \exp(34.218 - 31266/T) \quad [\mathrm{bar}^2]$$

Finally, the temperature-dependent adsorption coefficients are given below:

$$K_{CH_4} = 6.65 \cdot 10^{-4}\exp(38280/\mathrm{RT}) \quad [\mathrm{bar}^{-1}]$$
$$K_{H_2} = 6.12 \cdot 10^{-9}\exp(82900/\mathrm{RT}) \quad [\mathrm{bar}^{-1}]$$
$$K_{CO} = 8.23 \cdot 10^{-5}\exp(70650/RT) \quad [\mathrm{bar}^{-1}]$$
$$K_{H_2} = 1.77 \cdot 10^{5}\exp(82900/T) \quad [-]$$

Table 5 Pre-exponential terms and activation energies for the kinetics formulation of SMR

$A_1\left[\frac{\mathrm{kmol \cdot bar^{0.5}}}{\mathrm{kg \cdot h}}\right]$	4.22×10^{15}
$A_2\left[\frac{\mathrm{kmol}}{\mathrm{kg \cdot h \cdot bar}}\right]$	1.96×10^{6}
$A_3\left[\frac{\mathrm{kmol \cdot bar^{0.5}}}{\mathrm{kg \cdot h}}\right]$	1.02×10^{15}
$E_1\left[\frac{\mathrm{kJ}}{\mathrm{mol}}\right]$	240.10
$E_2\left[\frac{\mathrm{kJ}}{\mathrm{mol}}\right]$	67.13
$E_3\left[\frac{\mathrm{kJ}}{\mathrm{mol}}\right]$	243.9

Recently, a new study based on the same model developed by Xu and Froment was performed by Oliveira et al. (2009). The new model was applied to a different kind of catalyst (small extrudates with a diameter of 1.6 mm and a height of 4 mm) in order to obtain new model constants (activation energies and pre-exponential factors in the Arrhenius expression). The parameters obtained by Oliveira et al. are listed in the Table 6.

Equilibrium constants and adsorption expressions are the same used by Xu and Froment (1989).

Thermal balance of the fuel processor. For POX and autothermal conditions, no external heat is required. For steam- and dry- reforming conditions instead, a certain amount of heat is required to sustain the endothermic reactions in the fuel processor. Heat is often recovered from the afterburner exhaust gas or from the SOFC itself, whose heat production is generally sufficient to sustain the endothermic reforming reactions provided that FU is higher than 55–60 % (see the simulation depicted in Fig. 16).

Energy model of the SOFC stack. The SOFC performance for system modeling applications and preliminary evaluation of the performance of different process design configurations and operating conditions can be predicted by means of 0-D lumped volume models of the SOFC power module. In order to build a

Table 6 Kinetic parameters for the model of Oliveira et al

$A_1\left[\frac{\text{mol kPa}^{0.5}}{\text{kg}}\right]$	5.79×10^{13}
$A_2\left[\frac{\text{mol}}{\text{kg kPa}}\right]$	9.33×10^{4}
$A_3\left[\frac{\text{mol kPa}^{0.5}}{\text{kg}}\right]$	1.29×10^{14}
$E_1\left[\frac{\text{kJ}}{\text{mol}}\right]$	217.01
$E_2\left[\frac{\text{kJ}}{\text{mol}}\right]$	68.20
$E_3\left[\frac{\text{kJ}}{\text{mol}}\right]$	215.84

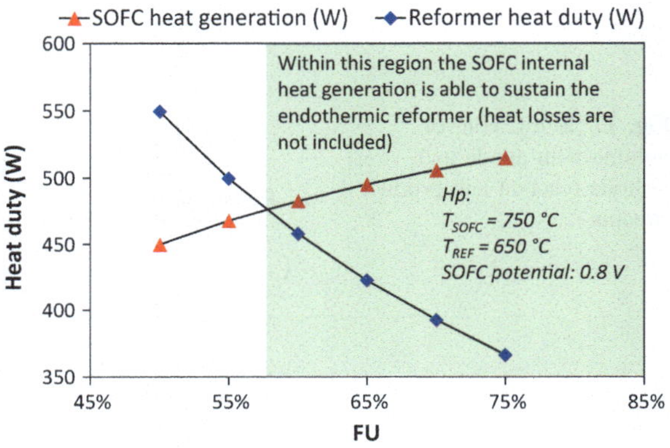

Fig. 16 Operating conditions for the SOFC to sustain the steam reforming of CH_4 (S/C ratio of 2) in term of thermal balance

thermodynamically consistent model, the following equations are used to calculate the amount of waste heat rate that is removed by the cooling cathode air within the SOFC stack (Eq. 39) refers to the anode side, while Eq. (40) refers to the cathode side). Heat generation occurs in the SOFC solid due to electrochemical reactions (reversible heat rate production) and polarization effects (irreversible heat rate production). The better are the performance of the SOFC (i.e., lower area specific resistance), the lower are the irreversibilities, and thus, less internal heat generation (i.e., waste heat) is involved.

$$\phi - W_{\text{el}} = \sum_i \pm \bar{n}_{\text{an},i} \bar{h}_{\text{an},i} \tag{39}$$

$$\phi = \sum_i \pm \bar{n}_{\text{cat},i} \bar{h}_{\text{cat},i} \tag{40}$$

Looking at the overall stack control volume, the energy balance equation reads as the following:

$$-W_{\text{el}} = \sum_i \pm \bar{n}_i \bar{h}_i \tag{41}$$

Equation (41) shows how the anode feed can also have a cooling effect on the SOFC stack as long as the reformate temperature is lower than the average stack temperature (Figure 17).

The amount of fuel required by the fuel cell is strictly depending on the overall current produced in the stack. Assuming for simplicity n_{CELLS} single cells series connected, the total stack current is as follows:

$$I_{TOT} = n_{CELLS} I \tag{42}$$

The fuel flow rate to supply to the SOFC is thus as follows:

$$\bar{n}_{\text{fuel}} = \frac{I_{\text{TOT}}}{n_{\text{el}} F} \cdot \frac{1}{\text{FU}} \tag{43}$$

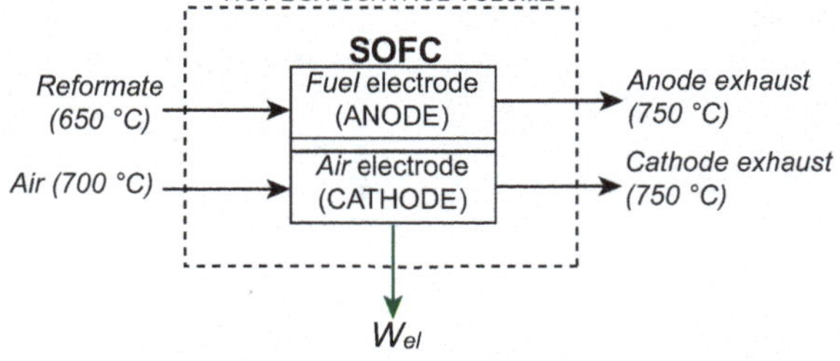

Fig. 17 SOFC control volume with anode and cathode reactant and product streams

The total electric power produced by the SOFC is calculated as follows:

$$W_{\mathrm{el}} = V_{\mathrm{op}} I_{\mathrm{TOT}} \tag{44}$$

The cathode airflow rate that is required to control the SOFC temperature at the desired value is calculated using Eqs. (39) and (41).

Finally, the overall plant electrical efficiency should take into account the losses in the power conditioning (PC) unit that serves for the DC/DC conversion of the SOFC electrical output and conversion to AC power through a DC/AC inverter. Also, the parasitic consumption of auxiliaries (e.g., air blower) must be accounted for. Hence, the final expression for the net electrical efficiency is as follows:

$$\eta_{\mathrm{el}}^{\mathrm{tot}} = \frac{V_{\mathrm{op}} I_{\mathrm{TOT}} \eta_{\mathrm{PC}} - W_{\mathrm{aux}}}{\bar{n}_{\mathrm{fuel}} \cdot \overline{\mathrm{LHV}}} \tag{45}$$

The SOFC polarization (and so the stack operating voltage) is generally a complex function of operating temperature and pressure, FU, stack interconnector flow fields, electrodes materials, etc.

Nonetheless, for system applications, the SOFC performance is often described in terms of its overall area specific resistance (ASR) that is measured in Ω cm^2. The *ASR* accounts for the overall polarization of the stack including cell polarization and the contact resistances between cell and interconnector. The operating voltage is then calculated as following:

$$V_{\mathrm{op}} = \bar{V}_{\mathrm{Nernst}}(T, p, \mathbf{y}) - j \cdot \mathrm{ASR} \tag{46}$$

The average Nernst voltage is calculated using the anode and cathode gas compositions evaluated at inlet and outlet channels of the fuel cell, respectively.

An alternative to the use of the Nernst voltage is the so-called Gibbs voltage. The Gibbs voltage denotes the maximum thermodynamic work that can be extracted from a fuel cell that operates without generating irreversibilities. The formulation of the Gibbs voltage is similar to that of Nernst voltage. Nonetheless, while the Nernst voltage is consistently determined only at open-circuit condition (OCV), the Gibbs voltage is the ideal reversible voltage that could be observed during stack operation (i.e., $I > 0$).

$$V_G = \frac{-\Delta G^r(T, p)}{I} = \frac{-\sum_i \pm \bar{n}_i \cdot g_i}{\bar{n}_{an,in} \cdot n^{\mathrm{el}} \cdot F \cdot \mathrm{FU}} \tag{47}$$

The Gibbs voltage thus takes into account the actual Gibbs free energy change across the SOFC stack. The mathematical derivation of the Gibbs voltage requires to apply both the first and second principles of thermodynamics to the stack control volume (that is assumed isothermal[4]):

$$\Phi - W_{el} = \Delta H = \sum_{i}^{N} \bar{n}_i \bar{h}_i \tag{48}$$

$$\frac{\Phi}{T} - \Sigma_{irr} = \Delta S = \sum_{i}^{N} \bar{n}_i \bar{s}_i \tag{49}$$

If no irreversibilities occur in the SOFC (i.e., $\Sigma_{irr} = 0$), then the previous two equations combined together yield the following expression:

$$-I \cdot V_G = \Delta H - T\Delta S = \Delta G \tag{50}$$

The previous equation finally leads to (51).

Finally, Eqs. (48) and (49) show the two contributions (sources) to the SOFC internal heat generation:

$$\Phi - V_{op} I = \Delta G - T\Delta S \tag{51}$$

The previous equation is rewritten as follows:

$$\Phi = (V_G - V_{op})I - T\Delta S = \eta_{pol} I - T\Delta S \tag{52}$$

The term $\eta_{pol} I$ indicates the internal heat source due to the irreversible stack polarization, while the term $-T\Delta S$ denotes that reversible heat connected to the electrochemical oxidation reactions.

When chemical reactions also take place inside the SOFC stack, the internal heat generation is changed. In case of the endothermic reforming reactions (e.g., this the case of direct internal reforming), the heat sink Φ_{REF} must be included in the SOFC thermal balance that will result in a reduced internal heat generation in the SOFC stack. Note that the energy balance expression given in Eq. 1.40 is still consistent as the heat source/sink terms are simply included as the enthalpy balance term.

Heat Exchanger Network. The heat exchanger network (HEN) includes the heat exchanger devices that are required to match hot and cold streams within the plant. The methodology of pinch analysis is generally used to identify the minimum external energy requirement of the analyzed process. The first step to develop the pinch analysis is to extract from the process flow sheet thermal data for each stream.

---

[4]Note that the stack control volume considered here is different from the previous one through which the SOFC waste heat rate was calculated. The stack control volume described here is isothermal and thus it does not include the cooling effect of cathode air. This is pretty much like looking at a restricted stack volume in which the cooling effect of reactants/products is not accounted for yet.

Table 7 Hot/cold stream in the SOFC-GT hybrid plant

Stream ID	Type	Start temperature (°C)	Target temperature (°C)
1 (fuel feed)	Cold	182	700
2 (water feed)	Cold	15	700
3 (air feed)	Cold	255	700
4 (SOFC exhaust)	Hot	771	174

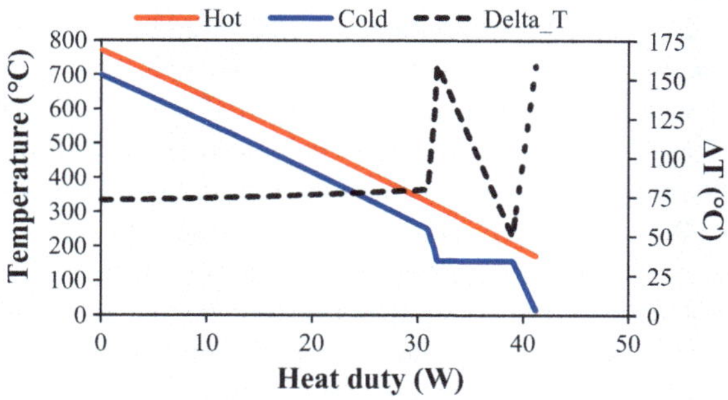

Fig. 18 Hot and cold composite curves for the SOFC-GT plant

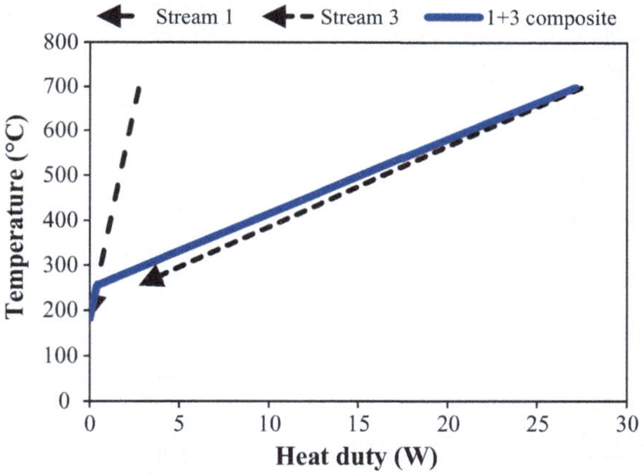

Fig. 19 Example on how to build a composite curve starting from two cold streams

As an example, hot and cold streams for the SOFC-GT plant used as example in this chapter are given in Table 7. 'Hot steams' refer to those streams that need cooling (i.e., heat sources), while 'cold streams' are the streams that require heating (i.e., heat sinks). Once the heat/cold streams are known, the composite curve is built (see Fig. 18). Composite curves consist of temperature–enthalpy (T-H) profiles of heat availability in the process (the 'hot composite curve') and heat demands in the process (the 'cold composite curve') together in a graphical representation. In Fig. 19, an example on how to build the composite curve of streams 1 and 3 given in Table 1 is shown. The two streams share the temperature range between 255 and 700 °C. Therefore, in this range, the heat duty of the two streams is added together.

To build the a generic hot/cold composite curve, this same procedure is applied to all hot/cold streams after having identified all the temperature ranges where one or more streams are being cooled/heated.

Since the water feed is changing phase and large temperature differences apply for each stream, a 1D discretization of the heating/cooling processes in several steps is beneficial to have accuracy of the results.

Reactants' recirculation. In this paragraph, the effect of the recirculation of the anodic and cathodic exhaust on the SOFC reversible voltage is discussed.

Hot anodic recirculation is mostly used to feed the required steam to the reformer reactor. In this way, not an external water source is avoided but also the latent heat of vaporization is recovered compared to cold recirculation systems.

Hot cathodic recirculation is instead employed to reduce the size of the cathode air preheater at the expenses of a more O_2-depleted reactant at the cathode side.

When exhaust recirculation is employed, it is always useful to distinguish between local parameters (e.g., local fuel utilization and local air excess ratio) and global ones. Local parameters express what is actually experienced by the SOFC stack, while global values refer to the same parameters, but they are experienced by the overall system. In Fig. 20, the control volumes defining both local and global parameters are shown, respectively.

It is particularly important to establish an analytical interrelation between local and global parameters as a function of the reactant recirculation ratios, to identify to what extent the recirculation of the SOFC exhaust modifies the baseline case, i.e., without any hot recirculation.

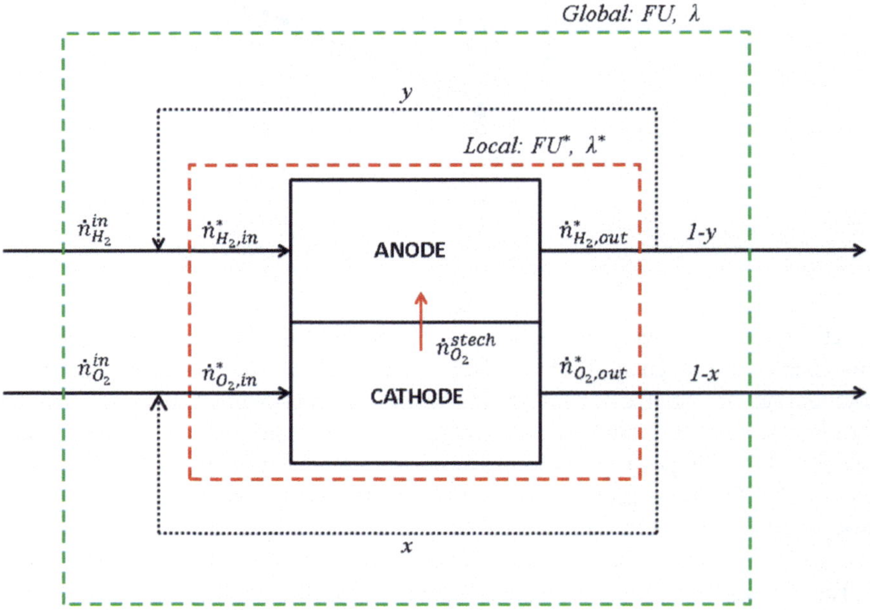

Fig. 20 Anodic and cathodic recirculation in an SOFC stack

From an electrochemical point of view, the exhaust recirculation always brings a dilution of the inlet reactants. So a recirculation (either on the anode or on the cathode side) always entails a decrease in the inlet Nernst voltage of the SOFC stack.

Anodic recirculation

For the anodic recirculation, we can define the FU inside (local FU) and outside (global FU) generator as the following:

$$\mathrm{FU} = \frac{\frac{I}{2F}}{\dot{n}_{\mathrm{H_2,in}}} \tag{53}$$

$$\mathrm{FU}^* = \frac{\frac{I}{2F}}{\dot{n}^*_{\mathrm{H_2,in}}} \tag{54}$$

The relation between FU and FU* as a function of the anodic recirculation ratio is simply obtained by solving a H_2 molar balance across the SOFC anode side.

The H_2 molar flow rate inlet and outlet of the SOFC, respectively, is given by:

$$\dot{n}^*_{\mathrm{H_2,in}} = \dot{n}_{\mathrm{H_2,in}} + y \cdot \dot{n}^*_{\mathrm{H_2O,out}} = \dot{n}_{\mathrm{H_2,in}} + y \cdot \left(\dot{n}^*_{\mathrm{H_2,in}} - \mathrm{FU} \cdot \dot{n}^{\mathrm{in}}_{\mathrm{H_2}}\right) \tag{55}$$

$$\dot{n}^*_{\mathrm{H_2,out}} = \dot{n}^*_{\mathrm{H_2,in}} - \frac{I}{2F} = \dot{n}^*_{\mathrm{H_2,in}} - \mathrm{FU} \cdot \dot{n}^{\mathrm{in}}_{\mathrm{H_2}} \tag{56}$$

By combining together Eqs. (54)–(56), we obtain the H_2 flow rate that feeds the stack as a function of the anodic recirculation ratio and the fresh inlet fuel (coming from the reformer):

$$\dot{n}^*_{\mathrm{H_2,in}} = \frac{\dot{n}_{\mathrm{H_2,in}}(1 - y \cdot \mathrm{FU})}{1 - y} \tag{57}$$

By inserting Eq. (57) into Eq. (53), we finally obtain an expression (Eq. 58) which provides the relation between local FU and global FU as depending on the anode recirculation ratio:

$$\mathrm{FU}^* = \frac{\mathrm{FU} \cdot \dot{n}^{\mathrm{in}}_{\mathrm{H_2}}}{\dot{n}^*_{\mathrm{H_2,in}}} = \frac{\mathrm{FU} \cdot (1 - y)}{1 - y \cdot \mathrm{FU}} \tag{58}$$

Cathodic recirculation

In analogy with what we did in the case of anode recirculation, we can also define local and global air excess ratio values:

$$\lambda = \frac{\dot{n}_{\mathrm{O_2,in}}}{\dot{n}^{\mathrm{stoich}}_{\mathrm{O_2}}}; \quad \lambda^* = \frac{\dot{n}^*_{\mathrm{O_2,in}}}{\dot{n}^{\mathrm{stoich}}_{\mathrm{O_2}}} \tag{59}$$

This time, we need solve for the O_2 molar balance across the SOFC cathode chamber to establish an analytic correlation between global and local λ values, as function of the cathodic recirculation ratio.

The O_2 molar flow entering and exiting the SOFC, respectively, is given by:

$$\dot{n}^*_{O_{2,\mathrm{in}}} = \dot{n}_{O_2,\mathrm{in}} + x \cdot \dot{n}^*_{O_2,\mathrm{out}} \tag{60}$$

$$\dot{n}^*_{O_2,\mathrm{out}} = \dot{n}^*_{O_{2,\mathrm{in}}} - \dot{n}^{\mathrm{stoich}}_{O_2} \tag{61}$$

By combining together Eqs. (60) and (61), we obtain the following expression:

$$\dot{n}^*_{O_{2,\mathrm{in}}}(1 - x) = \dot{n}_{O_2,in} - x \cdot \dot{n}^{\mathrm{stoich}}_{O_2} \tag{62}$$

Equation (62) can be rearranged to obtain an expression of the fresh O_2 molar flow at the inlet of the generator (global inlet), as function of the cathodic recirculation ratio, and the local O_2 molar flow at inlet inside the SOFC cathode (local inlet); the flowing expression is obtained by the following:

$$\dot{n}_{O_2,\mathrm{in}} = \dot{n}^*_{O_2,\mathrm{in}}(1 - x) + \dot{n}^*_{O_2,\mathrm{stoich}} \cdot x \tag{63}$$

By inserting Eq. (63) in Eq. (59), we finally obtain the expression which provides the analytic interdependency relation between local and global λ, as a function of the cathodic recirculation ratio:

$$\lambda^* = \frac{\dot{n}_{O_2,\mathrm{in}} - x \cdot \dot{n}^{\mathrm{stoich}}_{O_2}}{(1 - x) \cdot \dot{n}^{\mathrm{stoich}}_{O_2}} = \frac{\lambda - x}{1 - x} \tag{64}$$

Most of the SOFC systems do not employ cathode recirculation. For such systems, a quite common indication is that λ should be equal or greater than 2. This is equivalent to say the outlet oxygen partial pressure has to equal or greater than 0.12 roughly.

However, in the air-recirculated system, the inlet oxygen partial pressure in the cathode chamber is less than 0.21 (this value corresponds to a fresh air cathode stream, i.e., without any recirculation). For this reason, it is important to determine to which extent the amount of recirculated air affects both the inlet and outlet oxygen partial pressures of the fuel cell cathode.

The oxygen inlet partial pressure of SOFC cathode under a recirculation's regime is expressed as follows:

$$y^*_{O_2,in} = \frac{\dot{n}^*_{O_{2,in}}}{\dot{n}^*_{air,in}} \tag{65}$$

where

$$\dot{n}^{*}_{O_{2,in}} = \dot{n}_{O_2,in} + x\left(\dot{n}^{*}_{O_{2,in}} - \dot{n}^{stoich}_{O_2}\right) \tag{66}$$

Equation (66) can be rewritten the as follows:

$$\dot{n}^{*}_{O_{2,in}} = \frac{\dot{n}_{O_2,in} - x \cdot \dot{n}^{stoich}_{O_2}}{1 - x} \tag{67}$$

To calculate the oxygen molar fraction at the SOFC cathode, we need to solve the air molar balance across the cathode electrode. For the air balance, the two following equations hold:

$$\dot{n}^{*}_{air,in} = \dot{n}_{air,in} + x \cdot \dot{n}^{*}_{air,out} \tag{68}$$

$$\dot{n}^{*}_{air,out} = \dot{n}^{*}_{air,in} - \dot{n}^{stoich}_{O_2} \tag{69}$$

Combined together, they produce the following expression:

$$\dot{n}^{*}_{air,in} = \dot{n}_{air,in} + x \cdot (\dot{n}^{*}_{air,in} - \dot{n}^{stoich}_{O_2}) \tag{70}$$

which can be rewritten as follows:

$$\dot{n}^{*}_{air,in} = \frac{\dot{n}_{air,in} - x \cdot \dot{n}^{stoich}_{O_2}}{1 - x} \tag{71}$$

By substituting Eq. (71) into Eq. (65), we obtain an expression for the molar fraction of oxygen at the cathode inlet (local value):

$$y^{*}_{O_2,in} = \frac{\dot{n}_{O_2,in} - x \cdot \dot{n}^{stoich}_{O_2}}{\frac{\lambda \cdot \dot{n}^{stoich}_{O_2}}{y_{O_2,in}} - x \cdot \dot{n}^{stoich}_{O_2}} \tag{72}$$

Considering also the following equation:

$$\dot{n}_{air,in} = \frac{\lambda \cdot \dot{n}^{stoich}_{O_2}}{y_{O_2,in}} \tag{73}$$

we can rearrange Eq. (72) in the following form:

$$y^{*}_{O_2,in} = \frac{\lambda - x}{\frac{\lambda}{y_{O_2,in}} - x} \tag{74}$$

The oxygen molar fraction at the cathode outlet in the presence of cathodic recirculation can be derived considering again the usual molar/mass balances across the SOFC (one for the O_2 and one for air, respectively).

For the oxygen mass balance, we have the following:

$$\dot{n}^{*}_{O_2,out} = \dot{n}_{O_2,in} + x \cdot \dot{n}^{*}_{O_2,out} - \dot{n}^{stech}_{O_2} \quad (75)$$

that we can rewritten as follows:

$$\dot{n}^{*}_{O_2,out} = \frac{\dot{n}_{O_2,in} - \dot{n}^{stoich}_{O_2}}{1 - x} \quad (76)$$

Similarly, the air mass balance gives:

$$\dot{n}^{*}_{air,out} = \dot{n}_{air,in} + x \cdot \dot{n}^{*}_{air,out} - \dot{n}^{stoich}_{O_2} \quad (77)$$

that we can rewritten as follws:

$$\dot{n}^{*}_{air,out} = \frac{\dot{n}_{air,in} - \dot{n}^{stoich}_{O_2}}{1 - x} \quad (78)$$

Finally, we obtain the following expression for the oxygen molar fraction at the cathode outlet:

$$y^{*}_{O_2,out} = \frac{\dot{n}^{*}_{O_2,out}}{\dot{n}^{*}_{air,out}} = \frac{\dot{n}_{O_2,in} - \dot{n}^{stoich}_{O_2}}{\dot{n}_{air,in} - \dot{n}^{stoich}_{O_2}} = \frac{\dot{n}^{stoich}_{O_2}(\lambda - 1)}{\dot{n}^{stoich}_{O_2}\left(\frac{\lambda}{y_{O_2,in}} - 1\right)} \quad (79)$$

From Figs. 21 , 22 and 23, graphs concerning the quantities described in this paragraph are given. In particular, Figs. 22 and 23 show how critical the cathodic recirculation becomes for the Nernst voltage if a too vitiated air stream travels along the cathode channel. In other words, there exists a threshold above which an increase of air recirculation starts, leading to a significant drop in the Nernst voltage.

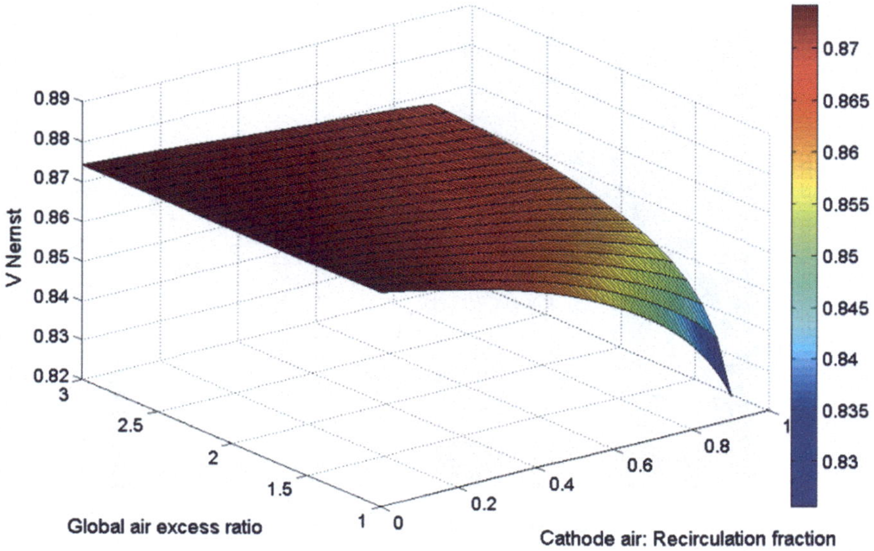

Fig. 21 Oxygen molar fraction as a function of the global air excess ratio and the cathodic recirculation fraction

Fig. 22 Nernst voltage with the outlet composition as a function of the global air excess ratio and the cathodic recirculation fraction (contour plot)

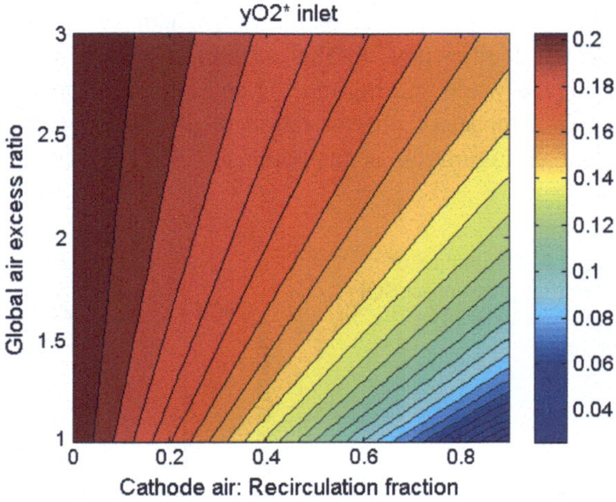

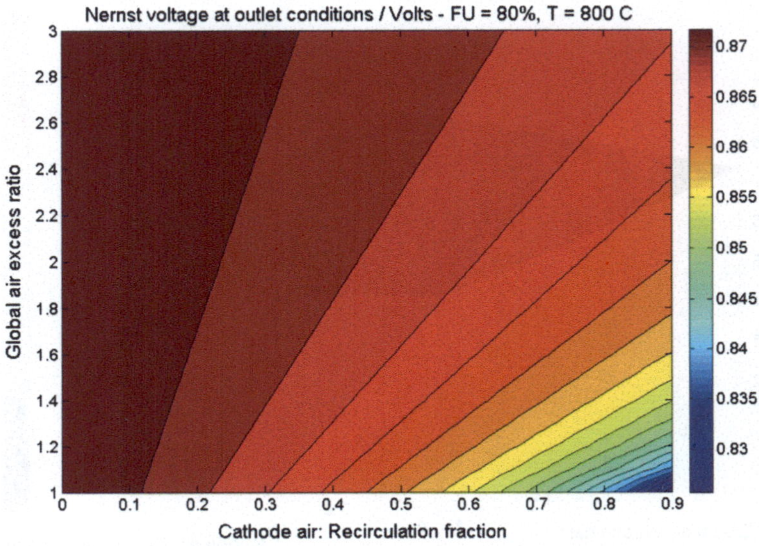

Fig. 23 Nernst voltage with the outlet composition as a function of the global air excess ratio and the cathodic recirculation fraction (surface plot)

5 SOFC Plant Optimization

The design parameters of the SOFC can be optimized in order to maximize the electrical efficiency. In this section, the performance of the SOFC-GT plant is analyzed while varying the FU and operating pressure.

The impact of pressurization on the SOFC-GT plant is given in Fig. 24. The plant configuration is the same as Fig. 12, with the only differences that a hot

Fig. 24 The impact of operating pressure on SOFC performance at FU = 70, 75, and 80 %

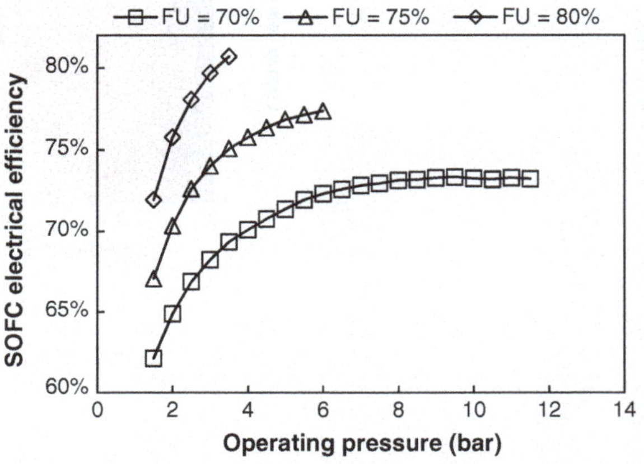

recirculator is used instead of the ejector-based system and that a single multistream heat exchanger is used to preheat anode and cathode feeds from the hot expanded exhaust. Increasing the operating pressure is beneficial to the SOFC performance. It is also observed that the larger efficiency increase is obtained for pressures below 6–7 bar. The trends shown in Fig. 24 resemble that of the efficiency of a conventional GT when varying the pressure ratio of the compressor. Maintaining this analogy, the SOFC acts as the combustor of the Joule cycle. However, the pressurization is also beneficial to the SOFC stack performance because of the higher Nernst voltage.

The curves in Fig. 24 are plotted in different pressure ranges because some point results unfeasible. For instance, all the simulation cases that require external heating (i.e., the waste heat recovery from the SOFC exhaust gas is not sufficient to preheat the anode and cathode feeds) are disregarded.

The optimal plant performance has been calculated introducing a constraint on the pinch point of the heat exchanger that is used to preheat the inlet feed gases. The optimization problem is solved using a sequential quadratic processing (SQP) algorithm.

Hence, the objective function is to maximize the electrical efficiency while varying the operating pressure and the FU of the SOFC with a minimum internal temperature difference of 50 °C in the heat exchanger that recover from the exhaust gas. The resulting optimal FU is 72.1 %, while the operating pressure is 7.4 bar. Under these operating conditions, the calculated SOFC-GT electrical efficiency is 74.84 %. In Fig. 25, the impact of FU is shown for the SOFC-GT cycle operated at 7.5 bar. Above FU = 72 %, the minimum internal temperature difference in the heat exchanger drops below 50 °C, thus violating the constraint.

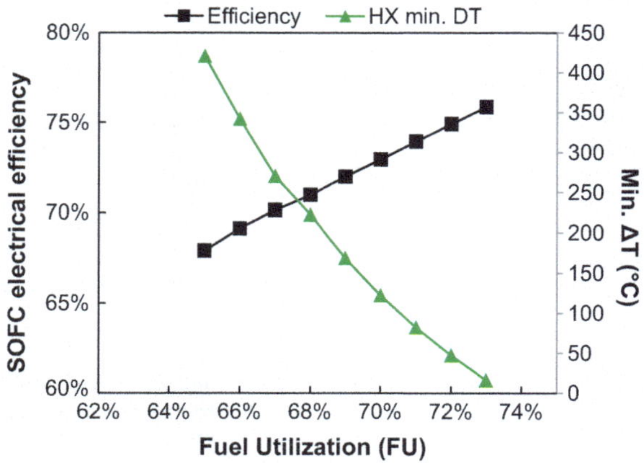

Fig. 25 Impactof FU on the SOFC-GT cycle pressurized at 7.5 bar

Economics of SOFC systems. The trade-off between energy efficiency and economic profit is crucial for every power-producing system. Schematic on the interplay between the fuel cell operating conditions and the resulting active area (for instance, an higher voltage entails better fuel cell efficiency at the expense of a larger required active area, which translates in larger capital costs) is shown in Figure 26. The SOFC electrical efficiency is enhanced by reducing the current density as less irreversibilities are generated that result in higher operating voltage.

In fact, the electrical efficiency of the SOFC is written as follows:

$$\eta_{\mathrm{el,SOFC}} = \frac{W_{\mathrm{el}}}{\mathrm{LHV}_f \bar{n}_f} = \frac{V_{\mathrm{op}} I_{\mathrm{TOT}}}{\mathrm{LHV}_f \frac{I_{\mathrm{TOT}}}{n_{\mathrm{el}} F} \cdot \frac{1}{\mathrm{FU}}} = \frac{n_{\mathrm{el}} F}{\mathrm{LHV}_f} V_{\mathrm{op}} \mathrm{FU} \tag{80}$$

Eq. (80) shows how the SOFC efficiency is directly proportional to both the operating voltage and FU.

To better clarify the impact of the operating point on the active area requirement, the graphs in Fig. 27 should be looked at. SOFC cells are generally operated at 0.8 V because a drastic increase in the required active area results for higher operating voltage.

The economic trade-off is therefore between operating and capital costs. A higher installed active area mostly increases the capital expenditure of the plant. However, the larger is the active, the higher is the electrical efficiency that results in lower operating costs due to reduced fuel consumption per electricity unit delivered. For the profitability of SOFC, maintenance costs are also important. Since the unit cost expressed in \$/kW for fuel cell systems is generally higher than that for conventional systems (e.g., ICES and gas turbines), operating costs must be lower to compensate for the higher initial capital cost. It has been already shown that the better fuel efficiency is beneficial for reducing operating costs. However, the

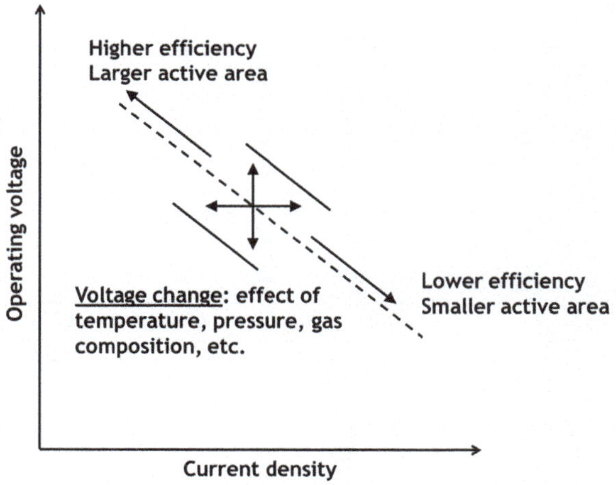

Fig. 26 The impact of SOFC performance on energy efficiency and capital investment

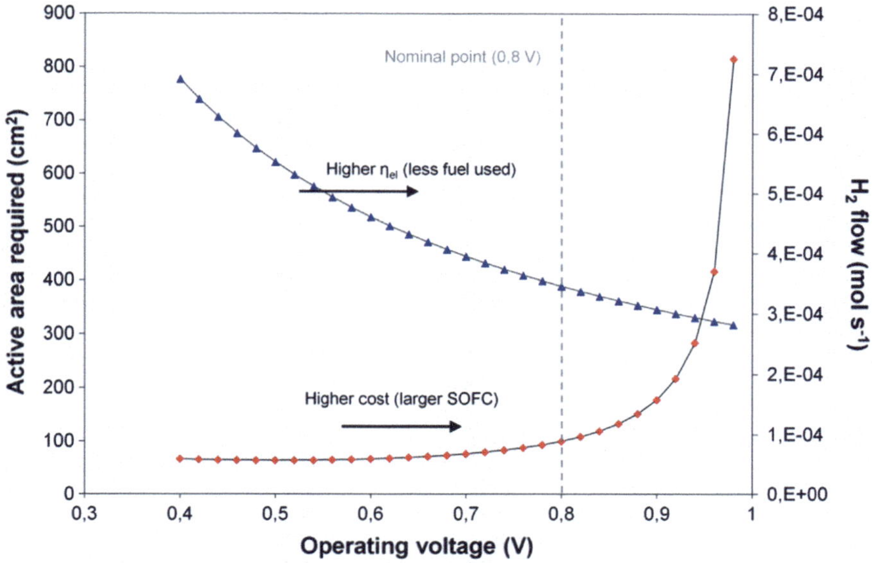

Fig. 27 SOFC efficiency versus active area required. (The calculation refers to the active area and hydrogen fuel that are required to produce 40 W of electric power. The fuel cell OCV is taken as 1.0 V and the ASR = 0.4 Ω cm^2. At the nominal condition of 0.5 A/cm^2, the operating voltage is 0.8 V and 40 W of electricity are delivered)

lifetime of the SOFC stack is also important. Ideally, SOFC systems require low maintenance compared to engines since no moving parts are present in the power core. However, the assumption here is that the degradation rate of the SOFC is almost negligible (i.e., below 0.1–0.2 % per 1000 h).

References

Campanari, S., Chiesa, P., & Manzolini, G. (2010). CO_2 capture from combined cycles integrated with molten carbonate fuel cells. *International Journal of Greenhouse Gas Control, 4*, 441–451.

Discepoli, G., Cinti, G., Desideri, U., Penchini, D., & Proietti, S. (2012). Carbon capture with molten carbonate fuel cells: Experimental tests and fuel cell performance assessment. *International Journal of Greenhouse Gas Control, 9*, 372–384.

Franzoni, A., Magistri, L., Traverso, A., & Massardo, A. F. (2008). Thermoeconomic analysis of pressurized hybrid SOFC systems with CO_2 separation. *Energy, 33*, 311–320.

Grol, E. (2009). *Systems analysis of an integrated gasification fuel cell combined cycle.* DOE/NETL-40/080609 report

Hansen, J. B. (2008). Correlating sulfur poisoning of SOFC nickel anodes by a Temkin isotherm. *Electrochemical and Solid-State Letters, 11*, B178.

Hassmann, K. (2001). SOFC power plants, the Siemens-Westinghouse approach. *Fuel Cells, 1*, 78–84.

International Energy Agency, IEA. (2012). *Energy technology perspective* 2012.

Lanzini, A. (2011). *Study of solid oxides fuel cell systems running on poly-fuel mixtures*. Ph.D. Dissertation, Politecnico di Torino, Italy.

Li, M., Rao, A. D., Brouwer, J., & Samuelsen, G. S. (2010). Design of highly efficient coal-based integrated gasification fuel cell power plants. *Journal of Power Sources, 195*, 5707–5718.

Madi, H., Lanzini, A., Diethelm, S., Papurello, P., Van herle, J., Lualdi, M., et al. (2015). Solid oxide fuel cell anode degradation by the effect of siloxanes. *Journal of Power Sources, 279*, 460–471.

NETL (2011a). *Cost and performance baseline for fossil energy plants volume 1: Bituminous coal and natural gas to electricity*. DOE/NETL-2010/1397 report.

NETL (2011b). *Analysis of natural gas fuel cell plant configurations*. DOE/NETL-2011/1486 report.

NETL (2011c). *Analysis of integrated gasification fuel cell plant configurations*. DOE/NETL-2011-1482 report.

Oliveira, E. L. G., Grande, C., & Rodrigues, A. E. (2009). Steam methane reforming in a Ni/Al_2O_3 catalyst: Kinetics and diffusional limitations in extrudates. *Canadian Journal of Chemical Engineering, 87*, 945–956.

Park, S. K., Kim, T. S., Sohn, J. L., & Lee, Y. D. (2011). An integrated power generation system combining solid oxide fuel cell and oxy-fuel combustion for high performance and CO_2 capture. *Applied Energy, 88*, 1187–1196.

Rao, A. D. (1991). *Reactor expander topping cycle*. U.S. Patent No. 4,999,993.

Romano, M. C., Spallina, V., & Campanari, S. (2011). Integrating IT-SOFC and gasification combined cycle with methanation reactor and hydrogen firing for near zero-emission power generation from coal. *Energy Procedia, 4*, 1168–1175.

Seidler, S., Henke, M., Kallo, J., Bessler, W. G., Maier, U., & Friedrich, K. A. (2011). Pressurized solid oxide fuel cells: Experimental studies and modeling. *Journal of Power Sources, 196*, 7195–7202.

Singhal, S. C. (1999). Progress in tubular solid oxide fuel cell technology. In *Proceedings of the SOFC-VI Symposium, ECS Proceedings Volume.*

Spallina, V., Romano, M. C., Campanari, S., & Lozza, G. (2011a). *A SOFC-based integrated gasification fuel cell cycle with CO_2capture*. Gas Turbines Power: J. Eng. 133.

Spallina, V., Romano, M. C., Campanari, S., & Lozza, G. (2011b). *Thermodynamic analysis and optimization of IT-SOFC-based integrated coal gasification*. Tech: J. Fuel Cell Sci. 8.

Trasino, F., Bozzolo, M., Magistri, L., & Massardo, A.F. (2011). Modeling and performance analysis of the Rolls-Royce Fuel Cell Systems Limited: 1 MW plant. *Journal of Engineering for Gas Turbines and Power-Transactions of the ASME, 133*.

Traverso, A., Magistri, L., & Massardo, A. F. (2010). Turbomachinery for the air management and energy recovery in fuel cell gas turbine hybrid systems. *Energy, 35*, 764–777.

Trembly, J. P., Gemmen, R. S., & Bayless, D. J. (2007). The effect of coal syngas containing HCl on the performance of solid oxide fuel cells: Investigations into the effect of operational temperature and HCl concentration. *Journal of Power Sources, 169*, 347–354.

Vora, S. D. (2005). SECA program at Siemens Westinghouse. In *Presentation at the Sixth Annual SECA Workshop Pacific Grove, CA,* April 18–21, 2005.

Xu, J., & Froment, G. F. (1989). Methane steam reforming, methanation and water-gas shift: I intrinsic kinetics. *AIChE Journal, 35*, 88–96.

DOE Methodologies for Analysis of Large SOFC Systems

Domenico Ferrero, Andrea Lanzini, Pierlugi Leone and Massimo Santarelli

Abstract The experimental analysis of large systems in operation is a complex task, due to the large number of variables which affect their operation, the limited number of measurements points, and the necessity to avoid malfunctions in the real operation. To make a good use of the collected experimental data, and to reduce the efforts of the test session, the experimental analysis of large systems can be effectively developed with the methods of Design of Experiments (DoE), and with a statistical analysis of the collected data. The design of experiments approach allows one to study the main effect of the factors and also their interaction effect. The approach also allows obtaining analytical relations which express the dependent variables as a function of the examined control factors through first (simple factorial) or second-order (e.g. spherical CCD) regression models. The results can be analysed in terms of sensitivity analysis and response surface analysis. Finally, optimization procedures can be applied to the regression models in order to get the best behaviour of the system analysed. The utilization of the DoE methods has been particular useful in experimental approach to large solid oxide fuel cells (SOFC) plants. As a real example, the experimental analysis of the CHP-100 SOFC Field Unit has been developed with the methods of Design of Experiments, and with a statistical analysis of the collected data.

1 Design of Experiment Methodologies for Analysis of Large SOFC Systems: Introduction

The experimental analysis of a real large SOFC generator in operation is a complex task. The generator is a big plant, with many variables which affect its operation; therefore, the experimental environment is completely different compared to a laboratory, because it is difficult to control the many variables involved during the development of the experiment; moreover, the data acquired in the experimental

D. Ferrero · A. Lanzini · P. Leone · M. Santarelli (✉)
Energy Department, Polytechnic of Turin, Turin, Italy
e-mail: massimo.santarelli@polito.it

M. Boaro and A.S. Aricò (eds.), *Advances in Medium and High Temperature Solid Oxide Fuel Cell Technology*, CISM International Centre for Mechanical Sciences 574,
DOI 10.1007/978-3-319-46146-5_7

session have to be carefully analyzed, in order to isolate the main effects which have to be detected, to take care of the interactions and to quantify the significance of every effect. This causes the necessity of a careful design of the experiment coupled with a consequent statistical analysis of the collected data in order to be able to outline the significance of every observed effect.

Moreover, the large plant is characterized by a not uniform distribution of its physical and chemical variables (e.g., temperatures, chemical composition of mass flows), and at the same time the measurement points are discrete; therefore, in many parts of the plant, the data are not available. As a consequence, a combination of some experimental techniques and of analytical models could be used to deduce the distribution, inside the generator volume, of some important variables.

Finally, the generator is in real operating conditions, and therefore, its experimental analysis has to avoid malfunctions and dangerous operations; usually, one possibility is the perturbation of some independent variables in a safety experimental range, and the analysis of the sensitivity of the other variables (the dependent variables) to the imposed perturbation.

As a conclusion, the experimental analysis of a large SOFC system can be developed with methods of design of experiments and with a statistical analysis of the collected data.

In these pages, the methods of design of experiments have been applied to an example of large SOFC generator installed in CCHP (combined cooling, heat and power) configuration: the CHP-100 SOFC Field Unit built by Siemens Power Generation Stationary Fuel Cells (SPG-SFC). The generator has been installed in a SOFC laboratory at TurboCare S.p.A. (a subsidiary of SPG-SFC) since June 19, 2005, and to date has shown the record availability of 99.5 %. The SOFC CHP-100 is the first to utilize the commercial prototype air electrode-supported cells and in-stack reformers. The generator is fed with natural gas from the grid. The stack is composed of 1152 single tubes, arranged in 48 cell bundles (4 cell bundles are connected in series to form a bundle row, and 12 bundle rows are aligned side by side, interconnected with an in-stack reformer between each bundle row).

A design of experiments procedure has been used to study in a rigorous manner the collected experimental data and also to relate, through analytical expressions, the investigated dependent variables with the control factors (independent variables). The design of experiments approach allows one to study the main effect of the factors and also their interaction effects. The aim is the description of the analytical relations which express the dependent variables as a function of the examined control factors through first- (simple factorial) or second-order (spherical CCD) regression models.

The experimental sessions have been designed in order to investigate the effect of two important operation factors: the overall fuel consumption (FC) and the air stoichs (λ_{ox}). The main expectation has been the characterization of the operation of the single sectors of the SOFC generator, pointing out the analysis on the distribution of the local voltages (and temperatures). Particular attention has been addressed in the description of the operation in terms of sensitivity maps of the main investigated dependent variables (voltages).

2 Description of the Plant

The CHP-100 kW_e SOFC Field Unit (Siemens Power Generation-Stationary Fuel Cells) is the first to utilize the commercial prototype air electrode-supported cells (22 mm diameter, 150 cm active length, 834 cm^2 active area) and in-stack reformers. The generator is fed with natural gas from the grid distribution. In Fig. 1, the picture representing the CHP 100 test site in TurboCare (Torino, Italy) is shown.

The upper level of the system hierarchy after the single cell is the cell bundle, which consists of a 24-cell array arranged as 8 cells in electrical series by 3 cells in electrical parallel. Two bundles in series form a sector; two sectors are connected in series to form a row, and 12 rows are aligned side by side, interconnected in serpentine fashion with an in-stack reformer between each row (for a total of 1152 single tubes). The schematic of the cell stack arrangement is shown in Fig. 2 (Kabs 2001), outlining the position of the fuel ejectors and the power leads.

In this chapter, the results will be discussed dividing ideally the primary generator in four main zones (north; south; power leads side, and ejectors side), and also in the central zone (in the middle between the power leads and the ejectors sides). Supply of NG arrives as liquefied NG and thus contains no sulfur. For instance in Italy, the gas pipeline network operator specifies for H_2S an upper limit of 6.6 mg/m^3, which corresponds to less than 5 ppm (v).

Cell AES Siemens:

- ELECTROLYTE: ZrO_2 + Y_2O_3 0.04 mm

Fig. 1 Picture of the SOFC CHP-100 test site in TurboCare (Torino, Italy) Walpole and Myers (1993)

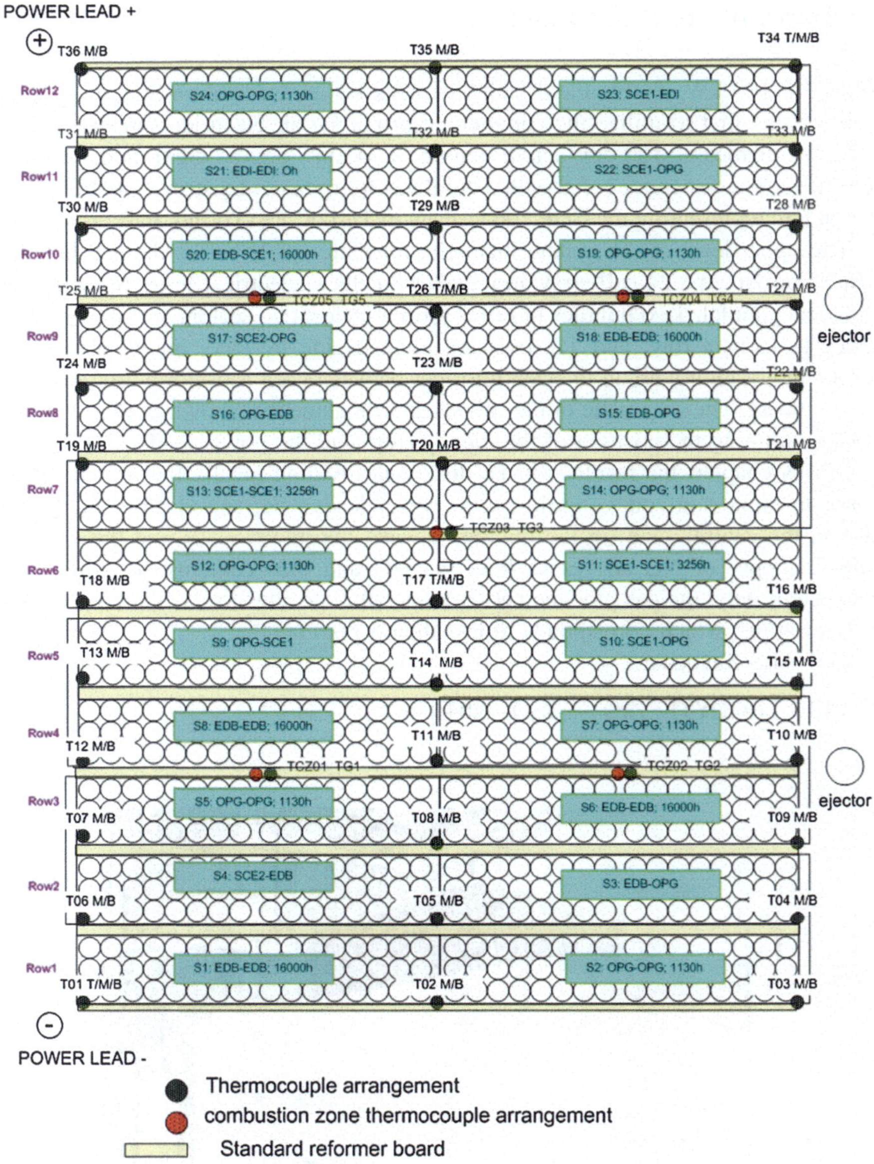

Fig. 2 Schematic of the stack arrangement

- ANODE: Ni/ZrO_2, 0.10 mm
- CATHODE: $LaMnO_3$, 2.20 mm

Cell bundle: 24-cell array arranged as 8 cells in electrical series by 3 cells in electrical parallel.

The stack is defined by 12 bundle rows aligned side by side, interconnected in serpentine fashion with an in-stack reformer between each bundle row (for a total of 1152 single tubes).

The system is completed by the balance of plant (BoP), with five major skids: generator module, electrical control system, fuel supply system (FSS), thermal management system (TMS), and heat export system (HES) (George 1998; Singhal 2004).

From the electrical point of view, the module is made up of four parts: (a) the electrochemical generator; (b) the power conditioning system (PCS): an inverter, which operates the DC/AC conversion; (c) the SOFC auxiliaries (blowers); (d) the board fuel cell (QFC), planned from Politecnico of Torino, which allows the electrical connection of the SOFC CHP-100 to the net. From the thermal point of view, the exhausts from the stack, passing in a cross-finned tubes gas-water exchanger, provide, in nominal conditions, approximately 60 kW_t of thermal energy, used for the winter and summer conditioning (through a absorption refrigerator cycle water–lithium bromide fed with warm water) of some offices of the TurboCare factory.

The simplified flow schematic of the SOFC CHP-100 BoP is shown in Fig. 3, while in Fig. 4 the simplified schematic of the primary generator structure is reported.

Data acquisition:

- Electric data:
  - generator terminal voltage and
    sector (two cell bundles) voltage, for a total of 24 measurements (around 10 V);

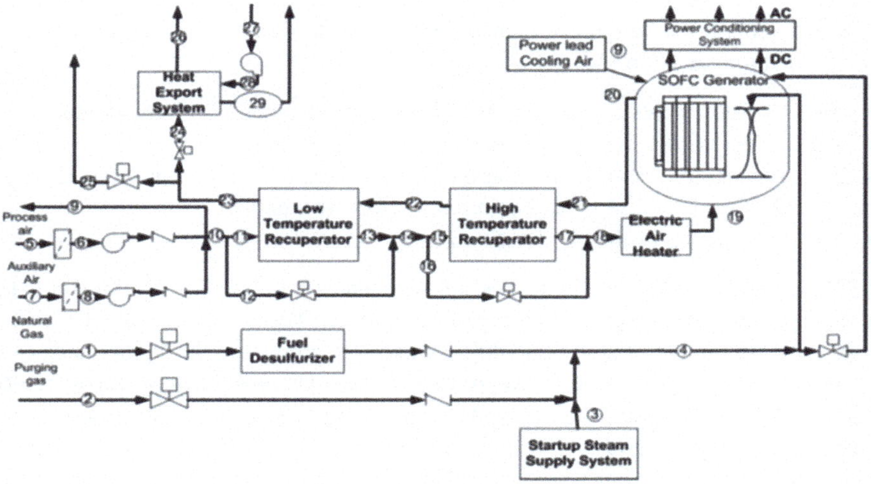

Fig. 3 Simplified flow schematic of the SOFC CHP-100 BoP

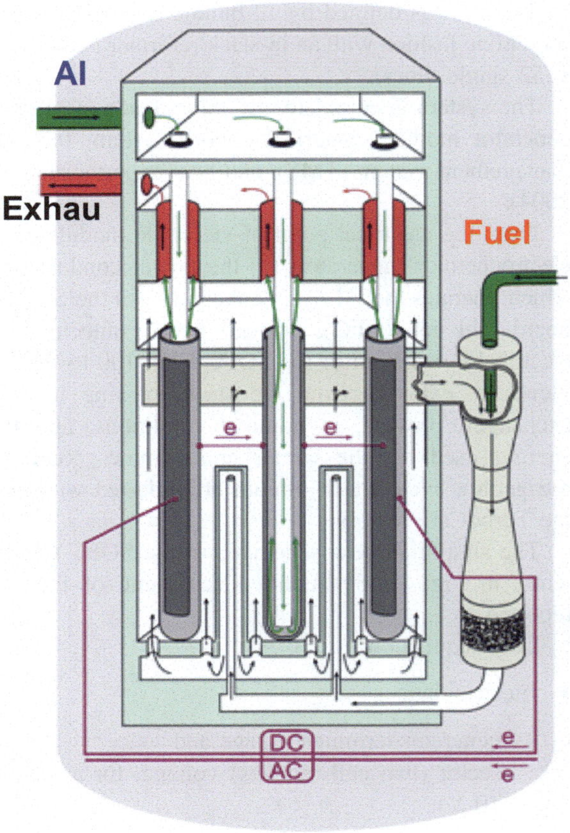

Fig. 4 Simplified schematic of the primary generator structure

- Thermal data:
  - 3 vertical measurement planes corresponding three cell lengths: bottom, medium, and top plane,
- 36 S-thermocouples at the bottom and medium plane (72 S-thermocouples),
- 4 S-thermocouples at the top plane,
- 5 S-thermocouples for the regulation of the generator (TG1 (row 3), TG2 (row 4), TG3 (row 7), TG4 (row 9), and TG5 (row 10), and
- 5 thermocouples in the combustion zone.

The commissioning date in TurboCare was June 19, 2005. The operational data of @ December 2006 report a number of run hours in TurboCare of around 12,000, with a total run hours of around 32,000 (including run hours in the Netherlands and Germany); a average stack temperature of 954 °C; a DC-generated power of 123.6 kW_e, at 246.1 VDC and 502.3 A; a AC-generated power of 113 kW_e; a power to TurboCare workshop grid of 103 kW_e (20 % of the workshop requirement); a heat generation of 60 kW_t (hot water @85 °C). Other characteristics are as follows: a reliable operation also with significant LHV changes of the natural gas; no measurable voltage or power degradation; a very high availability of 99.5 % (actual, on annual basis); 4 stops (one operation error, three inverter failures); 24 successful operation transition to power dissipator due to utility (AEMD, Azienda Energetica Metropolitana Torino Distribuzione) or TurboCare grid failures; one complete thermal cycle (August 2006); a very dependable automatic operation (no operator in control room); a remote control capability via modem; a easy maintenance (replacement of air filters and of reactant desulphurization).

Politecnico di Torino has developed several activities, such as the design of the SOFC thermal plant (Saroglia et al. 2005), the modeling of the generator and of the BoP (Leone 2005, 2006; Calì 2005a, 2007), and the safety design and analysis (Calì 2005b). In (Calì 2005c), a model of the operation of the SOFC CHP-100 has been developed using a 0-D approach and validated through a first session of experimental tests. In (Calì 2006a, b) the effect of the setpoint generator temperature and fuel consumption factors on several dependent variables (i.e., DC and AC electric power, recovered heat, electric and global efficiency, efficiency of the pre-reforming process) is analyzed in the form of screening tests, and the process responses are treated in form of response surface plots. In (Saltarelli 2006), the regression models have been used in constrained optimization procedures to maximize different objective functions (AC electric power and recovered heat).

3 ANOVA of the Experimental Data, the Regression Models, and RSP (Response Surface Plots)

The procedures of ANOVA to obtain suitable regression models for some dependent variables of the system (so, an analytical form in which the dependent variable is linked with an analytical function to the selected independent variables of the system) starting from the analysis of the experimental session are here applied to the case study of the SOFC CHP 100 system.

In the example, the independent variables (factors) that have been selected are among the most important control variables of such kind of systems:

- Setup temperature (TGEN)
- Fuel consumption (FC)

Having selected the independent variables of the system, is now possible to organize the experimental test session in order to obtain suitable regression models for the dependent variables.

In the example, two different designs of experiments are described:

- Simple 2^2 factorial: to obtain first-order regression models of the analyzed dependent variables;
- 2^2 central composite design factorial: to obtain second-order regression models of the analyzed dependent variables (Fig. 5).

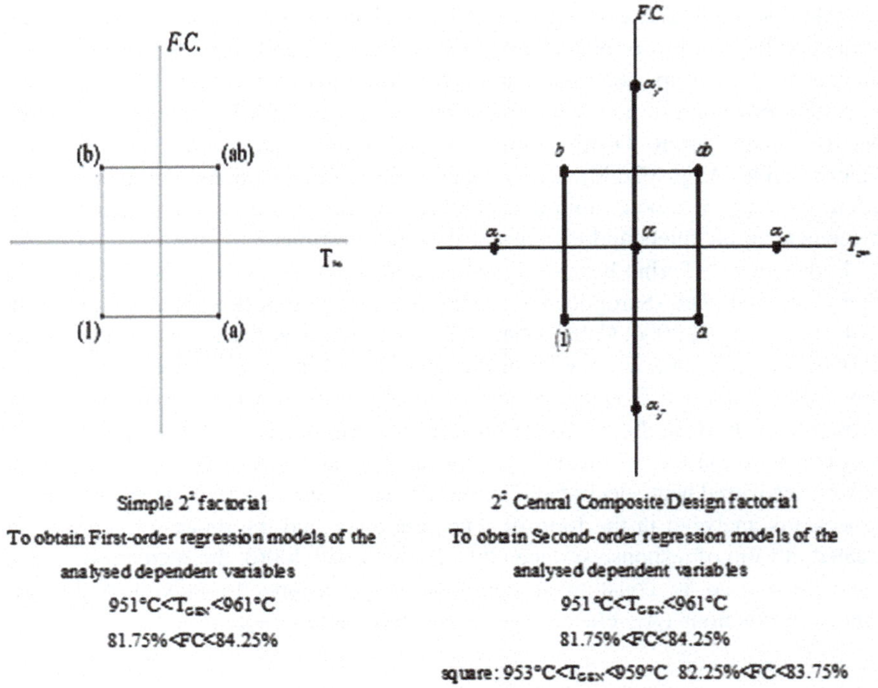

Fig. 5 Two different designs of experiments for the SOFC CHP100 system

Among the several dependent variables that can be analyzed with an ANOVA, and then expressed in form of a regression model, one of the most important is the generator voltage.

Therefore, the ANOVA on the experimental data of the generator voltage is described in Table 1.

From the ANOVA of the experimental data of the generator voltage versus the setup temperature (TGEN) and the fuel consumption (FC), it is possible to check that:

- The main effect of the Factor A (the setup temperature) over the generator voltage is significant.
- The main effect of the Factor B (the fuel consumption) over the generator voltage is significant.
- The main effect of the interaction AB of the two factors over the generator voltage is not significant.

Therefore, it is possible to consider that the crossing effect of the two independent variables has no significant effect on the dependent variable "generator voltage." As a consequence, the regression model of the generator voltage that can

Table 1 ANOVA on the experimental data of the generator voltage

Source of variation	Sum of squares	Degrees of freedom	Mean square	F_0	Percent contribution
Factor A: T_{GEN}	SSA: 0.609	$a-1$: 1	MS_A: 0.609	$F_{0,A}$: 42.8237	4.1
Factor B: FC	SSB: 14.2013	$b-1$: 1	MSB: 14.2013	$F_{0,B}$: 998.6598	95.52
Interact. AB	SS_{AB}: 0.00022898	$(a-1)(b-1)$: 1	MSAB: 0.00022898	F_0,AB: 0.0161	0.00154
Error	SS_E: 0.0569	$ab(n-1)$: 4	MS_E: 0.0142		
Total	SS_T: 14.8673	$abn-1$: 7			
Significance test					
$F_{0.05,\ DOFA,DOFE}$: 771		$F_{0,A} > F_{0.05DOFA,DOFE}$: *the main effect of the factor A is significant*			
$F_{0.05,\ DOFB,DOFE}$: 7.71		$F_{0,B} > F_{0.05,dofb,dofe}$: *the main effect of the factor B is significant*			
$F_{0.\ 05,DOFAB,DOFE}$: 7.71		$F_{0,AB} < F_{0.05,DOFAB,DOFE}$: the main effect of the interaction AB is NOT significant			

Table 2 ANOVA on the first-order regression model of the generator voltage

Source of variation	Sum of squares	Degrees of freedom	Mean square	F_0
Regression	SS_R: 14.8102	*k*: 2	MS_R: 7.4051	$F_{0,\ R}$: 648.3174
$^{SS}r(\beta_1\|\beta_1)$	(0.609)	(1)		$F_{0,\ \beta 1}$: 53.315
$SSr(\beta_2\|\beta_0,\ \beta_1)$	(14.2013)	(1)		$F_{0,\ \beta 2}$: 1243.3
Error	SS_E: 0.0571	$n-p$: 5	MS_E: 0.0114	
SS_{LOF}	(0.00022898)	$m-p$: (1)		
SS_{PE}	(0.0569)	$n-m$: (4)		
Total	Syy: 14.8673	$n-1$: 7		
R^2	0.9962	R^2adj	0.9946	
$\hat{\sigma}^2$	0.0114			
Significance test for the regression model				
*F*0.05, *k*, $n-k-1$: 5.79		*F*0, *R* > *F*0.05, *k*, $n-k-1$: At least one of the regressors contributes to the model		
*F*0.05, *r*, $n-p$: 6.61		*F*0, *b* 1 > *F*0.05, *r*, $n-p$: The T_{GEN} variable contributes to the regression model		
*F*0.05, *r*, $n-p$: 6.61		*F*0, *b* 2 > *F*0.05, *r*, $n-p$: The f.c. variable contributes to the regression model		
Coded variables: $V_{GEN} = 246.3321 + 0.2759\ t_{GEN} - 1.3323$ f.c.				
Physical variables: $V_{GEN} = 282.0447 + 0.05518\ T_{GEN} - 1.06608$ f.c.				

Significance test for the regression model

Table 3 2^2 central composite design factorial and second-order regression models with coded variables

Dependent variable	2° order regression models—CCD face centered	R^2
(1) Generator voltage (V)	$V_{GEN} = 471.1213 + 1.1516\ t_{GEN} - 18.3075$ f.c. $- 0.00096\ t_{GEN}^2 + 0.0584$ f.c.$^2 + 0.0084\ t_{GEN}$ f.c.	0.8974
(2) DC power (kW)	$W_{el.DC} = 327.9443 + 0.3166\ t_{GEN} - 8.3765$ f.c. $- 0.000312\ t_{GEN}^2 + 0.0289$ f.c.$^2 - 0.0035\ t_{GEN}$ f.c.	0.9029
(3) AC power (kW)	$W_{el.AC} = -578.2861 + 22.1368t_{GEN} - 8.64$ f.c. $- 0\ 0012\ t_{GEN}^2 + 0.0498$ f.c.$^2 - 0.000064\ t_{GEN}$ f.c.	0.6446
(4) Thermal power (kW)	$W_{TH} = 3176.3 - 16.4995\ t_{GEN} + 115.0163$ f.c. $+ 0.0147\ t_{GEN}^2 + 0.1024$ f.c$^2 - 0.1395\ t_{GEN}$ f.c.	0.198
(5) DC efficiency (%)	$\eta_{el,DC} = 288.4456 - 0.1471\ t_{GEN} - 4.391$ f.c. $- 0.00006\ t_{GEN}^2 + 0.0098$ f.c.$^2 + 0.0032\ t_{GEN}$ f.c.	0.9379
(6) AC efficiency (%)	$\eta_{el,AC} = -100.1377 + 0\ 6467\ t_{GEN} - 4.2018$ f.c. $- 0.000404\ t_{GEN}^2 - 0.0175$ f.c.$^2 + 0.0015\ t_{GEN}$ f.c.	0.5243
(7) Thermal efficiency (%)	$\eta_{TH} = 1327.2 - 6.6378\ t_{GEN} + 44.893$ f.c. $+ 0\ 0058\ t_{GEN}^2 + 0.036$ f.c.$^2 - 0.0536\ t_{GEN}$ f.c.	0.1223
(8) Global efficiency (%)	$\eta_g = 1234.4 - 6.0064\ t_{GEN} + 40.691$ f.c. $+ 0.0054\ t_{GEN.}^2 + 0.0535$ f.c.$^2 - 0.0521\ t_{GEN}$ f.c.	0.0765
(9) Exhaust temperature at HES [°C]	$T_{LEXGAS} = 10{,}637 - 8.208\ t_{GEN} - 149.2843$ f.c. $+ 0.0036\ t_{GEN}^2 + 0.7806$ f.c.$^2 + 0.015\ t_{GEN}$ f.c.	0.8536
(10) Hot Spot temperature (°C)	$T_{hot\ spot} = -12{,}892 + 29{,}864 \cdot t_{GEN} - 15{,}616 \cdot$f.c. $- 0.016 \cdot t_{GEN}^2 + 0.016 \cdot t_{GEN} \cdot$f.c.	0.592

be constructed from the ANOVA can neglect the crossed term represented by the product of the two dependent variables.

Thus, when using the simple 2^2 factorial, a first-order regression model of the generator voltage has been obtained in the form:

$$V = \beta_0 + \beta_1 T_{GEN} + \beta_2 FC$$

The ANOVA on the first-order regression model of the generator voltage is described in Table 2.

From the ANOVA over the regression model, it is possible to see that the regressors of both the independent variables have a significant effect on the correct analytical expression of the dependent variable "generator voltage."

The same activity has been performed using 2^2 central composite design factorial to obtain second-order regression models; the complete list of variables modeled is reported in Table 3.

When considering, as a comparison, only the generator voltage of the SOFC as dependent variable, the response surface and contour plot of the generator voltage with first- and second-order regression models are shown Figs. 6 and 7.

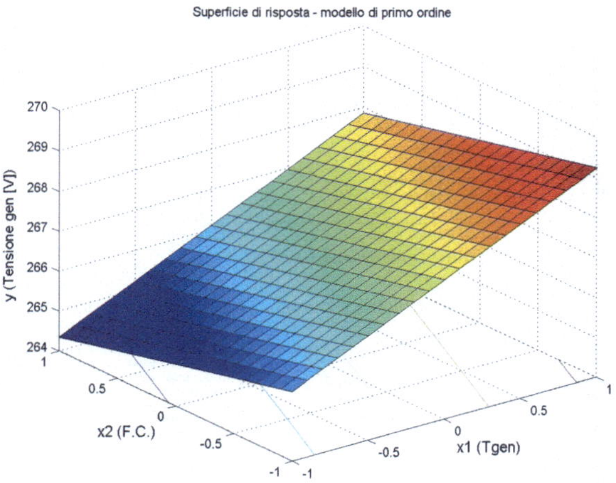

Fig. 6 Response surface plot of the generator voltage with first-order regression model

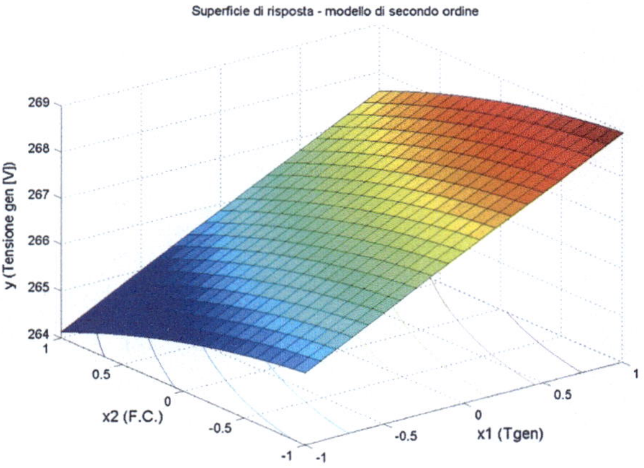

Fig. 7 Response surface plot of the generator voltage with second-order regression model

4 Operation Maps (Voltage, Heat–Power Ratio, DC and AC Electric Power)

From the DoE methods, and especially when obtained the regression models of the dependent variables, it is possible to obtain the operation maps of these dependent variables. In Fig. 8, the operation maps are shown for the generator voltage, the heat-to-power ratio, the DC and AC electric power, all as functions of the independent variables setup temperature (TGEN) and fuel consumption (FC). In Fig. 9, instead, the response surface and the contour plots of the temperatures of the anodic recirculation and of the pre-reformer outlet are reported:

- **High terminal voltage** is obtained at low fuel consumption and high temperature operating conditions because respectively: increase of average Nernst

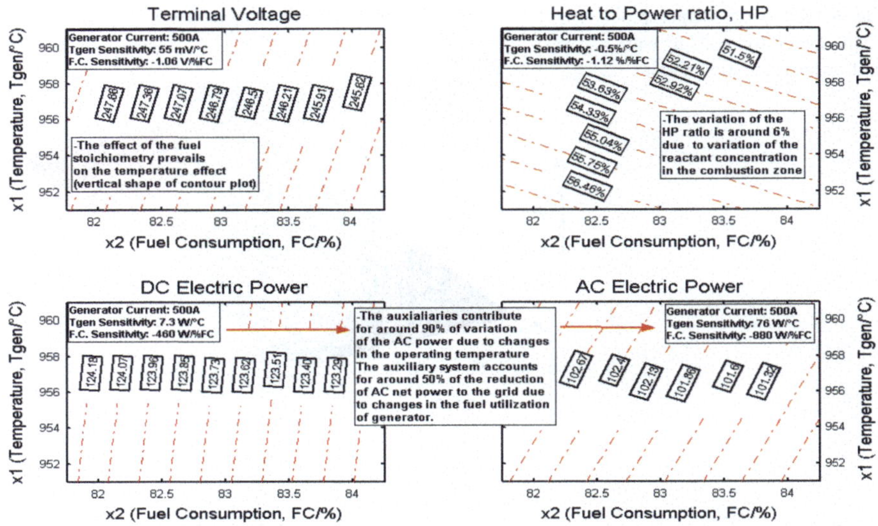

Fig. 8 Operation maps for the generator voltage, the heat-to-power ratio, the DC and AC electric power

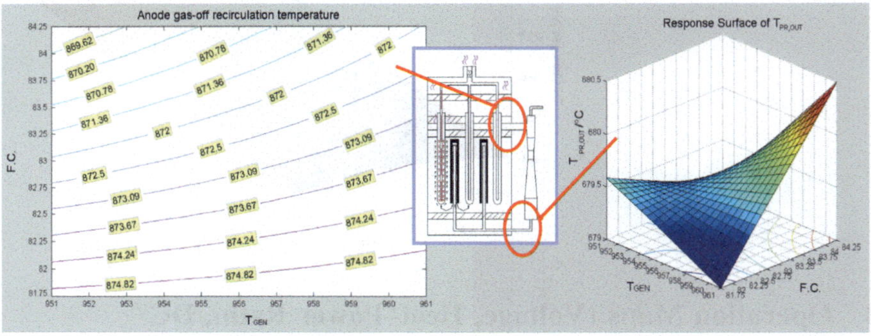

Fig. 9 Response surface and the contour plots of the temperatures of the anodic recirculation and of the pre-reformer outlet

potential and reduction of diffusion overpotential, decrease of the main cell overpotential (ohmic contribution).

- The **variation in the HP ratio is around 6 %** due to the operation of the afterburners.
- The **temperature** modification becomes **significant in the AC power** because its reduction determines an increase in air flow, and thus an increase in power consumed by the **blowers**.

Also, it is possible to obtain the response surface and the contour plots of the temperatures of the anodic recirculation and of the pre-reformer outlet:

- Anodic recirculation temperature: The increase in TGEN leads to a slight increase in the variable; the increase in FC has a negative effect due to the reduced temperature in the combustion chamber. Factor interaction: The anodic recirculation temperature shows a higher sensitivity to TGEN when the operation is carried out at high FC.
- Temperature of the pre-reformer outlet flow slightly changes during the session, and the factor interaction has a positive effect.

5 DoE on the Analytical Model (Regression Models and Optimization)

The methodology used for the planning of the experimental campaign can also been applied to a computer model simulation. In previous works (Leone 2005, 2006), we already applied the factorial analysis to the same problem: We considered a 2^5 factorial analysis performed at three current levels, useful to perform a first screening test. In fact, it is possible to obtain regression models and consequently to analyze the operation by treating with simple polynomial structures. Three independent variables have been chosen: fuel utilization factor (U_F), air utilization factor (U_{ox}), and the air temperature at the inlet of the generator (T_{ox}). The data are analyzed by applying the response surface methodology (RSM) analysis. Finally, constrained optimization methods have been applied to investigate the optimal operation points of the generator in terms of DC-generated power or in terms of exhausts' recovered heat. A comparison with the actual operation point of the SOFC generator is also shown.

5.1 Design for Fitting Regression Models to the Process Responses

A full factorial design of experiments has been employed in this work. The well-known Yates' technique has been applied (Torchio 2005; Walpole and Myers 1993; Montgomery 2005). This technique allows to perform a parametric analysis on the generator operation by evaluating whether a factor (independent variable) has a significant effect on the generator physical processes (dependent variables) or its effect is negligible. A positive main effect of a factor means that, when the factor is fixed at its upper level, it causes an increase in the dependent variable. In this work, the computer model simulation has been designed using a 2^3 factorial problem. The three factors chosen are air utilization factor, U_{ox}; fuel utilization factor, U_F; air pre-heating temperature, T_{ox}. The eight treatments (2^3 factorial) can be displayed geometrically as a cube, as shown in Fig. 3. In the figure, the lower and upper values of the factors are shown, with the randomized uncertainties used to modify the factors values: This uncertainty simulates the uncertainties of the measurement instruments. The designed treatments have been performed with

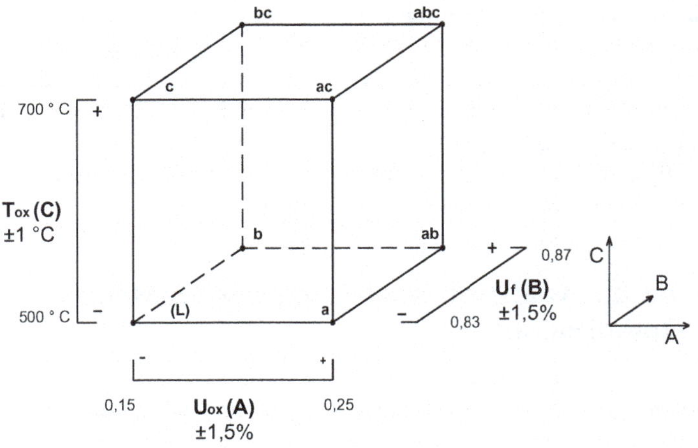

Fig. 10 Cube plot of Yate's treatment for NG feeding

completely randomized design and with three repetitions for each treatment in order to analyze the simulated data with an analysis of variance (ANOVA). The treatments have been performed for three different current levels of the generator: 400 A (0.16 A cm^{-2}); 500 A (0.2 A cm^{-2}); and 600 A (0.24 A cm^{-2}).

With this design of experiments, it is possible to obtain first-order regression models for the investigated dependent variables in terms of the significant factors. If higher-order regression models would be obtained, then the experimental campaign should be designed by using a different approach (i.e., 2^k factorial analysis with central composite design CCD for the second-order regression models or 3^k factorial analysis). The aim of this sequential approach is to obtain analytical relations between the dependent variables (i.e., average generator temperature, combustion zone temperature, exhaust's temperature, and generator voltage) and the analyzed independent variables (fuel utilization factor, air utilization factor, and air pre-heating temperature). The regression models allow to represent multiple responses of the generator operation by plotting contour plots and response surfaces. Moreover, they allow to apply constrained optimization methods in order to maximize one or more dependent variables (i.e., DC-generated power or heat recovered) at different current levels.

In the regression models, the coefficient linking the independent and dependent variables is not one dimension, because they represent the sensitivity coefficient linking these variables when the regression models are expressed in the physical form. Therefore, the unit measures are consistent in a regression model due to the procedure applied to obtain them.

In tables and figures presented here, the regression models are written in coded forms, where the independent variables varies between the values −1 (corresponding to the lower bound of its experimental domain) and +1 (corresponding to the upper bound of its experimental domain). The models can be expressed in physical form with a simple manipulation of the equation (Fig. 10).

5.2 Response Surface Plots

Several variables were investigated during the simulation work. The attention was focused mainly on two types of dependent variables. The first class is related to two physical performance data of the SOFC CHP operation: the generator voltage and the generator exhausts' temperature. The second class involves physical data related to potentially dangerous operation of the generator: Such dependent variables should be constrained in order to avoid dangerous operation. The analysis was carried out first by applying a Yates's analysis (estimation of the significance of the independent variables), then regression models, and finally by observing the response surface plots. In this paragraph, the responses of the average generator voltage and of the exhaust temperature are shown at the three considered current levels.

Generator Voltage. In Fig. 11, the contour plots of the average generator voltage are shown for the three different current levels of 400, 500, and 600 A. The three analyzed factors have main significant effects on the average generator voltage. The air utilization factor at high level ($U_{\text{ox}} = 0.25$) has a positive main effect on the generator voltage; this is due to an increase in the generator average temperature and subsequently to a reduction in the cell overvoltages; the fuel utilization factor has a negative main effect on the voltage when it is at the upper level ($U_{\text{F}} = 0.87$), this is due to an increase in the steam partial pressure at the

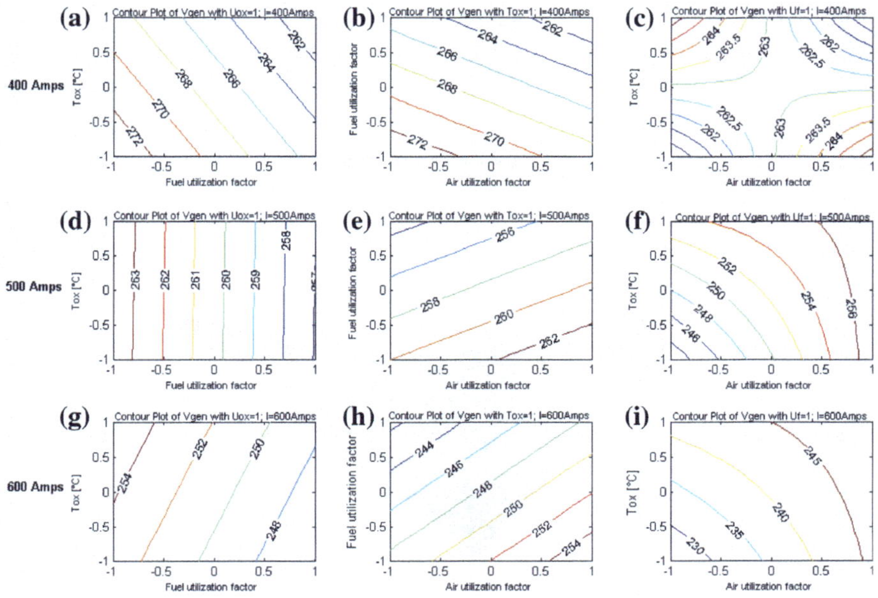

Fig. 11 Contour Plot of the average terminal generator voltage for the three analyzed current levels

anode side with a reduction in the Nernst voltage, and moreover with an increase in the activation and concentration losses; the air temperature at the inlet canister at high level ($T_{ox} = 700$ °C) leads to an increase in the generator operating temperature with a reduction in the ohmic overpotentials (positive main effect on the voltage). The contour plots shown in Fig. 11 express the interaction effect between the factors. At fixed air utilization factor $U_{ox} = 0.25$ (Fig. 11a, d, g), (1) the fuel utilization factor has a negative effect on the voltage if it is at high level for the three current levels; (2) the effect of the air temperature at the inlet canister changes for the three considered current levels, and it is related to the values of the equilibrium temperature reached in the generator and to the consequent effect on the polarization behavior: If the generator works at high temperature, the effect of the reduction on the ohmic overvoltages cannot balance the reduction of the Gibbs free energy with the temperature at low current, and in Fig. 11A, at the fixed value of U_{ox} of 0.25 (small air-cooling effect) and at low currents, the increase of the air temperature causes a reduction in the generator voltage. At the current levels of 500 and 600 Amps and fixing $U_{ox} = 0.25$ (Fig. 11D,G), the effect on the voltage of the air temperature is nearly negligible, while the fuel utilization factor shows a significant effect.

In Fig. 11b, e, h, the contour plots of the voltage are shown at fixed air temperature at the inlet canister, $T_{ox} = 700$ °C: (1) The effect of the U_{ox} changes at different current levels due to the already described behavior of the polarization with tube temperature; (2) an increase in the fuel utilization factor has a negative main effect on the voltage at every current level because of the increasing of the steam partial pressure and the reduction in the Nernst voltage.

In Figs. 12, 13, and 14, the response surface plots are shown at the fixed fuel utilization factor of $U_F = 0.87$: (1) The effect of the cell tube temperature is quite clear in Fig. 4c where the interaction of U_{ox} and T_{ox} is well represented: It is

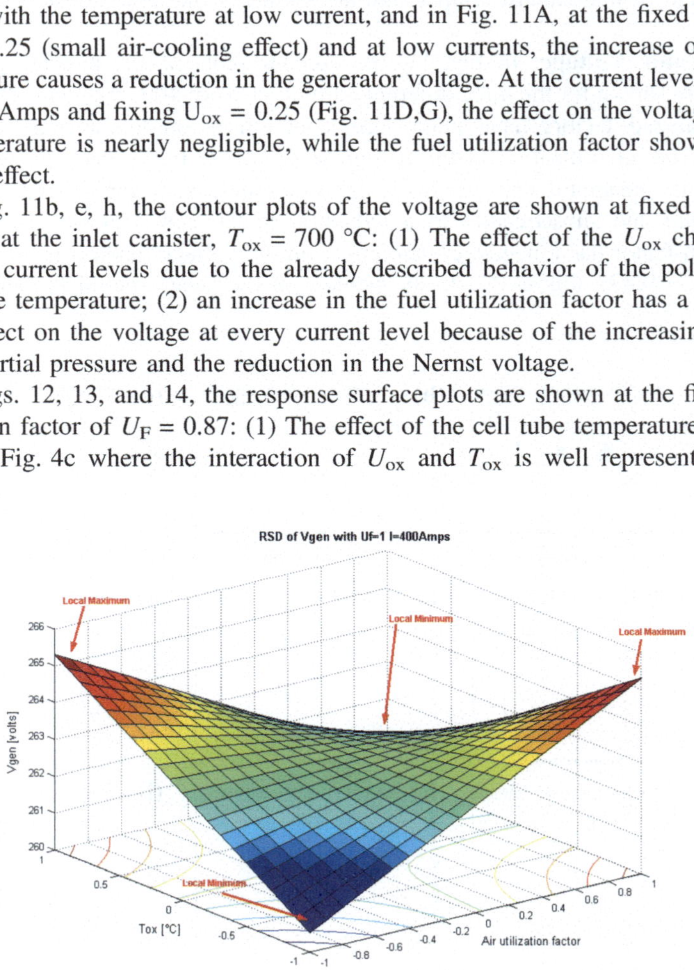

Fig. 12 Response surface at the fixed fuel utilization factor of 0.87 for the average terminal generator voltage at 400 A

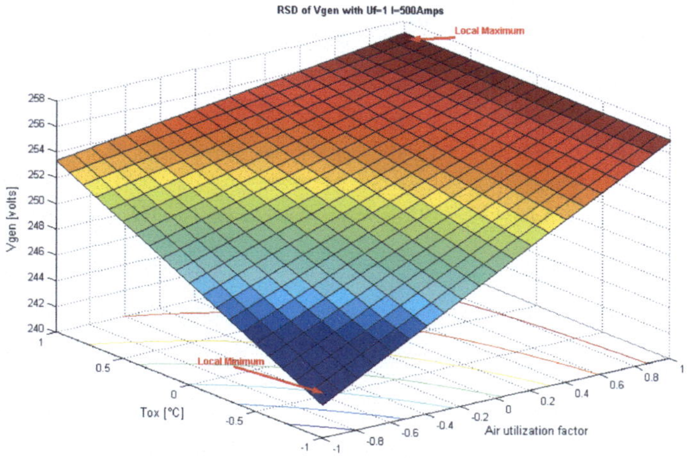

Fig. 13 Response surface at the fixed fuel utilization factor of 0.87 for the average terminal generator voltage at 500 A

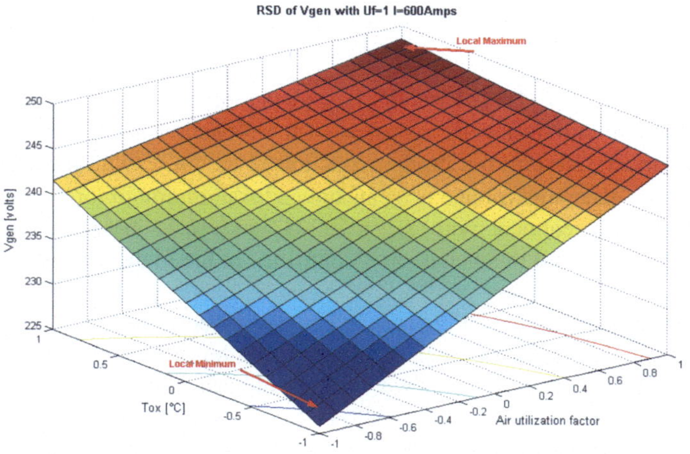

Fig. 14 Response surface at the fixed fuel utilization factor of 0.87 for the average terminal generator voltage at 600 A

interesting to notice that working at high temperatures (treatment with T_{ox} = 700 °C and U_{ox} = 0.25) causes a local minimum for the average generator voltage (Fig. 12) when operating at low currents; in Fig. 11c, it is also interesting to notice that the average voltage presents a second local minimum when operating at low temperature (treatment with T_{ox} = 500 °C and U_{ox} = 0.15): therefore, at low currents, the optimal design point should be found between these two conditions (see Fig. 12); (2) the minimum due to high temperature operations disappears when increasing the

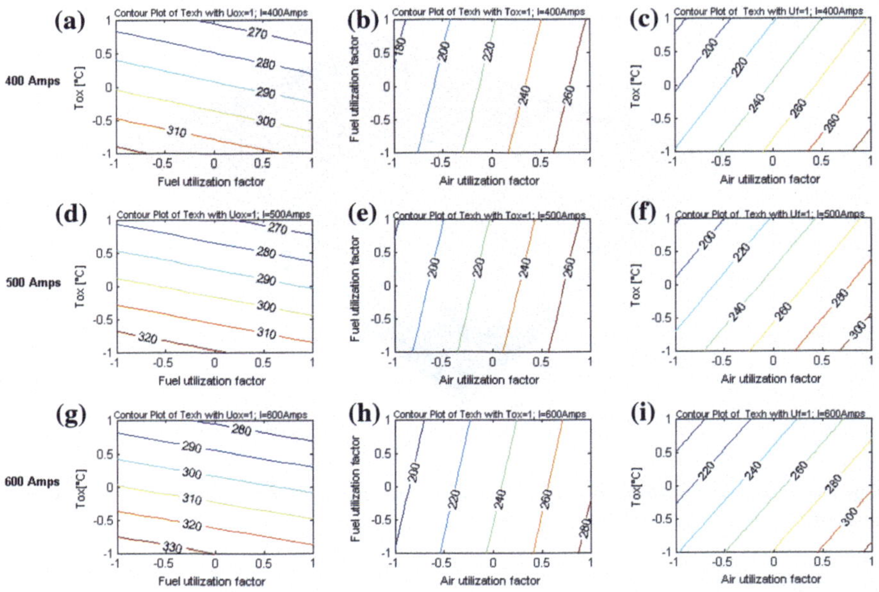

Fig. 15 Contour plot of the generator exhaust temperature for the three analyzed current levels

current demand (Figs. 11f, i, 13 and 14), and the effect of T_{ox} is positive on the voltage: At these current levels, the effect of the reduction in the ohmic overvoltages with temperature prevails on the reduction in the OCV with temperature; the response of the generator voltage presents only a local minimum for the operation at low temperature (treatment with T_{ox} = 500 °C and U_{ox} = 0.15) at the nominal condition (Fig. 13), and this behavior is confirmed at the highest current (Figs. 11i and 14).

Exhaust temperature. In Fig. 15, the contour plots of the generator exhausts' temperature are shown for the three different current levels of 400, 500, and 600 A. All the three considered factors have main significant effects: (1) The main significant factor is the air utilization, with positive effect. The air utilization at high level (U_{ox} = 0.25) means to feed the generator with less air flow; the generator is less cooled and its temperature rises up, with the effect of an increase in the exhausts' temperature; working at high air utilization factor means also a lower cooling of the exhaust flow to pre-heat the process air because of the reduced air mass flow; (2) the air temperature at the inlet canister has negative effect when the factor is at high level (T_{ox} = 700 °C); if the air process is highly pre-heated by the stack exhausts, then their exit temperature is reduced; (3) the fuel utilization factor has a negative effect. At high fuel utilization (U_F = 0.87), less heat is recovered in the post combustion zone because a smaller amount of depleted fuel can be combusted, causing a decrease in the temperature of the exhausts; nevertheless, it has to be noticed that its effect is nearly negligible (only 2 % contribution of the total variability), as shown in the contour plots.

At different current levels, the tendency is similar, and even the contribution of each factor to the total variability of the dependent variable remains quite constant (the effect of the air utilization is slightly decreased and the air temperature slightly increased).

5.3 Results and Discussion: Optimization

Factorial analysis allows to define analytical relations among the analyzed dependent variables and the investigated factors, the regression models. In Table 4, the regressions model obtained by the factorial analysis for some of the analyzed dependent variables are shown (first-order regression models with interactions), where the regression models are written in coded forms.

In the regression models, the coefficient linking the independent and dependent variables is not one dimension, because they represent the sensitivity coefficient linking these variables. Therefore, the unit measures are consistent in a regression model due to the procedure applied to obtain them. This is true if the regression models are expressed in the physical form. For example, consider Eq. (1):

Table 4 First-order regression models (coded form) at different currents

Generator current (A)	Regression models
400	$T_{Gen} = 990.70 + 41.54 \times U_{ox} + 8.81 \times U_F + 29.23 \times T_{OX}$ $+ 2.79 \times U_{ox} \times U_F + 3.04 \times U_{ox} \times T_{OX}$ $T_{CZ} = 979.1337 + 61.55667 \times U_{ox} - 11.5589 \times U_F + 48.87205 \times T_{OX}$ $T_{EXH} = 248.5264 + 43.40135 \times U_{ox} - 7.32563 \times U_F - 23.0719 \times T_{OX}$ $V_{Gen} = 267.0624 - 4.1632 \times U_F - 2.3872 \times U_{ox} \times T_{OX}$ $W_{DC} = 107.23 - 1.41 \times U_F - 0.48 \times U_{ox} \times U_F - 0.60 \times U_{ox} \times T_{OX}$ $\Phi = 40.43 + 3.33 \times U_{ox} - 2.16 \times U_F - 7.45 \times T_{OX} + 2.49 \times U_{ox} \times T_{OX}$
500	$T_{Gen} = 976.97 + 39.65 \times U_{ox} + 5.40 \times U_F + 26.06 \times T_{OX}$ $T_{CZ} = 993.84 + 62.16 \times U_{ox} - 11.51 \times U_F + 45.18 \times T_{OX}$ $T_{EXH} = 252.91 + 43.24 \times U_{ox} - 7.10 \times U_F - 24.64 \times T_{OX}$ $V_{Gen} = 255.80 + 4.52 \times U_{ox} - 3.35 \times U_F + 2.70 \times T_{OX} - 2.65 \times U_{ox} \times T_{OX}$ $W_{DC} = 128.27 + 2.04 \times U_{ox} - 1.91 \times U_F + 1.40 \times T_{OX} - 1.33 \times U_{ox} \times T_O$ $\Phi = 52.24 + 3.73 \times U_{ox} - 2.86 \times U_F - 9.91 \times T_{OX} + 3.06 \times U_{ox} \times T_{OX}$
600	$T_{Gen} = 972.97 + 41.79 \times U_{ox} + 4.61 \times U_F + 25.50 \times T_{OX} + 2.20 \times U_{ox} \times T_{OX}$ $T_{CZ} = 1007.92 + 60.33 \times U_{ox} - 10.61 \times U_F + 43.74 \times T_{OX}$ $T_{EXH} = 261.85 + 42.42 \times U_{ox} - 6.55 \times U_F - 25.66 \times T_{OX}$ $V_{Gen} = 243.94 + 6.76 \times U_{ox} - 3.50 \times U_F + 4.55 \times T_{OX} - 3.32 \times U_{ox} \times T_{OX}$ $W_{DC} = 146.79 + 4.29 \times U_{ox} - 1.53 \times U_F + 2.68 \times T_{OX} - 2.19 \times U_{ox} \times T_{OX}$ $\Phi = 67.01 + 3.16 \times U_{ox} - 2.83 \times U_F - 12.58 \times T_{OX} + 3.69 \times U_{ox} \times T_{OX}$

$$V_{\mathrm{Gen}} = 427.495 - 227.6 \cdot U_{\mathrm{ox,phy}} - 167.5 \cdot U_{\mathrm{f,phy}} - 0.079 \cdot T_{\mathrm{ox,phy}} + 0.53 \cdot U_{\mathrm{ox,phy}} \cdot T_{\mathrm{ox,phy}} \quad (1)$$

The sensitivity of the terminal voltage to the fuel utilization is estimated in −1.675 V/%U_{f}; this value corresponds to a single cell sensitivity of around −4 mV/%U_{f}. In a recent experimental session, we estimated for the generator a fuel utilization sensitivity of around −1.06 V/%FC. The difference is due to the different tested experimental domains of the fuel utilization: In the model analysis, we tested 83 % < U_{f} < 87 %, while the experimental investigation was performed in the range 81.75 % < U_{c} < 84.25 % (77.5 % < U_{f} < 80 %). In the model analysis, we have found a higher sensitivity because the operation was simulated at lower fuel excess. The study of fuel utilization tests is actually more complicated because another concept of fuel utilization should be introduced, including also eventual cell leakages or fuel bypass in the stack. A deeper investigation on this aspect will be performed experimentally, but very rich information can be found in (Gopalan 2004).

Once the regression models are defined, optimization procedures can be implemented, with different objective functions. In this work, the optimization is performed separately on the DC-generated electric power (W_{DC}) and on the exhausts' heat recovered (Φ). There is no optimization for the electric efficiency or global plant efficiency. We decided to not optimize the efficiencies because the DC efficiency is not so useful, while the useful one would be the AC efficiency; but the regression model of the AC power has no negligible error due to the model of the auxiliaries; therefore, the optimization of the efficiencies (AC electric and global) has not been performed.

The procedure is repeated for the different current levels considered. In the optimization procedure, some constraints are imposed in order to find optimal operation points which have also technical feasibility and which ensure safe operation of the generator. The constraints are applied with particular attention to the working cell tube temperature and the combustion zone temperature. Also the considered factors (independent variables) have been constrained by defining an upper bound and a lower bound (experimental domain). Three optimization procedures have been considered: They differ for the constraints applied to the average generator temperature and the combustion zone temperature. They can be resumed in Table 5. These values come from the direct operational experience on the

Table 5 Performed optimization procedures

Optimization procedures	Case 1	Case 2	Case 3
Constraints on some variables	$0.15 < U_{ox} < 0.25$ $0.83 < U_F < 0.87$ $500 < T_{ox} < 700$ °C $940 < T_{Gen} < 970$ °C $T_{CZ} < 1020$ °C	$0.15 < U_{ox} < 0.25$ $0.83 < U_F < 0.87$ $500 < T_{ox} < 700$ °C $940 < T_{Gen} < 1000$ °C $T_{CZ} < 1020$ °C	$0.15 < U_{ox} < 0.25$ $0.83 < U_F < 0.87$ $500 < T_{ox} < 700$ °C $940 < T_{Gen} < 1020$ °C $T_{CZ} < 1050$ °C

SOFC CHP 100 and from previous technical reports. The various limits mean that there are normal operating conditions (applied to assure a high number of operation hours to the generator), but also more "extreme" operating conditions (which can be performed assuring the safety operation and which create no problems on the materials, but which are not suggested for a high number of operation hours). As an example, T_{GEN} has a suggested value of 970 °C, but the generator could operate in same conditions even at 1000 °C. Moreover, the evaluation is made in this simulation activity also to analyze which would occur in case of relaxing some of the limits in the control variables such as T_{GEN} and T_{CZ}.

DC-generated electric power. The obtained results of the optimization procedure performed, imposing the maximization of the DC-generated electric power, are shown in Table 6. The first case of the optimization procedure is characterized by a maximum constrained generator temperature of 970 °C (Table 5, Case 1), while in the second case the constraint on this variable is relaxed (T_{gen} < 1000 °C, Table 5, Case 2); in the third case also, the constraint on the combustion zone temperature is relaxed and is fixed at 1050 °C (T_{CZ} < 1050 °C, Table 5, Case 3). In the Case 1, the maximum of the DC power is obtained for the value of the cell tube temperature at the constrained upper bound (970 °C), while in the Cases 2 and 3 the value of the cell tube temperature is not at its constrained upper bound when the DC electric power is maximized (Table 6). In Case 3, the cell tube temperature rises up more than in Case 2, due to the relaxation imposed to the combustion zone temperature; this condition leads to a higher generator equilibrium temperature with a consequent increase in the cell tube temperature. The post-combustion temperature $\boldsymbol{T_{CZ}}$ considered in the Case 3 is too high for a safety operating condition, and the DC-generated electric power with respect to the Case 2 is only slightly higher: For this reason, the optimal working condition for the generator concerning the DC-generated electric power $\boldsymbol{W_{DC}}$ seems to be the one of the Case 2.

Table 6 Optimization of the plant in terms of DC-generated electric power

Case 1 (W_{DC} optimized)									
Generator Current	U_{ox}	U_f	T_{ox} (°c)	Power DC (KW)	Φ (KW)	T_{EXH} (°c)	V_{gen} (V)s	T_{gen} (° c)	T_{cz} (° c)
400 A (a)	0.224	0.83	500	109.18	50.45	300.0	272.38	970.0	971.7
500 A (b)	0.231	0.83	500	130.87	65.43	311.4	260.89	970.0	998.6
600 A (c)	0.234	0.83	500	150.08	82.08	323.1	249.80	970.0	1016.0
1.2 Case 2 (W_{DC} optimized)									
400 A (a)	0.250	0.83	500	109.74	50.89	322.3	273.61	988.4	1003.4
500 A (b)	0.248	0.83	500	130.04	65.66	326.3	263.36	983.0	1020.0
600 A (c)	0.238	0.83	500	150.49	82.04	325.8	250.44	972.5	1020.0
1 Case 3 (W_{DC} optimized)									
400 A (a)	0.250	0.83	500	109.74	50.89	322.3	273.61	988.4	1003.4
500 A (b)	0.250	0.83	561.2	132.21	61.49	312.8	263.67	1001.1	1050.0
600 A (c)	0.250	0.83	534	152.29	78.89	327.8	253.40	991.9	1050.0

In Table 6 also, the heat recoverable in the W_{DC} optimized operation points are shown. With the increasing of generator current, the heat recoverable increases due to the operation irreversibilities.

Exhausts' heat recovered. The obtained results of the optimization procedure performed imposing the maximization of the exhausts' heat recovered Φ are shown in Table 7. In the case of this dependent variable, the optimization procedure allows to find an absolute maximum for the three analyzed cases. In Case 1 (Table 7), the cell tube temperature is forced at its constrained upper bound, while this does not happen in other cases. The behavior is different at low and nominal currents with respect to the high current. At high current, the heat recovered is mainly due to the irreversibilities of the operation; the maximum refers to the operating condition at lower generator temperature where the single cell overvoltages are at their higher values; in this condition, the operation is characterized by low exhausts' temperatures but high mass flows (Table 7 Case 2(c): $T_{EXH} = 291$ °C and $G_{EXH} = 1504$ kg/h with $\lambda = 5$). At the low and nominal currents, the operation is characterized by high exhaust temperature and smaller exhaust's mass flow [Table 7 Case 2(a, b)].

It is interesting to notice that in Case 1(a, b), Case 2(a, b), Case 3(a) (Table 7), the maxima found for the heat recovered are the same points found in the optimization of the DC-generated power (Table 6): It means that there are operating conditions which maximizes both DC-generated power W_{DC} and exhausts' heat recovered Φ; thus, in order to optimize the CHP performances, the SOFC plant should work at the optimal points found in the optimization procedure for the DC power labeled with Case 2 (Table 6).

This fact does not happens at the highest current. In Case 2 (the suggested case), the operation point with maximum W_{DC} [Table 6 Case 2(c)] does not correspond to the operation point with maximum Φ [Table 7 Case 2(c)], but the heat recovered

Table 7 Optimization of the plant in terms of recovered heat

Generator current	U_{ox}	U_f	T_{ox} (°c)	Power DC (KW)	Φ (KW)	T_{EXH} (°c)	V_{gen} (V)s	T_{gen} (°c)	T_{cz} (°c)
1.3 Case 1 (Φ optimized)									
400 A (a)	0.224	0.83	500	109.180	50.46	300.0	272.37	970	971.7
500 A (b)	0.230	0.83	500	130.879	65.43	311.3	260.89	970	998.6
600 A (c)	0.196	0.83	500	145.170	82.47	291.0	242.16	940	970.4
1.4 Case 2 (Φ optimized)									
400 A (a)	0.250	0.83	500	109.740	50.89	322.3	273.61	988.3	1003.4
500 A (b)	0.248	0.83	500	132.044	65.66	326.2	263.36	983.0	1020.0
600 A (c)	0.196	0.83	500	145.170	82.47	291.0	242.16	940.0	970.4
Case 3 (Φ optimized)									
400 A (a)	0.250	0.83	500	109.740	50.89	322.3	273.61	988.3	1003.4
500 A (b)	0.250	0.83	561.2	132.170	65.69	327.9	263.63	985.1	1022.3
600 A (c)	0.250	0.83	534	145.170	82.47	2917.0	242.16	940.0	970.4

Table 8 Comparison between the actual operation point and the CHP optimized operation point

CHP 100 design point[a]						CHP 100 optimized operation				
1.5 Generator current	U_{Ox}	U_f	T_{ox} (°c)	Power DC (KW)	Φ (KW)	U_{Ox}	U_f	T_{ox}(°c)	Power DC (KW)	Φ (KW)
400 A (a)	0.163	0.85	655.41	107.859	33.000	0.25	0.83	500	109.74	50.89
500 A (b)	0.195	0.85	596.51	128.546	52.512	0.248	0.83	500	132.04	65.66
600 A (c)	0.231	0.85	537.62	149.883	74.884	0.237	0.83	500	150.49	82.04

[a]Model result

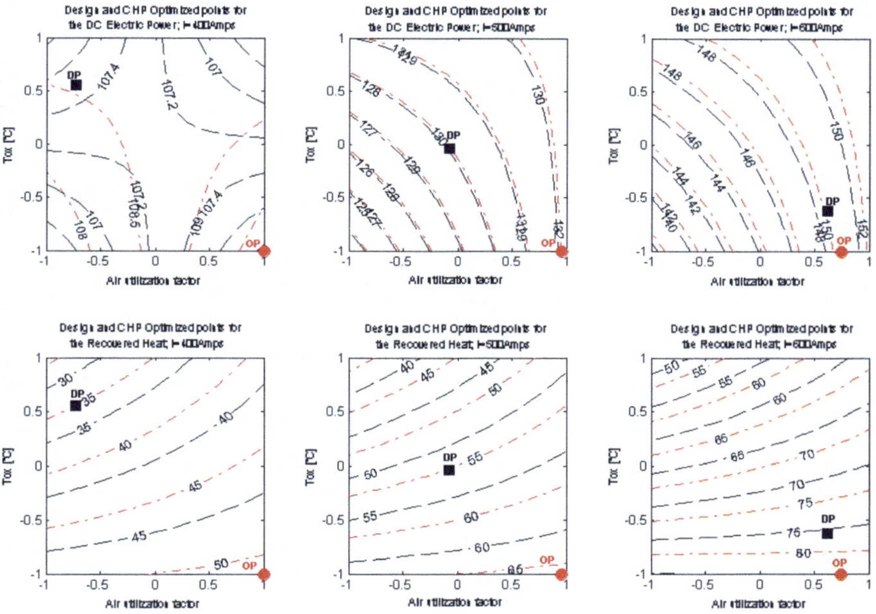

Fig. 16 Comparison between the actual operation and the optimized operation point with the respective values of the obtained DC power and recovered heat

values are quite similar, resulting in the opportunity of choosing the operation point which maximizes the DC-generated electric power [Table 6 Case 2(c)].

In Table 8, the comparison between the actual SOFC CHP 100 kW_e operation point and the CHP optimized operation point is shown.

The optimization procedure seems to suggest that the global CHP performance (sum of electric power and recovered heat) of the SOFC plant could be improved.

Finally, we can represent in a graph the actual operation points of the SOFC CHP 100 and the optimized operation (Fig. 9). The data refer to Table 8, and the contour plots are defined for two fixed values of the fuel utilization factor ($U_F = 0.85$ dashed contour plot and $U_F = 0.83$ dash-dot contour plot). In Fig. 16, this comparison between the actual operation point and the CHP optimized operation is shown in terms of DC-generated power and recovered heat, and the actual operation points are indicated with squares (DP: design point) while the optimized points with circles (OP: optimized point). The optimized operation is characterized by low air pre-heating (lower value of 500 °C) and small air excess; the results suggest that it is possible to improve the global CHP performance by modifying the thermal management of the generator and reducing the energy used to pre-heat the process air.

References

Calì, M. et al. (2005a). Comparison of the behavior of the CHP-100 SOFC field unit fed by natural gas or hydrogen through a computer experimental analysis. In *World Hydrogen Technology Convention, Singapore.*

Calì, M. et al. (2005b). The EOS project: A SOFC pilot plant in Italy—Safety aspects. In *HySafe—International Conference on Hydrogen Safety, Pisa (Italy).*

Calì, M. et al. (2005c). Benchmark characterization of a tubular SOFC CHP generator: Model description and experimental validation. In *ASME 1st European Fuel Cell Technology and Applications Conference, Rome (Italy),* December 2005.

Calì, M. et al. (2006a). Operation of the tubular SOFC CHP100 kWe field unit in Italy. General topics and operation description by means of regression models. In *World Hydrogen Energy Conference WHEC 2006, Lyon (France), June 2006.*

Calì, M. et al. (2006b). Experimental activity on the tubular SOFC CHP100 kWe Field Unit in Italy: Factor significance, effects and regression model analysis. In *Proceedings of ESDA2006, 8th Biennial ASME Conference on Engineering Systems Design and Analysis, Torino (Italy),* July 2006.

Calì, M., et al. (2007). Design of experiments for fitting regression models on the tubular SOFC CHP100 kWe: Screening test, response surface analysis and optimization. *International Journal of Hydrogen Energy, 32*, 343–358.

George, R. A. (1998). Status of tubular SOFC field unit demonstrations. *Journal of Power Sources, 73*, 251–256.

Gopalan, S., et al. (2004). Fuel sensitivity tests in tubular solid oxide fuel cells. *Journal of Power Sources, 125*, 183–188.

Kabs, H. (2001). Operational experience with Siemens-Westinghouse SOFC cogeneration systems. In *Proceedings of Lucerne Fuel Cell Forum 2001*, Lucerne (Switzerland).

Leone, P. et al. (2005) Computer experimental analysis of a tubular SOFC CHP to evaluate factors effects on performances and S/C ratio, Eco-efficiency 2005—HYSYDays. In *1st World Congress of Young Scientists on Hydrogen, 18-20 May 2005, Torino, Italy.*

Leone, P., et al. (2006). Computer experimental analysis of the CHP performance of a SOFC stack by a factorial design. *Journal of Power Sources, 156*, 400–413.

Montgomery, D. C. (2005). *Design and analysis of experiments*. INC: Wiley.

Santarelli, M. et al. (2006) Experimental activity on the tubular SOFC CHP100 generator in Italy: Regression models analysis and optimization, *Lucerne Fuel Cell Forum 2006, Lucerne (Switzerland), July 2006.*

Saroglia, S. et al. (2005) Design and development of a cogenerative system for a SOFC CHP100, eco-efficiency 2005—HYSYDays. In *1st World Congress of Young Scientists on Hydrogen, 18-20 May 2005, Torino, Italy.*

Singhal, S. C., & Kendall, K. (2004). *High temperature solid oxide fuel cells: fundamentals.* Design and Applications: Elsevier.

Torchio, M. F., et al. (2005). Experimental analysis of the CHP performance of a PEMFC stack by a 2^4 factorial design. *Jornal of Power Sources, 149*, 33–43.

Walpole, R. E., & Myers, R. H. (1993). *Probability and statistics for engineers and scientists.* New Jersey: Prentice-Hall International Inc.

Solid Oxide Fuel Cells Modeling

Domenico Ferrero, Andrea Lanzini and Massimo Santarelli

Abstract A solid oxide fuel cell (SOFC) is a complex system consisting of different components, in which interconnected physical phenomena occur simultaneously and contribute to determine the global thermo-electrochemical response of the system. The simulation and prediction of the response of an SOFC are of paramount importance for the analysis of possible applications without resorting to extensive experimental investigations. Simulating the SOFC response requires to develop reliable models that can describe the significant phenomena occurring in the system. Different approaches can be followed for the SOFC modeling, depending on the goals of the model. This chapter will provide an introduction to SOFC modeling focusing on a macroscopic, physically based approach.

Abbreviations

List of Symbols

a	Thermodynamic activity
B_p	Permeability (m^2)
C_f	Drag constant
C_p	Specific heat at constant pressure (J kg^{-1} K^{-1})
d	Molecule, particle, pore diameter (m, μm)
D	Diffusion coefficient (m^2 s^{-1}, cm^2 s^{-1})
D_T	Thermal diffusion coefficient (kg m^{-1} s^{-1})
E	Equilibrium, electrode potential (V)
E_{act}	Activation energy (J mol^{-1})
E_b	Emissive power of black body (W m^{-2})
f	Volume fraction of ionic/electronic phase in the electrode
$\boldsymbol{f}$	Body forces acting on the fluid (m s^{-2})
F	Faraday's constant (C mol^{-1})
$\boldsymbol{F}$	Volume force (N m^{-3})
F_{i-j}	View factor between i and j surface elements

D. Ferrero · A. Lanzini · M. Santarelli (✉)
Energy Department, Polytechnic of Turin, Turin, Italy
e-mail: massimo.santarelli@polito.it

M. Boaro and A.S. Aricò (eds.), *Advances in Medium and High Temperature Solid Oxide Fuel Cell Technology*, CISM International Centre for Mechanical Sciences 574,
DOI 10.1007/978-3-319-46146-5_8

$\bar{g}$	Molar Gibbs free energy variation (J mol^{-1})
$\bar{h}$	Molar enthalpy (J mol^{-1})
H_0	Incident irradiation (W m^{-2})
i	Current density (A m^{-2})
i_0	Exchange current density (A m^{-2})
i_v	Volumetric current density (A m^{-3})
i_{TPB}	Current per unit of TPB length (A m^{-1})
I	Current (A)
$\vec{j}$	Mass flux (kg m^{-2} s^{-1})
k	Thermal conductivity (W m^{-1} K^{-1})
k_B	Boltzmann constant (J K^{-1})
K_r	Equilibrium constant of r reaction
L_p	Characteristic size of the pore (m)
M_n	Molecular weight (kg mol^{-1})
n	Number of electrons involved in redox reactions
p	Pressure (Pa, bar)
P	Percolation probability
q	Rate of charge-transfer reaction (mol m^{-1} s^{-1})
$\vec{q}$	Heat flux (W m^{-2})
Q	Volumetric heat source (W m^{-3})
r	Reaction rate (mol m^{-3} s^{-1})
R	Ideal gas constant (J mol^{-1} K^{-1})
R_{con}	Contact resistance (Ω cm^2)
$\bar{s}$	Molar entropy (J mol^{-1} K^{-1})
$\dot{s}_k$	Molar rate of k species (mol cm^{-2} s^{-1})
S	Mass source term (kg m^{-3} s^{-1})
t	Time (s)
T	Temperature (K)
$\boldsymbol{u}$	Fluid velocity vector (m s^{-1})
$\bar{\boldsymbol{u}}$	Superficial velocity (m s^{-1})
V	Cell voltage (V)
V_a	Atomic diffusion volumes (cm^3 mol^{-1})
x	Mass fraction
$[X]$	Molar concentration (mol m^{-3}, mol m^{-2})
y	Molar fraction

Greek Symbols

α	Symmetry coefficient of Butler–Volmer equation
β	Symmetry coefficient of charge-transfer reaction
β_e	Extinction coefficient of the medium (m^{-1})
γ	Pre-exponential activation parameter (A cm^{-2})
γ_s	Scaling factor (1 or m^{-1})
γ_{0i}	Sticking coefficient of i-reaction

Γ	Surface site density (mol cm^{-2})
ε	Porosity
η	Overpotential (V)
θ_k	Surface coverage of k species
λ	Mean free path (m)
λ_{TPB}	Volumetric TPB density (m^{-2})
μ	Gas viscosity (Pa s)
ν	Stoichiometric coefficient
ξ	Surface emissivity
ρ	Density (kg m^{-3})
σ	Electronic, ionic conductivity (S m^{-1})
σ_B	Stefan–Boltzmann constant (W m^{-2} K^{-4})
σ_k	Coordination number of k species
$\sigma_{\alpha\beta}$	Average collision diameter (Å)
τ_g	Tortuosity
$\vec{\tau}$	Stress tensor (Pa)
ϕ	Electronic, ionic potential (V)
ϕ_v	Viscous dissipation (kg m^{-1} s^{-3})
χ	Volumetric charge density (C m^{-3})
ψ	Volumetric charge source (C s^{-1} m^{-3})
$\Omega_{\alpha\beta}$	Collision integral

List of Subscripts and Superscripts

act	Activation
adv	Advection
an	Anode
cat	Cathode
chem	Chemical
con	Contact
conc	Concentration
diff	Diffusion
eff	Effective
eq	Equilibrium
irr	Irreversible
mol	Molecular
oc	Open-circuit
ohm	Ohmic
rad	Radiative
react	Reaction
res	Resistance
rev	Reversible

List of Acronyms

BV	Butler–Volmer
DGM	Dusty gas model
SMM	Stefan–Maxwell model
SOFC	Solid oxide fuel cell
SRU	Stack repeating unit
TPB	Three phase boundary

The scope of this chapter is to provide an introduction to solid oxide fuel cell (SOFC) modeling. A SOFC is a complex system consisting of three main components (electrolyte and porous electrodes, i.e., anode and cathode), each one composed of peculiar materials in which interconnected physical phenomena occur simultaneously involving gas and solid phases.

In SOFC applications, the individual cells are stacked together to increase the generated power, and if we consider the single unit of a stack of cells, which is called SRU (i.e., stack repeating unit), other components as interconnects, seals, and gas channels must be taken into account. Finally, if we look at the entire stack of cells, gas manifolds, insulation, and current collection plates have to be considered. Therefore, the modeling SOFC cells, SRUs, and stacks are the challenging tasks due to the wide variety of the components involved.

SOFC modeling can be carried out following different approaches, techniques, and levels of details depending on the objective of the model (e.g., cell performance simulation, study of the degradation of materials, and optimization of fluid distribution) and on the particular component or group of components on which the model is focused.

From a general point of view, SOFC systems (whether they are single cells, SRUs or stacks) can be considered as nonlinear dynamic systems with multiple inputs and outputs in which mass, momentum, energy, and charge transfer take place together with chemical and catalytic reactions. The goal of modeling is to develop mathematical tools capable of simulating the response of the system, with the purpose to provide models that can be applied to the design, analysis, control, or diagnostic of SOFC systems.

SOFC models can be generally classified into two main categories: experimentally based and physically based models (Wang et al. 2011a).

Experimentally based models of SOFC systems are developed using statistical data-driven approaches without applying equations derived from the knowledge of the involved physics. Regression-based and artificial neutral network techniques are applied to experimental databases in order to identify the relationship between inputs and outputs of the system that are implemented into predictive models which are mostly applied in the design of SOFC control strategies.

In the SOFC literature, most of the models are physically based. These models range from microscale (atomic or molecular level) to macroscale due to the fact that physical processes of SOFC systems have characteristic length and time scales from

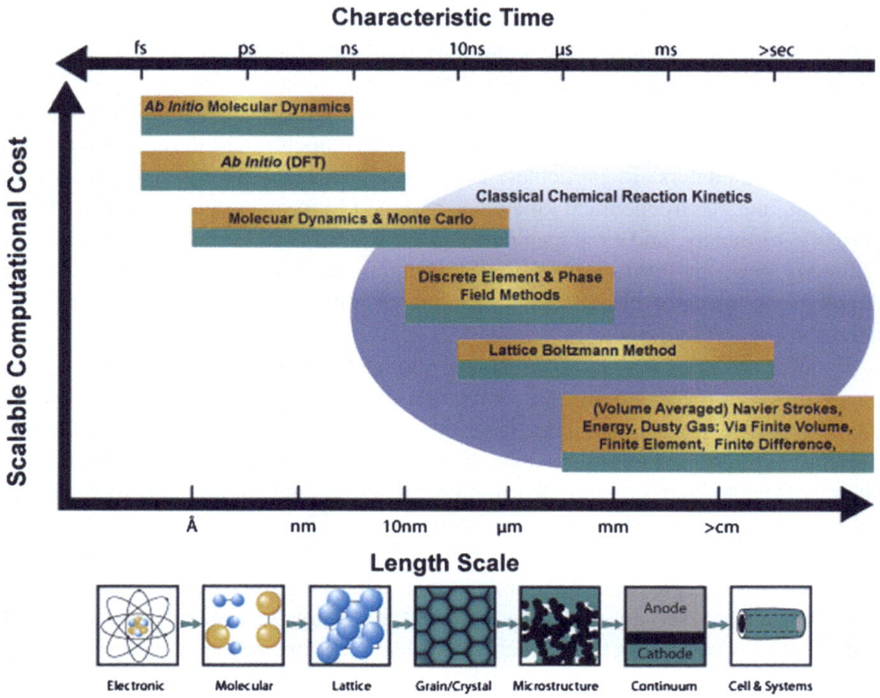

Fig. 1 Multidimensional SOFC modeling. [Reprinted with permission from Grew and Chiu (2012)]

angstrom and femtoseconds to centimeters and seconds. Consequently, the modeling techniques adopted strictly depend on the length and time scale of the described phenomena (Fig. 1).

At the scale of the electronic structure of matter, ab initio methods are used for the study of atomic interactions, followed by molecular dynamics and Monte Carlo techniques at the level of molecular structures, while discrete elements and phase field methods are adopted at the characteristic length of grain and crystals. Lattice Boltzmann Methods can be applied for the study of the fluid transport within microstructures, and finally the modeling methods that follow the continuum approach and use volume-averaged equations can be applied from the length scale of micrometers onwards to describe the physics of SOFC systems from the macroscopic point of view (Grew and Chiu 2012).

The modeling approaches presented in this chapter are physically based and focus on the macroscopic description of the phenomena.

In general, macroscopic models describe SOFC systems by using conservation laws and governing equations of the involved physics and range from 0-D to 3-D depending on the model objectives. Multidimensional models take into account the spatial distribution of the physical variables (temperature, species concentration, etc.) and are typically aimed at simulate cell/stack for design or optimization purposes. The prediction of the steady state and transient response at cell, stack, and system level for diagnostic and control is frequently addressed by 0-D and 1-D models, due to their low computational cost.

When a macroscopic approach is adopted, many physical phenomena that are implicitly derived in the atomistic and molecular modeling are described by using empirical parameters. In particular, the representation of microscopic structures, chemical, and electrochemical kinetics is assigned to macroscopic parameters (e.g., porosity, tortuosity, and exchange current density) that can be estimated directly or indirectly (i.e., by fitting) from experimental measurements. Thus, in order to develop a physically based model, a representative set of experimental data is necessary.

Finally, physically based macroscopic models of SOFCs can follow two main goals: they can be oriented to the simulation of the performance or they can study the degradation processes occurring in the materials. In the first case, the models calculate system responses mainly in terms of voltage, current, temperature, chemical species, and pressure distributions; in the second, simulations focus on the calculation of thermal stresses, strain, and stress fields.

In the following sections, the discussion will be focused on the mathematical modeling of the physical phenomena that determine the SOFC performance. Transport and conservation of mass, momentum, charge, and energy are described and the basics of electrochemical and chemical reactions modeling are given.

1 Modeling the Mass Transfer in SOFCs

In SOFC systems, mass transfer takes place both in the gas phase (i.e., stack manifolds, gas channels, and porous electrodes) and in the solid phase (i.e., transport of ions in the electrolyte). The mass transfer of gases is studied in this section, while the transport of ions in the electrolyte is addressed in Sect. 4.

Mass transfer in the gas phase occurs by advection and diffusion. In general, mass transport and conservation can be expressed by using the continuity equation in the advection–diffusion form:

$$\frac{\partial \rho}{\partial t} + \nabla \cdot \left(\vec{j}_{\text{adv}} + \vec{j}_{\text{diff}}\right) = S \tag{1}$$

where ρ is the fluid density, t is the time, $\vec{j}_{\text{adv}}$ is the advective flux of mass due to the motion of the fluid, $\vec{j}_{\text{diff}}$ is the total diffusive flux of mass related to local gradients of temperature and partial pressures, and S accounts for the volumetric mass sources or sinks.

At the high operating temperatures of SOFC systems, fluids can be considered as ideal gases with a good approximation, thus the ideal gas law can be applied for the calculation of the fluid density:

$$\rho = \frac{pM_n}{RT} \tag{2}$$

where p is the total pressure of the gas mixture, M_n is the molecular weight of the mixture, T is the temperature, and R is the ideal gas constant.

The composition of gas mixtures in SOFC systems is not spatially homogeneous due to chemical and electrochemical reactions; moreover, a gas that moves in the electrodes can occupy only the void fraction of the porous domains. Hence, the mass transport Eq. (1) has to be re-elaborated in order to formulate a species balance that takes into account the porosity of materials and the mass fractions of chemical species in the gas mixture. For each component, the mass balance can be written as:

$$\frac{\partial(\varepsilon \rho x_\alpha)}{\partial t} + \nabla \cdot (\vec{j}_{\mathrm{adv},\alpha} + \vec{j}_{\mathrm{diff},\alpha}) = S_\alpha \tag{3}$$

where ε is the porosity and x_α is the mass fraction of the α component. Equation (3) is also valid in non-porous domains, where the porosity assumes the value of 1. For non-porous domains, the advective term of Eq. (3) can be expressed as:

$$\vec{j}_{\mathrm{adv},\alpha} = \rho x_\alpha \mathbf{u} \tag{4}$$

where $\mathbf{u}$ is the fluid velocity field. In porous media, the advection term can be written as:

$$\vec{j}_{\mathrm{adv},\alpha} = \rho x_\alpha \bar{\mathbf{u}} \tag{5}$$

where $\bar{\mathbf{u}}$ is the superficial velocity of the fluid in the porous domain (also called velocity of permeation, filtration or Darcy velocity). The superficial velocity is given by the Dupuit–Forchheimer relationship: $\bar{\mathbf{u}} = \varepsilon \mathbf{U}$, being $\mathbf{U}$ the mean velocity of the fluid through the pore space of the electrode, obtained by averaging the fluid velocity over a macroscopic volume of the electrode (Nield and Bejan 2006).

The source term of Eq. (3) accounts for the net volumetric production of the α species due to the electrochemical and chemical reactions. In general, S_α is given by:

$$S_\alpha = M_\alpha \sum_i \gamma_{si} v_{\alpha i} r_i \tag{6}$$

where r_i is the molar rate of the i-reaction, $v_{\alpha i}$ is the net stoichiometric coefficient of α species in the i-reaction, and γ_{si} is a scaling factor that assumes either a unitary value if the i-reaction rate is given per unit volume or it has the unit of inverse length (i.e., surface area per unit volume) if the i-reaction has a rate expressed per unit surface.

The molar rates of electrochemical reactions are given by the Faraday's law:

$$r_i = \frac{|i_\mathrm{v}|}{nF} \tag{7}$$

where F is Faraday's constant, n is the number of electrons released during the reaction of one fuel molecule, and i_v is the volumetric current density. The calculation of the current density requires an electrochemical model that describes the charge-transfer chemistry in the SOFC electrodes, this topic will be treated in Sect. 4.

The molar rates of chemical reactions can be calculated from the study of global or detailed reaction mechanisms, as described in Sect. 5.

The electrochemical reactions involve limited regions of electrodes where charge-transfer reactions occur on the electrochemically reactive sites characterized by the coexistence of electron and ion conductor phases in the presence of gas-phase reactants. These regions are called triple- or three-phase boundaries (TPBs) and spread from the electrode/electrolyte interface of SOFCs into the electrode volume. In anode-supported cell models, the three-phase boundary is frequently assumed as a layer of negligible thickness at the anode/electrolyte interface (Costamagna et al. 2004; Ni 2009; Laurencin et al. 2011; Ferrero et al. 2015) and the mass sources due to the electrochemical reactions are imposed as boundary conditions at the border between electrode and electrolyte, instead of being included in the source term S_α.

Chemical reactions take place within the gas phase (i.e., homogeneous reactions) or on the surface of the solid medium of the electrode that acts as a catalyst for the reactions (i.e., heterogeneous reactions). In both cases, the volume in which chemical reactions occur covers the entire domain of SOFCs electrodes, and the reaction rates must be included in the volumetric source term of Eq. (3).

As stated before, the total diffusive flux of mass is due to the presence of temperature and partial pressures gradients. The thermal diffusion can be easily highlighted:

$$\vec{j}_{\mathrm{diff},\alpha} = \vec{j}_{\mathrm{d},\alpha} - D_{\alpha,\mathrm{T}}\frac{\nabla T}{T} \tag{8}$$

where $D_{\alpha,\mathrm{T}}$ is the thermal diffusion coefficient. The thermal diffusion of mass is also referred to as the Soret effect, which occurs in mixtures with high temperature gradients and large variations in molecular weight of the species. This type of diffusion is always neglected in SOFC models. The diffusive flux $\vec{j}_{\mathrm{d},\alpha}$ related to partial pressure gradients of the species will be described in detail in the following sections, where different diffusion models are introduced.

The solution of the mass balances requires to combine Eqs. (3) and (2) with a momentum balance (see Sect. 2) for the determination of the fluid velocity field, with a gas diffusion model for the calculation of the diffusive fluxes of the species, with an energy balance (see Sect. 3) for the determination of the temperature distribution, and also with electrochemical and chemical models (see Sects. 4 and 5), which are required for the calculation of the species source term [see Eq. (6)].

As will be shown later, some gas diffusion models for the porous media [i.e., Fick's model in the advection–diffusion form (22) and dusty gas model (29)] already include the momentum balance. In these models, the mass flux determined by Eq. (5)—usually expressed as a function of the total pressure gradient by the application of Darcy's law (38) for the conservation of momentum—is included in the calculation of the total diffusive flux and referred to as a viscous flux.

The modeling of the gas diffusion depends on the medium where the diffusion occurs (i.e., porous or non-porous) and on the characteristics of the gas mixture (i.e., binary or multicomponent).

Before introducing the mathematical description of the diffusive flux $\vec{j}_{\mathrm{d},\alpha}$, a brief presentation of the diffusive models is given.

In the SOFC literature, three theoretical models are usually applied to describe the diffusive mass transport: Fick model, Stefan–Maxwell model, and dusty gas model.

Diffusion models based on Fick's law assume that the flux of a chemical species in a gas mixture is proportional to its concentration gradient. These models are often presented in the advective-diffusive form, in which molecular diffusion due to concentration gradients and viscous flow due to pressure gradients are linearly combined (Webb and Pruess 2003). Fick's law is rigorously valid only for binary mixtures or in the case of diffusion of dilute species in a multicomponent mixture (Krishna and Wesselingh 1997; He et al. 2014a, b), and its application to the diffusion in porous media is consistent in a very narrow range of conditions (Bertei and Nicolella 2015). Nevertheless, Fick-based models are widely employed not only in modeling the diffusion in binary mixtures, but also in the modeling of concentrated species diffusion in multicomponent mixtures in both porous and non-porous media, due to their simplicity (Ferguson et al. 1996; Ho et al. 2008, 2009; Goldin et al. 2009; Elizalde-Blancas et al. 2013).

The Stefan–Maxwell model is frequently used in the literature to overcome the limitations of Fick's law (Krishna and Wesselingh 1997; Suwanwarangkul et al. 2003). This model is derived from the kinetic theory and correctly describes the multicomponent diffusion in non-porous domains, but it does not include the interaction between pore walls and gas molecules (Webb and Pruess 2003; Suwanwarangkul et al. 2003). In some diffusion models, the Stefan–Maxwell equations have been modified to include the gas-pore interactions, as in the binary friction model of Kerkhof (1996) or in the work of Hussain et al. (2005).

The dusty gas model (Mason and Malinauskas 1983), which takes into account both the interactions between the different components of a gas mixture and the gas–wall collisions, has proven to be the most suitable and rigorous model for the description of multicomponent diffusion in porous media (Suwanwarangkul et al. 2003; Hernández-Pacheco et al. 2004). Even if it has the higher predictive capability, the dusty gas model is not widely applied as the Fick's one due to its complexity. In particular, Fick and Stefan–Maxwell models can be solved analytically by deriving explicit expressions for the diffusion fluxes, while the dusty gas model requires a numerical solution (Suwanwarangkul et al. 2003). For this reason, the dusty gas model is frequently presented in simplified forms in the SOFC literature, usually by assuming uniform pressure in the electrodes (Jiang and Virkar 2003; Hernández-Pacheco et al. 2004; Janardhanan and Deutschmann 2006; Matsuzaki et al. 2011; Geisler et al. 2014; Ferrero et al. 2015), and in the work of Kong et al. (2012) it has been reformulated in the form of a Fickian model in order to facilitate its implementation.

Modeling diffusion in non-porous domains: Fick and Stefan–Maxwell models
In the non-porous domains (i.e., gas channels and manifolds), $\vec{j}_{\mathrm{d},\alpha}$ is the mass flux originated by the molecular diffusion of the species α in the gas mixture that can be binary (i.e., typically air in the cathode channels) or multicomponent. The molecular diffusion (also called continuum or ordinary diffusion) is due to the relative motion of the different species of the gas mixture driven by partial pressure gradients. For the diffusion modeling in non-porous media, Fick and Stefan–Maxwell models are usually applied in the literature.

The simplest diffusion model is the Fick's one. The model is given by (Bird et al. 2006):

$$\vec{j}_{\mathrm{d},\alpha} = -\rho D_{\alpha\mathrm{m}} \nabla x_\alpha \tag{9}$$

where $D_{\alpha\mathrm{m}}$ is the diffusivity of the species α in the gas mixture. For a binary mixture, $D_{\alpha\mathrm{m}}$ coincides with the ordinary diffusion coefficient of the gas phase, $D_{\alpha\beta}$, which is independent of the gas mixture composition and can be calculated using the theoretical correlation of Chapman–Enskog (Poling et al. 2001):

$$D_{\alpha\beta} = \frac{0.00266}{\sqrt{2}} \left(\frac{1}{M_\alpha} + \frac{1}{M_\beta} \right)^{1/2} \frac{T^{3/2}}{p\sigma_{\alpha\beta}^2 \Omega_{\alpha\beta}} \tag{10}$$

where M_α and M_β are molecular weights, $\sigma_{\alpha\beta}$ is the average collision diameter, $\Omega_{\alpha\beta}$ denotes the collision integral, and p is the total pressure of the mixture (bar). In the literature, the empirical correlations of Fuller et al. (1966) are frequently used for the calculation of the binary diffusion coefficients:

$$D_{\alpha\beta} = \frac{0.00143 \cdot T^{1.75}}{p\left[\frac{2}{(1/M_\alpha + 1/M_\beta)}\right]^{1/2} \left[(\Sigma V_{aA})^{1/3} + (\Sigma V_{aB})^{1/3} \right]^2} \tag{11}$$

where ΣV_{ai} are the sums of the atomic diffusion volumes, p is the total pressure of the mixture (bar), and $D_{\alpha\beta}$ is expressed in (cm^2 s^{-1}).

Many researchers have applied Fick's law to multicomponent diffusion modeling; in this case, the ordinary diffusivity of the species α in the gas mixture is usually given by the Wilke's formula (Yakabe et al. 2000; Wilke 1950a, b):

$$D_{\alpha m} = \frac{1 - y_\alpha}{\sum_{\beta \neq \alpha} \frac{y_\beta}{D_{\alpha\beta}}} \tag{12}$$

Equation (12) is strictly valid for the diffusion of gases in a stagnant multi-component mixture. When this assumption is not satisfied, as in the case of diffusion in SOFC channels, manifolds, and electrodes, the solution of the system of Eq. (9) leads to an intrinsic flux inconsistency, i.e., the sum of the diffusive fluxes is not zero (Désilets et al. 1997). In order to overcome this drawback of the model, it is necessary to replace one of the Eq. (9) with the consistency condition:

$$\sum_{\forall \alpha} \vec{j}_{\mathrm{d},\alpha} = 0 \tag{13}$$

In this way, one of the diffusion mass fluxes is "artificially" calculated so that their sum gives zero. A consistent method that combines Fick's law and flux consistency has been proposed by Ramshaw (1990):

$$\vec{j}_{\mathrm{d},\alpha} = -\rho D_{\alpha m} \nabla x_\alpha + \rho x_\alpha \sum_{\forall \beta} (D_{\beta m} \nabla x_\beta) \tag{14}$$

The Stefan–Maxwell model describes the multicomponent mass transport following a rigorous theoretical approach that allows to correctly describe the counterdiffusion effects of ternary mixtures of gases. Stefan–Maxwell equations are formulated as force balances on the chemical species of a gas mixture. The equations are written as a balance between the driving force of the motion of a species (i.e., the partial pressure gradient) and the friction between that species and each of the other species of the mixture (Krishna and Wesselingh 1997). The equations are given by:

$$\sum_{\beta \neq \alpha} \frac{y_\beta \vec{J}_{\mathrm{d},\alpha} - y_\alpha \vec{J}_{\mathrm{d},\beta}}{D_{\alpha\beta}} = -\frac{p}{RT} \nabla y_\alpha \tag{15}$$

where $\vec{J}_{\mathrm{d},\alpha}$ is the molar diffusive flux relative to concentration gradients, and the Stefan–Maxwell diffusion coefficient $D_{\alpha\beta}$ is equal to the binary diffusion coefficient used for Fick's law (11). Equation (15) can be rearranged to show the mass diffusive fluxes:

$$\sum_{\beta \neq \alpha} \frac{M_n}{M_\beta} \left(\frac{x_\beta \vec{j}_{\mathrm{d},\alpha} - x_\alpha \vec{j}_{\mathrm{d},\beta}}{D_{\alpha\beta}} \right) = -\rho \nabla x_\alpha \tag{16}$$

It is worth noting that Stefan–Maxwell expressed with Eqs. (15)–(16) does not take into account the effect of total pressure gradients, as well as the Fick's model (9).

Some authors have proposed different diffusion models based on Stefan–Maxwell equations that also include the viscous effects due to the presence of a total pressure gradient.

In the works of Andersson et al. (2010), Stefan–Maxwell equations are formulated with a different approach (Curtiss and Bird 1999) that includes the total pressure gradient in the forces balance on the gas species:

$$\vec{j}_{\mathrm{d},\alpha} = -\left(\rho x_\alpha \sum_i D_{\alpha i} \boldsymbol{d}_i\right) \tag{17}$$

where $\boldsymbol{d}_i$ is the diffusional driving force:

$$\boldsymbol{d}_i = \nabla y_i + \frac{1}{p}[(y_i - x_i)\nabla p] \tag{18}$$

In the work of Novaresio et al. (2012), Stefan–Maxwell equations are derived by using a thermodynamic approach in which partial pressure gradients are expressed as the sum of the pressure gradients due to diffusive and viscous effects. The equation obtained is given by:

$$\frac{1}{RT}\nabla p_\alpha = \sum_{\beta \neq \alpha} \frac{y_\alpha \vec{J}_{\mathrm{d},\beta} - y_\beta \vec{J}_{\mathrm{d},\alpha}}{D_{\alpha\beta}} + \frac{1}{RT} x_\alpha \nabla p \tag{19}$$

For a multicomponent mixture composed of n gases, the Stefan–Maxwell model expressed by (15), (17), or (19) is a system of n equations in the n flux unknowns. However, only $n - 1$ of the equations are linearly independent (Ramshaw 1990). Thus, the flux consistency (13) has to be imposed to close the system of equations.

Finally, it is worth noting that both the Fick and Stefan–Maxwell models presented in this section do not comprise a momentum balance; thus, in the formulation of the complete mass transport model for the fluid, the momentum conservation equations have to be added. A complete comparison of the performance between Fick's and Stefan–Maxwell models is given in the works of Krishna and Wesselingh (1997) and Suwanwarangkul et al. (2003).

Modeling diffusion in porous domains: Fick and dusty gas models The prediction of partial pressure profiles within porous electrodes is of paramount importance to allow the correct estimation of current density, electrode potential, and local reaction rates in a cell model. In order to correctly model the gas transport in SOFC electrodes, it is necessary to describe the motion of gas mixtures in porous media by taking into account the interactions of the gas species among themselves and with the walls of the pores.

The mass transport of gases in porous media is generally described by three mechanisms: viscous flow, molecular diffusion, and Knudsen diffusion. The viscous flow is related to total pressure gradients, the molecular diffusion to partial pressure gradients, and the Knudsen diffusion is produced by both type of gradients combined with molecule-pore wall collisions. The transport of adsorbed gas molecules on the solid surfaces of pores is another transport mechanism that takes place in porous structures; however, its contribution to the diffusivity is usually neglected (Froment et al. 1990; Kast and Hohenthanner 2000).

In order to identify the type of transport mechanism that is dominant in the porous electrode, the Knudsen number is usually adopted (He et al. 2014a, b):

$$Kn = \frac{\lambda}{L_{\mathrm{p}}} \tag{20}$$

where λ is mean free path of gas molecules and L_{p} is the characteristic size of the pore, typically its diameter if pores are assumed as spherical. The mean free path of a gas molecule can be directly calculated from the kinetic theory:

$$\lambda = \frac{k_{\mathrm{B}}T}{\sqrt{2}p\pi d_{\mathrm{mol}}^2} \tag{21}$$

where d_{mol} is the molecule diameter and k_{B} is the Boltzmann constant.

Depending on the value assumed by the Knudsen number, three different flow regimes can be identified: a continuum regime for Kn smaller than 0.01, a transition regime for Kn in the range of 0.01–10, and a Knudsen regime when Kn is larger than 10.

Molecular diffusion and viscous flow are the dominant mechanisms in the continuum regime; in this case, the momentum transfer occurs by collisions between molecules, which are more frequent than the surface collisions between molecules and pore walls. In the continuum regime, concentration gradients lead to mass transfer due to molecular diffusion and a total pressure gradient produces a viscous flow.

In the Knudsen regime, the molecule-pore collisions are more frequent than the intermolecular collisions, and the momentum transfer is determined by the interactions between molecules and pore walls. In this regime, a gradient of pressure or concentration leads to a mass transfer due to Knudsen flow, as there is no distinction between flow and diffusion in a non-continuum regime (Kast and Hohenthanner 2000).

In an SOFC electrode, the gas flow takes place in a transitional regime, as the mean pore diameter usually ranges between 0.4 and 2.6 μm (Funahashi et al. 2007; Greene et al. 2006; Hao et al. 2008; Jung et al. 2006; Lanzini et al. 2009; Moon et al. 2008; Park et al. 2009; Yakabe et al. 2000; Zhu and Kee 2003) and the mean free path for typical SOFC gases and operating conditions is about 0.2–0.5 μm (Hirschfelder et al. 1954); thus, in the presence of concentration and pressure gradients, all the three mechanisms must be taken into account.

For the diffusion modeling in porous media, Fick and dusty gas model are usually applied in the literature.

The Fick model given by Eq. (9) is also applied in the porous media. In order to consider the viscous diffusion, the model is frequently presented in the advective-diffusive form (Webb and Pruess 2003):

$$\vec{j}_{d+a,\alpha} = -\rho(D^{\text{eff}}_{\alpha mK}\nabla x_\alpha + \frac{B_p x_\alpha}{\mu_g}\nabla p) \tag{22}$$

where B_p is the permeability of the porous medium, μ_g is the gas viscosity, and $D^{\text{eff}}_{\alpha mK}$ is the effective diffusive coefficient which takes into account both the molecular and Knudsen diffusivity. The permeability can be expressed by thy Kozeny–Carman relationship (Bear 1972), which is based on the assumption that the porous electrode is formed by closely packed spherical particles:

$$B_p = \frac{\varepsilon^3 d_p^2}{72\tau_g(1-\varepsilon)^2} \tag{23}$$

where τ_g is the tortuosity parameter and d_p is the diameter of the particles (μm). Firstly introduced by Carman (1956), the tortuosity takes into account the complexity of the diffusion path of a fluid inside the porous media and can be defined as the ratio between the lengths of real diffusion path and straight path. There is a strong disagreement in the literature about the value of the tortuosity in fuel cell electrodes; the survey of Brus et al. (2014) shows values between 1 and 10 and indicates that the most precise estimations of tortuosity can be derived from the image analysis of real electrodes obtained by FIB–SEM methods (Lanzini et al. 2009; Wilson et al. 2011; Joos et al. 2011; Lee et al. 2013; Iwai et al. 2010; Kishimoto et al. 2011) (Fig. 2).

Using these techniques, the typical tortuosity of SOFC anodes is in the range of 1.5–4 (Brus et al. 2014).

The Knudsen diffusivity of the gas species α in a porous media is given by Lehnert et al. (2000):

$$D_{K,\alpha} = \frac{d_\varepsilon}{3}\sqrt{\frac{8RT}{\pi M_\alpha}} \tag{24}$$

where d_ε is the mean pore size of the porous media (μm).

The molecular and Knudsen diffusion coefficients can be combined in a global diffusion coefficient (Welty et al. 2001):

$$\frac{1}{D_{\alpha mK}} = \frac{1}{D_{K,\alpha}} + \frac{1-\Upsilon y_\alpha}{D_{\alpha m}} \tag{25}$$

The coefficient Y is usually assumed to be zero in the SOFC literature, leading to the Bosanquet formula (Suwanwarangkul et al. 2003; Pollard and Present 1948; Veldsink et al. 1995):

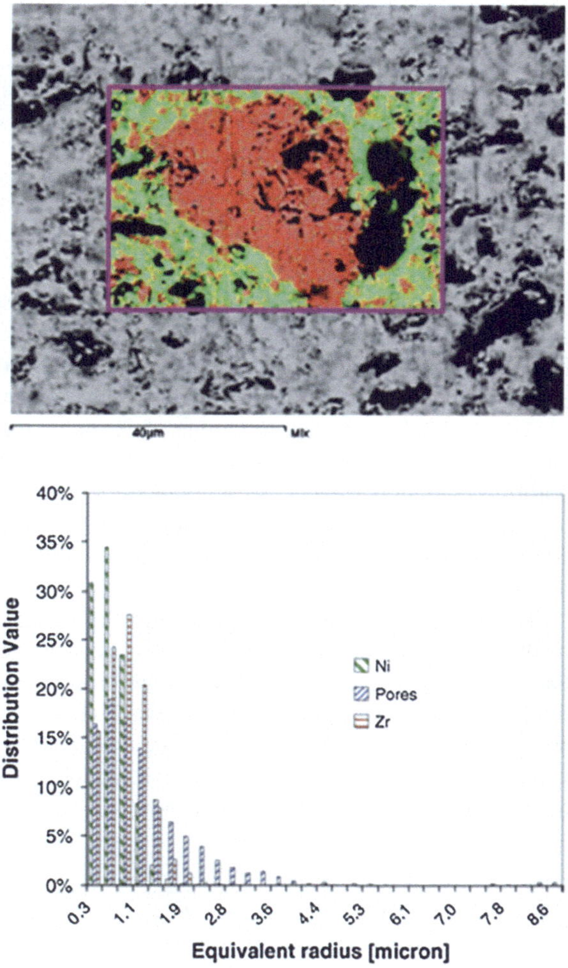

Fig. 2 Microstructural characterization of SOFC: in the *upper panel*, the EDS element mapping of Ni/YSZ anode regions (*red* Ni, *green* Zr, *black* pore); in the *lower panel*, the average phase distribution of the anode. [Reprinted with permission from Lanzini et al. (2009)]

$$D_{\alpha mK} = \left(\frac{1}{D_{K,\alpha}} + \frac{1}{D_{\alpha m}} \right)^{-1} \tag{26}$$

The Bosanquet relation is based on diffusion in aligned cylindrical pores, and it does not take into account the real geometry of the electrode structure in which gases diffuse through convoluted paths. The effective diffusivity has been studied by Bruggeman (1935) who analyzed the properties of various heterogeneous substances. If the porous electrode is assumed as composed of a solid phase made of spheres and the bed phase surrounding the spheres is treated as the void fraction of the electrode, the Bruggeman's equation for the diffusivity can be used:

$$D_{\alpha mK}^{\text{eff}} = \varepsilon^{1.5} D_{\alpha mK} \tag{27}$$

In the SOFC literature, the effective diffusivity is usually presented (Chan et al. 2001; Hajimolana et al. 2011) in a different form, which takes into account the tortuosity of the porous structure:

$$D_{\alpha mK}^{\text{eff}} = \frac{\varepsilon}{\tau_{\text{g}}} D_{\alpha mK} \tag{28}$$

In the work of Webb and Preuss (2003), the Knudsen diffusion is included in the advective-diffusive form of the Fick model through the use of the Klinkenberg factor to obtain an effective permeability, while in the first term of Eq. (22), the effective diffusivity used does not include the Knudsen effect.

As stated before, numerous studies have shown the limitations and drawbacks of the use of Fick's law to predict the diffusion fluxes for multicomponent mixtures in porous media. In particular, the work of Bertei and Nicolella (2015) has pointed out how the use of Bosanquet formulation, which is strictly valid for self-diffusion or equimolar counter transfer (Welty et al. 2001), can lead to inconsistent results in the diffusive flow calculations.

The dusty gas model is derived from the kinetic theory and treats the porous medium as one component of the gas mixture. The medium is assumed as a gaseous phase of giant molecules (the "dust") uniformly distributed in the porous domain, motionless and with infinite molar mass (Krishna and Wesselingh 1997). By applying the Stefan–Maxwell equations to this mixture, the transport of gases is described by an implicit expression that includes the effect of concentration and total pressure gradients (Mason and Malinauskas 1983):

$$-\frac{1}{RT}\nabla p_\alpha = \sum_{\beta \neq \alpha} \frac{y_\beta \vec{J}_\alpha - y_\alpha \vec{J}_\beta}{D_{\alpha\beta}^{\text{eff}}} + \frac{\vec{J}_\alpha}{D_{K\alpha}^{\text{eff}}} + \frac{1}{D_{K\alpha}^{\text{eff}}} \frac{p_\alpha}{RT} \frac{B_{\text{p}}}{\mu_{\text{g}}} \nabla p \tag{29}$$

where $D_{\alpha\beta}^{\text{eff}}$ and $D_{K\alpha}^{\text{eff}}$ are the effective multicomponent and Knudsen diffusivities, directly calculated from $D_{\alpha\beta}$ and $D_{K,\alpha}$ by using (27) or (28). The dusty gas model includes the momentum balance in the form of Darcy's Law, as can be seen from the last term of Eq. (29), which is the viscous flow, and intrinsically ensures the flux consistency. A large number of studies have shown the validity of the dusty gas model (Krishna and Wesselingh 1997; Veldsink et al. 1995; Tseronis et al. 2008; Wang et al. 2012) for multicomponent flows in porous media.

In the work of García-Camprubí et al. (2010), the total flux of a species in a multicomponent mixture obtained from the application of the dusty gas model has been expressed as the contribution of three terms:

$$\vec{J}_\alpha = -\Gamma_\alpha \nabla p_\alpha + \vec{v}_\alpha^{p} p_\alpha + \vec{v}_\alpha^{N} p_\alpha \tag{30}$$

where

$$\Gamma_{\alpha} = \frac{1}{RT\left[\sum_{\beta \neq \alpha}\left(\frac{p_{\beta}}{D_{\alpha\beta}^{\text{eff}}}\right) + \frac{1}{D_{K\alpha}^{\text{eff}}}\right]} \tag{31}$$

$$\vec{v}_{\alpha}^{p} = \frac{\Gamma_{\alpha}}{D_{K\alpha}^{\text{eff}}}\left[-\frac{B_{\text{p}}}{\mu_{\text{g}}}\nabla p\right] \tag{32}$$

$$\vec{v}_{\alpha}^{N} = \Gamma_{\alpha} RT \sum_{\beta \neq \alpha}\left[\frac{\vec{J}_{\beta}}{pD_{\alpha\beta}^{\text{eff}}}\right] \tag{33}$$

The first term represents the total diffusion of the species related to concentration gradients (i.e., a purely diffusive term), the second is the viscous flux due to pressure gradients, and the third is the flow of the species α induced by the motion of the other species. The last term, peculiar of the dusty gas model and totally neglected by Fick-based models, has a growing importance with the increasing of the current density. For this reason, the application of dusty gas model is recommended (Suwanwarangkul et al. 2003; Cayan et al. 2009) for predicting concentration overpotentials in SOFC models in the polarization regions where limiting currents occur due to the presence of high current densities and low concentration of electrochemical reactants.

The implicit formulation of fluxes and the presence of a term dependent on the pressure gradient make it difficult to solve the equations of dusty gas model. Therefore, many authors assume the viscous flow to be negligible and use the model in combination with the momentum equation given by Darcy's Law. However, the dusty gas model applied with the uniform pressure simplification no longer guarantees the flux consistency, and Eq. (13) must be applied to calculate the flux of one species. In particular, Bertei and Nicolella (2015) have shown that the uniform pressure assumption is equivalent to assume the Graham's law of effusion, which is rigorously valid only in a confined system in the absence of reactions and pressure gradients, conditions that are not satisfied in an SOFC electrode.

2 Momentum Conservation

The momentum conservation equations derive directly from the application of the second Newton's law of motion. Mathematically, the momentum conservation in the non-porous domains (i.e., gas channel and manifolds) is described by the Navier–Stokes equations for compressible fluids:

$$\frac{\partial(\rho \boldsymbol{u})}{\partial t} + \nabla \cdot (\rho \boldsymbol{u}\boldsymbol{u}) = -\nabla p + \nabla \cdot \vec{\tau} + \boldsymbol{F} \tag{34}$$

where $\boldsymbol{F}$ are the body forces (e.g., gravity and electromagnetic forces.) and $\vec{\tau}$ is the stress tensor. Equation (34) represents a force balance on the fluid particles; it states that the total force applied to the particles is the sum of three contributions: pressure, stress, and external forces. The external forces in SOFC channels are usually neglected. For a Newtonian compressible fluid, the stress tensor is given by:

$$\vec{\tau} = \left[\mu_g\left(\nabla \boldsymbol{u} + (\nabla \boldsymbol{u})^T\right) - \frac{2}{3}\mu_g(\nabla \cdot \boldsymbol{u})\right] \tag{35}$$

where μ_g is the dynamic viscosity of the fluid. This property can be estimated through the combination of the viscosities of single components by using the Wilke's formula (Wilke 1950a, b):

$$\mu = \sum_{\forall i}\left[\frac{y_i \mu_i}{\sum_{\forall j}(y_j \theta_{ij})}\right] \tag{36}$$

where θ_{ij} is given by the equation:

$$\theta_{ij} = \frac{\left[1 + \left(\frac{\mu_i}{\mu_j}\right)^{1/2}\left(\frac{M_j}{M_i}\right)^{1/2}\right]^2}{(4/\sqrt{2})\left[1 + \left(\frac{M_i}{M_j}\right)\right]^{1/2}} \tag{37}$$

The fluid flow through porous media is characterized by convoluted paths, and it is not possible to apply the classical laws of mechanics separately to fluid and solid phases, due to the complex configuration of the contact boundaries between the phases. A continuum approximation has to be applied in order to formulate the momentum balance, and the macroscopic equations are derived by using averaging methods.

Several approaches have been proposed to formulate the momentum balance through a porous media; in the simplest form, the momentum conservation is expressed by Darcy's law, which assumes a linear proportionality between the flow velocity and the applied pressure difference:

$$\bar{\boldsymbol{u}} = -\frac{B_p}{\mu_g}\nabla p \tag{38}$$

This equation describes the balance on the fluid between the force applied by the pressure gradient and the frictional resistance due to the presence of a porous medium. In the Darcy's equation, the inertia forces are neglected, and the validity of the model is limited to laminar flows in low-porosity media dominated by viscous forces. Typically, the linear relation of Darcy's equation is valid for Reynolds number of the flow in the order of unity or smaller (Nield and Bejan 2006).

An extension of the Darcy model to high velocity flows in porous media, or to high porosity media, is given by the Forchheimer's equation:

$$\nabla p = -\frac{\mu_g}{B_p}\bar{\boldsymbol{u}} - C_f \rho \frac{|\bar{\boldsymbol{u}}|\bar{\boldsymbol{u}}}{\sqrt{B_p}} \tag{39}$$

where C_f is a dimensionless drag constant. The last term of Eq. (39) is referred to as Forchheimer term and takes into account the inertia effects in the fluid flow. The C_f coefficient varies with the characteristics of the porous medium; a thorough discussion on the several different approaches adopted in the literature for the evaluation of this coefficient can be found in Nield and Bejan (2006).

The main limitation of the Darcy and Forchheimer equations is the impossibility to impose the no-slip boundary condition (Amhalhel and Furmański 1997). Hence, when Eqs. (38) or (39) is imposed in the porous medium, it is difficult to define interfacial conditions with an adjacent domain in which there is a free-flow and Navier–Stokes equations are applied, as typically happens at the electrode/channel interface of SOFC cells (Andersson et al. 2010).

The Brinkmann–Darcy flow model can be adopted to overcome the limitations of Darcy–Forchheimer equations. The Brinkmann–Darcy equations are given by Brinkman (1949a, b):

$$\nabla p = -\frac{\mu_g}{B_p}\bar{\boldsymbol{u}} + \tilde{\mu}_g \nabla^2 \bar{\boldsymbol{u}} \tag{40}$$

where $\tilde{\mu}_g$ is an effective dynamic viscosity that Brinkman set equal to the gas viscosity. More recent studies have shown that the effective viscosity is a function of the characteristic of the porous medium, in particular of the porosity (Amhalhel and Furmański 1997; Nield and Bejan 2006). The Brinkman–Darcy flow model allows to account for all boundary conditions at a solid or fluid interface.

A generalized flow model that includes the Forchheimer term into the Brinkman–Darcy equation has been derived by Hsu and Cheng (1990) starting from the Navier–Stokes equations and utilizing volume-averaging techniques:

$$\frac{1}{\varepsilon}\left[\frac{\partial(\rho\bar{\boldsymbol{u}})}{\partial t} + \nabla\cdot\left(\frac{\rho}{\varepsilon}\bar{\boldsymbol{u}}\bar{\boldsymbol{u}}\right)\right] = -\nabla p + \nabla\cdot\left(\frac{\vec{\boldsymbol{\tau}}}{\varepsilon}\right) + \mu_g\nabla^2\bar{\boldsymbol{u}} - \frac{\mu_g}{B_p}\bar{\boldsymbol{u}} - C_f\rho\frac{|\bar{\boldsymbol{u}}|\bar{\boldsymbol{u}}}{\sqrt{B_p}} + \frac{\boldsymbol{F}}{\varepsilon} \tag{41}$$

Equation (41) is known as Darcy–Brinkman–Forchheimer (DBF) flow model. In the work of Lage (1993), a complete study on the influence of each term of the DBF equation depending on the flow regime is presented.

The Darcy–Brinkman–Forchheimer equation is the most complete formulation of momentum conservation in the porous medium. When the free-flow approximation (i.e., infinite permeability and $\varepsilon = 1$) is applied in Eq. (41), the equation reduces to the Navier–Stokes form (34). Therefore, when modeling a fuel cell, it is possible to apply Eq. (41) in both free-flow and porous-medium domains. With this approach, the velocity field is continuous in the entire domain and coupling conditions between porous electrodes and free channels are not need. The Darcy–Brinkman flow model (with or without the Forchheimer term) is the standard model used by many CFD softwares (Fluent, COMSOL Multiphysics, OpenFOAM, etc.) to deal with fluid transport problems in porous media.

3 Energy Transport and Conservation

Modeling the heat transfer in SOFC systems allows the prediction of temperature distribution within cells and stacks, which is necessary for an accurate simulation of cell performance and for the prediction of thermo-mechanical degradation of cells and stack components.

Heat transport models must take into account the different heat transfer mechanisms, namely convective heat transfer between solid surfaces and gas mixtures, conductive heat transfer in gaseous and solid phases, and radiative heat transfer. Moreover, energy conservation equations must include the presence of heat sources (or sinks) due to chemical and electrochemical reactions, and the heat production due to the motion of electronic and ionic charges.

The conservation of energy can be implemented in a model by applying the first law of thermodynamics, which assumes different forms depending on the heat transfer phenomena that dominate the domain under investigation.

In the following part of the section, the equations describing heat transfer and conservation are introduced by type of domain from the non-porous fluid and solid domains to the porous domains of the electrodes.

Heat transfer in non-porous media: fluid domains The general form of the energy conservation equation for the heat transfer in a fluid domain is given by the enthalpy conservation equation, which can be expressed in terms of temperature, as follows:

$$\frac{\partial(\rho C_p T)}{\partial t} + \nabla \cdot (\rho C_p T \boldsymbol{u}) + \nabla \cdot \vec{q} = \frac{Dp}{Dt} + \phi_v + \rho \boldsymbol{f} \boldsymbol{u} + Q \tag{42}$$

where $\vec{q}$ is the heat flux by conduction, ϕ_v is the viscous dissipation, $\boldsymbol{f}$ are the specific body forces acting on the fluid (e.g., gravity), and Q represents the volumetric heat sources. The energy dissipation due to viscous forces is important for highly viscous fluids at high velocity but is negligible for gas flows under the typical laminar regimes of SOFCs. Also the pressure work may be neglected, since the pressure differences of fuel cells are very small. Moreover, it can be assumed with a good approximation that the body forces are irrelevant in the energy balance. Thus, Eq. (42) can be rewritten as follows:

$$\frac{\partial(\rho C_p T)}{\partial t} + \nabla \cdot (\rho C_p T \boldsymbol{u}) + \nabla \cdot \vec{q} = Q \tag{43}$$

The conductive heat transfer is given by the Fourier's law:

$$\vec{q} = -k_g \nabla T \tag{44}$$

where k_g is the thermal conductivity of the gas mixture. This property can be calculated by the Wilke's formula [see Eq. (36)], as in the case of the viscosity.

The volumetric heat sources in fluid domains are chemical reactions and radiative heating. The latter heating mechanism is related to the absorption, scattering, and emission of radiation by the fluid that occurs in the presence of participating gases.

The heat source term is negligible when considering the cathodic fluids of SOFCs, because they are composed of non polar molecules (i.e., oxygen and nitrogen) that do not react with each other and that can be considered as transparent gases non-interacting with thermal radiation at the conditions of SOFC applications.

Chemical reactions between the typical fuel mixture components (i.e., H_2, H_2O, CO, CO_2, and CH_4) at SOFC operating conditions occur when the gas flow comes in contact with suitable catalysts, as typically happens within the porous structure of the anode. The studies of Gupta et al. (2006) and Walters et al. (2003) have shown that homogeneous reactions cannot be ignored in non-catalytic regions of SOFCs only with particular fuel mixtures, specifically air/methane and dry natural gas. However, if we consider the typical fuel mixtures in stack manifolds and cell channels, chemical reactions between components can be neglected. Moreover, the calculated gas transmittance for a typical SOFC fuel stream composition at atmospheric pressure yields a value approaching unity (Damm and Fedorov 2005); thus, the fuel gas medium can be treated as transparent. Hence, the Q term is negligible for both anodic and cathodic gas mixtures in the non-porous domains of a SOFC system.

Once the conservation equation has been imposed in the fluid domains of the model, proper boundary conditions have to be chosen.

The conditions imposed at the boundaries of fluid domains strictly depend on the geometry and assumptions of the model. Most the models impose a fixed temperature at the fluid inlets and a convective flux at the outlets. Typical boundary conditions at the walls of fluid domains are thermal insulation, convective heat transfer, continuity of the temperature field across the boundary, or periodic boundary conditions, depending on the model. Even if heat transfer by radiation is not included into the energy conservation equation, the radiative exchange between the surfaces of channel and manifolds should be considered when defining the boundary conditions at the walls of the fluid domains. The surface-to-surface radiation is usually modeled using view-factor methods (Sánchez et al. 2007;

Damm and Fedorov 2005), and the radiative flux for an infinitesimal element of a surface is given by (García-Camprubí 2011):

$$\vec{q}_{\mathrm{rad},i} = \xi_i \left[E_{\mathrm{b},i} - \sum_j \left(F_{i-j} E_{\mathrm{b},j} \right) - H_{0,i} + \sum_j \left[\left(\frac{1}{\xi_j} - 1 \right) F_{i-j} q_{\mathrm{rad},j} \right] \right] \vec{n} \tag{45}$$

where ξ is the emissivity of the surface, E_{b} is the emissive power of a black body, H_0 is the incident radiation, and F_{i-j} is the view factor between two infinitesimal elements i and j. The view factor of Eq. (45) represents the fraction of radiation emitted by the surface element i that is directly incident on the element j. In the case of domains with high aspect ratio, typical of planar-type SOFC channels, the walls can be treated as black surfaces with unitary emissivity (i.e., $\xi_i = 1$). The radiative heat flux calculated at the surface can be then imposed as a boundary flux:

$$-\nabla \cdot (k \nabla T) = \nabla \cdot \vec{q}_{\mathrm{rad}} \tag{46}$$

The modeling study of Yakabe et al. (2001) on a planar SOFC has shown that when the surface-to-surface heat exchange is taken into account, the temperature distribution in the cell is flatter and the maximum temperature is 30 °C lower with respect to profiles obtained without considering the radiant heat exchange.

Heat transfer in non-porous media: solid domains In the solid domains of SOFC systems (i.e., electrolyte, interconnects, and other impervious stack components), conduction is the dominant heat transfer mechanism, and radiation can play a role, while the convection is negligible since the material is not moving. In the absence of convective terms, the energy conservation equation is given by:

$$\frac{\partial \left(\rho C_{\mathrm{p}} T \right)}{\partial t} + \nabla \cdot \vec{q} = Q_{\mathrm{v}} \tag{47}$$

In order to determine whether or not to include radiation in the heat transfer model of solid domains, it is necessary to evaluate the magnitude of the heat transfer by radiation and compare it with that of the conductive heat flux. A simple evaluation method is suggested by Damm and Fedorov (2005), which is based on the comparison of the maximum possible heat flux exchanged by radiation between two black walls separated by a transparent medium with the conductive flux calculated by the Fourier's law. If the magnitude of the radiation is not negligible, the heat flux exchanged by radiation should be included in Eq. (47) into the source or heat flux terms.

The radiative heat flux can be calculated by solving the radiative transfer equation (RTE) (Modest 2013). The RTE is an integro-differential equation, whose analytic solution exists only for few simple cases, and its numerical solution has a high computational cost; thus, the radiative flux is usually evaluated using approximate solutions of the RTE.

If the material is optically thin, radiation can be included in the energy conservation equation as a volumetric source that is accounted for in the Q_{v} term. In

Eq. (47), the total volumetric heat source Q_v can be expressed as the sum of the Joule heating, which occurs in electrically conductive materials traversed by ionic or electronic current, and the radiative heating. Electrochemical and chemical reactions take place in the electrode domains and should not be accounted for in the source term. The resulting total source is given by:

$$Q_v = Q_{ohm} + Q_{rad} = \frac{\left|i_{e/i}\right|^2}{\sigma_{e/i}} - \nabla \cdot q_r. \tag{48}$$

where $\sigma_{e/i}$ is the ionic/electronic conductivity of the material, $\left|i_{e/i}\right|$ is the local current density (see Sect. 4), and q_r is the radiative heat flux, which derives from approximate solutions of the RTE, for example the Schuster–Schwartzchild two-flux approximation in the case of 1-D models (Murthy and Fedorov 2003). The traditional YSZ electrolyte of SOFCs can be considered as an optically thin material (Damm and Fedorov 2004); thus, Eq. (48) is applicable in the YSZ domains. The work of Murthy and Fedorov (2003) has shown that the radiative heat flux strongly affects the temperature distribution in thick electrolytes (i.e., electrolyte supported cells), while the effect of radiation is negligible for thin electrolytes (i.e., anode-supported cells).

If the material is optically thick, the radiative heat flux can be calculated by the Rosseland diffusion approximation and included into the $\vec{q}$ term of Eq. (47) (Murthy and Fedorov 2003). The term $\vec{q}$ takes into account both the conductive and radiative heat fluxes and is given by:

$$\vec{q} = -(k_s + k_{rad})\nabla T \tag{49}$$

where k_s is thermal conductivity of the solid and k_{rad} expressed by the Rosseland approximation is given by:

$$k_r = \frac{16 n_r^2 \sigma_B T^3}{3\beta_e} \tag{50}$$

where n_r is the refractive index of the medium, σ_B is Stefan–Boltzmann constant, and β_e is the spectrally averaged mean extinction coefficient of the medium.

Heat transfer in porous media The problem of modeling the heat transfer in the porous domains of SOFCs, which are composed of mixed solid and gas phases, is usually addressed employing a local thermal equilibrium (LTE) approach that locally assumes the same temperature for gas species and solid structure. The LTE assumption is very common in thermal modeling of SOFCs and allows to use only one energy conservation equation for both the phases in the computational domain of the porous medium (Andersson et al. 2013; Haberman and Young 2004; Ferrero et al. 2015). The conservation equation is given by:

$$(\rho C_p)_{eff}\frac{\partial T}{\partial t} + \rho_g C_{pg}\boldsymbol{u}\cdot\nabla T = \nabla\cdot(k_{eff}\nabla T) + Q_v \tag{51}$$

where the terms $(\rho C_p)_{eff}$ and k_{eff} are effective transport parameters, namely the energy stored per unit volume and the effective heat conduction flux, both obtained as volume averages of the quantities defined for the gas and solid phases. The effective properties are given by:

$$(\rho C_p)_{eff} = \varepsilon\rho_g C_{pg} + (1-\varepsilon)\rho_s C_{ps} \tag{52}$$

$$k_{eff} = \varepsilon k_g + (1-\varepsilon)k_s \tag{53}$$

where the subscripts "g" and "s" stand for gas and solid.

The validity of the LTE approach has been discussed by Damm and Fedorov for hydrogen fueled SOFCs (Damm and Fedorov 2006) and subsequently by Zheng et al. (2013) who investigated the local thermal non-equilibrium (LTNE) effects in SOFCs electrodes in the presence of methane reforming and ammonia thermal cracking. Both studies indicates that LTNE effects within SOFC electrodes lead to insignificant local temperature differences between gas and solid phases of the order 10^{-2}–10^{-3} K; thus, the LTE assumption can be safely adopted in the thermal modeling of SOFCs.

The heat transfer by radiation is not included in Eq. (51), since SOFC electrodes are opaque to radiation and have a negligible radiative conductivity (Damm and Fedorov 2005). However, the radiative heat flux should be considered when defining the boundary conditions by using the same approach described for the heat transfer in fluid domains. In this case, surface-to-surface radiation is imposed on the boundary between electrode and gas channel.

The volumetric heat source term of Eq. (51) must include all the sources (or sinks) related to the phenomena that occur within the SOFC electrode in both the solid and gas phases. Heat generation is related to three different phenomena: (1) electrochemical reactions, (2) chemical reactions, and (3) Ohmic losses due to the resistance of the materials to the charge flow (i.e., Joule effect).

The heat generation due to the electrochemical reactions can be divided in reversible and irreversible; the first one account for the thermodynamic heat released by the ideal reactions and the second one takes into account the heat released for the activation of the charge-transfer reactions. These source terms are given by:

$$Q_{el,rev} = -\left(\frac{|i_v|}{nF}\right)T\Delta\bar{s} \tag{54}$$

$$Q_{el,irr} = \eta_{act}|i_v| \tag{55}$$

where $\Delta\bar{s}$ is the molar entropy change of the electrochemical reactions, the term $(|i_v|/nF)$ is the molar volumetric flow of reacted molecules steaming from the Faraday's Law, and η_{act} is the activation overpotential (see Sect. 4). As previously stated, anode-supported cells models frequently assume the electrochemical reactions to be confined at the interface between the fuel electrode and the electrolyte. If this assumption is adopted, the electrochemical heat generation should be imposed as a boundary condition at the electrode/electrolyte border instead of being included in the source term of Eq. (51).

In typical SOFC systems, chemical reactions between the gas species occur in the anode, where gas streams typically containing H_2, H_2O, CO, CO_2, and CH_4 come in contact with the metal phase of the porous structure—usually nickel—which promotes the heterogeneous chemical reactions (see Sect. 5). The chemical reactions can be endothermic or exothermic, thus the chemical source term of Eq. (51) can be positive or negative depending on the reaction. The chemical heat source is given by:

$$Q_{\text{chem}} = -\sum_i \gamma_{si} r_i \Delta\bar{h}_i \tag{56}$$

where $\Delta\bar{h}_i$ is the molar enthalpy of reaction.

The ohmic losses are due to electronic and ionic resistivities of the solid structure of the electrode. Under operating conditions, SOFC electrodes are traversed by ionic and electronic currents—which depend on electrode morphology, temperature, and reactant distributions—that can be determined by using an electrochemical model of the cell (see Sect. 4). The heat released in the electrode volume due to the ohmic losses is calculated as in Eq. (48) by using the effective conductivity of the electronic/ionic phase in the porous media, whose expression is given in Eq. (81).

Finally, ohmic heating due to contact resistance between electrodes and interconnects should also be taken into account when defining the thermal boundary conditions.

$$-\boldsymbol{n} \cdot (-k_{\text{eff}} \nabla T) = Q_{\text{ohm,res}} \tag{57}$$

This type of resistance is usually expressed in terms of an area-specific resistance (i.e., Ω cm^2) that depends on the contact method, the interface area between the materials and their resistivities (Wu et al. 2013). The heating source is given by:

$$Q_{\text{ohm,res}} = R_{\text{con}} \left|\vec{i}_{\text{b}}\right|^2 \tag{58}$$

where $\left|\vec{i}_{\text{b}}\right|$ is the absolute loss of the local current per unit surface that is crossing the boundary interface.

4 Electrochemical Modeling

The goal of an electrochemical SOFC model is to provide a mathematical formulation of the relation between the electrical variables of the fuel cell (i.e., current and voltage) and the thermo-fluidic and chemical ones (i.e., temperature and species partial pressures). In order to do this, it is necessary to implement in the model equations that give a suitable description of the electrochemical phenomena occurring in a fuel cell.

Before going into the details of the physics-based modeling, a brief introduction is given to highlight the main phenomena that an electrochemical model must address.

A SOFC is an electrochemical device that performs the direct conversion of the chemical energy of a fuel into electricity through redox reactions. The reactants (fuel and oxygen) are supplied in the gaseous form to the electrodes of the cell, where the electrochemical reactions take place, and an ion conductive layer—the electrolyte—ensures that the charged molecules produced in the redox processes can move between the electrodes. A potential difference arises between electrodes when reactants are supplied to them; if the electrodes are not electrically connected, this potential difference is exactly the electromotive force due to redox reactions and represents the maximum potential difference that the cell could achieve with these reactants. Instead, if the electrodes are connected through an external electrical circuit, the electrons move from an electrode to the other driven by the potential difference and irreversible phenomena connected to electrochemical reactions and charge transport occur and reduce the available potential difference.

A complete electrochemical model should encompass all these phenomena and give a mathematical description of the following: (1) generation of potential difference between electrodes, (2) electrochemical reactions, and (3) charge transport.

In the following sections, the description of the three aforementioned phenomena is given by presenting an overview of the approaches usually followed in the physical-based SOFC modeling.

Modeling the Equilibrium Potential The generation of a potential difference between the anode and cathode of a fuel cell depends on the redox reactions occurring at the electrodes. In general, the overall half-cell reactions consist in the oxidation of fuel molecules at the anode and the reduction of oxygen (or other oxidizing agents) at the cathode.

Typical fuels that are oxidized at the anode are hydrogen and carbon monoxide, but solid carbon, hydrogen sulfide, methane, and other higher hydrocarbons can participate directly to the electrochemical oxidation. In this chapter, the analysis will be limited to reactions involving H_2 and CO, as they are the electrochemical fuels in the vast majority of SOFC applications. The methane is frequently provided directly to the SOFC; however, it plays a role more as a reactant into the internal reforming reactions in which H_2 and CO are produced (see Sect. 5) rather than being the principal fuel in the electrochemical oxidation processes.

The half-cell oxidation reactions for H_2 and CO in SOFCs with oxygen-conductive electrolytes are given by:

$$H_2(g) + O^{2-}(el) \leftrightharpoons H_2O(g) + 2e^- \tag{59}$$

$$CO(g) + O^{2-}(el) \leftrightharpoons CO_2(g) + 2e^- \tag{60}$$

The oxygen reduction at cathode is given by:

$$O_2(g) + 4e^- \leftrightharpoons 2O^{2-}(el) \tag{61}$$

The reversible potential of the cell generated by the redox reactions—which is also called equilibrium, open-circuit, Nernst potential, or voltage—is the theoretical maximum potential difference that a fuel cell can produce between the electrodes for a given reactants composition. The reversible potential can be expressed as the difference between the equilibrium potentials of the reactions occurring at the anode and the cathode (Bagotsky 2005):

$$E_{\mathrm{rev}} = E_{\mathrm{eq},c} - E_{\mathrm{eq},a} \tag{62}$$

The equilibrium potentials are functions of the Gibbs free energy of reaction, which depends on the temperature and activities of reactants and products. The equilibrium potential of the single electrode is given by:

$$E_{\mathrm{eq}} = \frac{1}{n_e F}\left[\Delta\bar{g}_e^o + RTln\prod_k a_k^{v_k}\right] \tag{63}$$

where n_e is the number of electrons exchanged per molecule of fuel/oxidant in the electrode reaction, $\Delta\bar{g}_e^o$ is the molar standard-state free energy change of the reaction, a_k is the activity of species k, and v_k is the stoichiometric coefficient of the species (negative for reactants). In the case of gaseous reactants/products at low pressure—typical conditions of SOFC electrodes—the activity can be expressed as the ratio of the partial pressure of the gas over the standard pressure. For the coupled reactions (59)–(61) and (60)–(61), Eqs. (62) and (63) (84) result in the Nernst's formulation of the reversible potential:

$$E_{\mathrm{rev},H_2-H_2O} = \frac{RT}{4F}ln\left(\frac{p_{O_2}^{\mathrm{TPB}}}{p^o}\right) - \frac{1}{2F}\left[\Delta\bar{g}_{H_2-H_2O}^o + RTln\left(\frac{p_{H_2O}^{\mathrm{TPB}}}{p_{H_2}^{\mathrm{TPB}}}\right)\right] \tag{64}$$

$$E_{\mathrm{rev},CO-CO_2} = \frac{RT}{4F}ln\left(\frac{p_{O_2}^{\mathrm{TPB}}}{p^o}\right) - \frac{1}{2F}\left[\Delta\bar{g}_{CO-CO_2}^o + RTln\left(\frac{p_{CO_2}^{\mathrm{TPB}}}{p_{CO}^{\mathrm{TPB}}}\right)\right] \tag{65}$$

where the partial pressures are those of the species on the reacting surfaces, i.e., the TPB.

In most of the SOFCs models, the potentials given by Eqs. (64) or (65) are rearranged to show the species concentration in the bulk of the feeding gases:

$$E_{\mathrm{rev,H_2-H_2O}} = -\frac{\Delta \bar{g}^o_{\mathrm{H_2-H_2O}}}{2F} + \frac{RT}{2F} ln \left[\frac{p^{\mathrm{b}}_{\mathrm{H_2}}}{p^{\mathrm{b}}_{\mathrm{H_2O}}} \left(\frac{p^{\mathrm{b}}_{\mathrm{O_2}}}{p^o} \right)^{1/2} \right] - \eta_{\mathrm{conc},a} - \eta_{\mathrm{conc},c} \quad (66)$$

where the first two terms on the right side grouped together are referred to as reversible voltage under open-circuit conditions—usually named V_{oc}—and η_{conc} is the concentration overpotential due to the variation of the partial pressure of the species from the bulk of feeding streams to the reacting regions of the electrodes. The concentration overpotential is given by:

$$\eta_{\mathrm{conc}} = \frac{RT}{n_e F} \ln \left(\prod_k \frac{p_k^{b^{\nu_k}}}{p_k^{\mathrm{TPB}^{\nu_k}}} \right) \quad (67)$$

The difference in species concentration between the feeding flow and the reacting zone is usually attributed to the depletion of fuel/oxidizer due to the electrochemical reactions and to the mass transfer limitations in the electrodes that determine lower partial pressure of reactants in the TPB. Thus, the concentration overpotentials are frequently neglected at open-circuit, when electrochemical reactions do not occur. However, this is an oversimplification if gas streams contain components that can chemically react among themselves; in this case, the species concentrations also vary because of the chemical reactions even if the cell is at open-circuit.

A large part of SOFC models use Eq. (66) in combination with the activation and ohmic overpotentials to obtain the voltage of the cell:

$$V_{\mathrm{cell}} = V_{\mathrm{oc}} - \eta_{\mathrm{conc},a} - \eta_{\mathrm{conc},c} - \eta_{\mathrm{act},a} - \eta_{\mathrm{act},a} - \eta_{\mathrm{ohm}} \quad (68)$$

This is the standard expression of the cell polarization, usually adopted in the models to impose a constraint to the sum of the overpotentials of the cell (Fig. 3).

Given certain temperature and species distributions calculated from the thermo-fluidic models, the dependent variable of the polarization equation is the current density, which is given by the solution of Eq. (68). It is worth noting that the use of the Nernst's equation for the calculation of the reversible potential is not rigorously valid or applicable in all the models, unless proper assumptions are made. In particular, the Nernst's potential is a singular scalar value, whose calculation requires to evaluate the difference between the equilibrium potentials of anode and cathode; however, these two potentials are defined in different domains and the subtraction can be made if and only if each of them assumes a single scalar value. This is always true in 0-D models, but for higher dimension models the equilibrium potentials in general are not constant within the electrodes. Common assumptions adopted in 1-D and 2-D models impose the electrochemical reactions

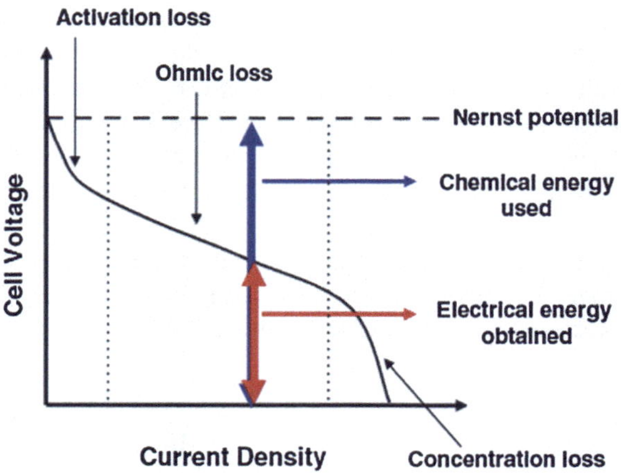

Fig. 3 Polarization curve. [Reprinted with permission from Kim et al. (2009)]

at the electrode/electrolyte interface and assume the electrodes as ideal electron conductors on which the electronic potential is constant (Ferrero et al. 2015; Ni 2009; Janardhanan and Deutschmann 2007). In this way, the TPBs are treated as lines on which the equilibrium potential varies with the position along the length of the electrode, and the Nernst potential can be calculated as the difference between the anodic and cathodic equilibrium potentials at each position of the electrode length. Another further assumption can be made by considering the equilibrium potential of the cathode as a constant, because in SOFC applications the composition of the air electrode mixture is less variable than the anodic one during the operations.

The use of Nernst's equation has no meaning in 3-D models or 2-D models where electrochemical reactions are imposed in the volume instead of being assumed at the electrolyte/electrode interface. In this case, a more general modeling approach based on the implementation of equations which contain the electronic and ionic potentials of the cell as dependent variables should be used. With this approach, it is not necessary to calculate a reversible potential to be introduced in a polarization equation, but it is sufficient to calculate the local equilibrium potential with Eq. (66) or by using the potential expression dependent on oxygen partial pressure in the gas mixture at equilibrium, as shown later in the case of multi-fuel mixtures. The current density produced in the cell can be then calculated as a function of the difference between electronic, ionic, and equilibrium potentials in the TPBs regions by implementing a Butler–Volmer equation for the charge-transfer reactions, as will be shown in the next pages, and the transport of charge is the result of the gradients of the potentials. In the general approach based on the potentials, the voltage of the cell is imposed on the surface of one of the electrodes as a boundary condition for the electrical potential, while on the surface of the other the ground potential (i.e., zero potential) can be assumed.

Some considerations have to be spent in the case of multi-fuel mixtures. In principle, when two or more different electrochemical reactants (i.e., H_2 and CO) are present at the anode at the same time, each oxidation reaction has a different value of equilibrium potential, and thus, it seems not possible to define a unique value of reversible potential. However, it is possible to demonstrate that, under the assumption of having the fuel mixture in equilibrium conditions on the TPB surfaces of the electrode, the equilibrium potential is the same for all the fuels, and consequently, it is possible to define unambiguously the reversible potential of the cell.

When the chemical equilibrium of the gaseous species within the electrode is assumed, the Nernst's voltage can be also re-written in terms of the oxygen partial pressures in the anode and cathode TPBs:

$$E_{\text{rev}} = \frac{RT}{4F} ln \left(\frac{p_{O_2}^{\text{TPB,cat}}}{p_{O_2}^{\text{TPB,an}}} \right) \tag{69}$$

where the partial pressure of oxygen at the anode is directly evaluated by calculating the equilibrium composition of the gas mixture. Expression (69) is useful to evaluate the ideal reversible potential of multi-fuel mixtures inside the anode just from the equilibrium composition of the fuel mixture.

It is worth noting that the equality between the equilibrium potentials of different reactions in multi-fuel mixtures is not valid when the gas mixture within the anode is far from equilibrium. If we consider anode-supported SOFCs with Ni/YSZ anodes, the equilibrium assumption can be considered applicable because the fuel gas has to cross a large volume of electrode before reaching the TPB, and thus, it has a sufficient contact time with the nickel catalyst that allow gas-shift and other reactions to reach the equilibrium. However, the gas equilibrium assumption has to be carefully verified case by case before applying it into a model.

If the gas mixture is not in equilibrium, it is not possible to define a single value of equilibrium potential within the electrode and both the Nernst- and potentials-based approach are not applicable. In this case, an elementary mass-action formulation based on the modeling of the rates of the single charge-transfer reactions is needed (Goodwin et al. 2009). As will be shown in the next section, with this approach, it is possible to avoid the calculation of the equilibrium potentials because the current generated by the electrochemical reactions is computed directly from the rates of the single charge-transfer reactions and the use of Butler–Volmer formalism is not required.

Electrochemical Reactions Modeling The electrochemical reactions can occur only where electron conductive, ion conductive, and gas phase coexist. The simultaneous presence of the three phases allows the conduction of electrons, the migration of ions, and the transport of gas molecules to/from the reaction sites. As stated before, these are the TPB regions of the electrodes.

The mere presence of the TPB is not sufficient to ensure the electrochemical reactions to happen, but the TPB must be connected to the rest of the structure. If either the electronic, ionic, or pore network is interrupted or badly interconnected, the electrochemical reactions cannot take place properly because electrons, ions, or gaseous reactants/products cannot reach or leave the TPB surfaces. Most of the electrode materials are predominantly electronic conductors (e.g., Ni metal and La–Sr–Mn oxides), and when these materials are used, the TPB is limited to the contact region between the electrode and the electrolyte. In particular, the commonly used anode Ni–YSZ cermets have a TPB extension that several studies have estimated in the order of 5–20 μm (Cai et al. 2011; Zhu and Kee 2008). The TPB length of SOFC cathodes can be typically higher, when mixed ionic and electronic conductors are used as cathode materials.

The reaction mechanisms that occur on the TPB consist of complex chains of intertwined physiochemical phenomena, which include adsorption/desorption of gas molecules on/from the electrode surface, dissociation, surface transport and solid-state diffusion of adsorbed species, and charge-transfer reactions. The study of electrochemical reaction mechanisms of SOFC electrodes has been addressed by countless works, and it is out of the scope of this chapter, a thorough review of the literature pertaining these mechanisms has been presented by Hanna et al. (2014) and Li et al. (2010).

The problem of modeling the electrochemical reaction mechanisms is usually addressed by following two different approaches: (1) by assuming a charge-transfer step to be the rate-determining step of the entire reaction mechanism and using Butler–Volmer expressions for the calculation of the current electrochemically generated in SOFC (Noren and Hoffman 2005) or (2) by using fundamental mass-action kinetics to describe each elementary reaction of the entire mechanism and calculating the electrochemical current from the rates of the elementary charge-transfer reactions (Goodwin et al. 2009).

In the following paragraphs, these two different modeling approaches are described.

In both approaches, it is necessary to define what the elementary steps of the reactions are, but in the first one, the total rate of the electrochemical reaction is assumed to be controlled only by the transfer of the electric charge at the TPB and not by the transport of the species. Most of the SOFC models adopt this approach and use the Butler–Volmer equation either combined with a polarization equation that couples the reversible cell potential—usually calculated by the Nernst's equation—with the voltage losses (i.e., overpotentials) related to the irreversible phenomena, namely the activation; ohmic and concentration overpotentials are expressed as a function of the electric, ionic, and equilibrium potentials. Some authors have also adopted a mixed approach where the description of elementary charge-transfer reactions through mass-action kinetics is introduced into the Butler–Volmer formulation by adopting simplifying assumptions (Zhu et al. 2005; Menon et al. 2013).

When the global electrochemical reaction is assumed to be controlled by a charge-transfer step, the Butler–Volmer equation can be used to calculate the net current density generated in the electrode by the reaction (Bagotsky 2005):

$$i = i_0 \left[\exp\left(\alpha_f \frac{n_{BV} F \eta_{act}}{RT} \right) - \exp\left(-\alpha_b \frac{n_{BV} F \eta_{act}}{RT} \right) \right] \tag{70}$$

where i_0 is the exchange current density of the electrode reaction, α_f and α_b are symmetry parameters of forward and backward reactions, η_{act} is the activation overpotential of the reaction, and n_{BV} is the number of electrons transferred in the charge-transfer step. It is worth noting that only if the electrochemistry is represented with a single, global charge-transfer process that corresponds with the half-cell reaction (Shi et al. 2011), then n_{BV} is the number of electrons transferred in the half-cell reaction.

If the Bulter–Volmer equation is describing a global charge-transfer reaction, the coefficients α_f and α_b have no constraints, while for elementary reactions—in which only one electron is transferred—these factors take on values between 0 and 1 and their sum is constrained to 1 (Goodwin et al. 2009).

The activation overpotential arises because the electric charge cannot move directly between the ionic and electronic conductive phases of a cell. Both the phases have free charge carriers and are globally neutral, however an excess charge is distributed on their surfaces. Therefore, at the interface between the phases—the TPB—an electric double layer is formed, with the charged surfaces behaving as the plates of a capacitor. During the charge-transfer reaction, the electrons are transferred across the double layer moving against the potential difference existing between the ionic and electronic phases. When the net current crossing the double layer is zero, the potential difference between the phases is equal to the equilibrium potential of the electrode, while if a potential difference higher than the equilibrium is needed to allow a non-zero net current to exist. The activation overpotential measures the disequilibrium between the potential difference in the phases and the equilibrium potential:

$$\eta_{act} = \phi_e - \phi_i - E_{eq} \tag{71}$$

where ϕ_e is the electronic potential of the electrode, ϕ_i is the ionic potential and E_{eq} is the equilibrium potential of the electrode that can be expressed by using Eq. (63).

In most of models that use a Butler–Volmer approach, the activation overpotential is related to the current density by Eq. (71) that is coupled to the polarization Eq. (68) to solve the electrochemical problem and obtain the current density distribution without explicitly introduce the ionic and electronic potentials in the equations (Chan et al. 2001; Ni et al. 2007; Ferrero et al. 2015). However, for the distributed charge-transfer modeling in 2-D and 3-D models, the electronic and ionic potentials are frequently used as dependent variables instead of the current density; in this case, the activation overpotential is expressed by using Eq. (71) and the Butler–Volmer equation is coupled with the charge transport equations to define the current density distribution within the electrodes (Zhu et al. 2005; Klein et al. 2007; Shi et al. 2007; Andersson et al. 2012).

The exchange current density provides a quantitative measure of the electro-catalytic activity of the electrode for a certain electrochemical reaction. Its value depends on the charge-transfer kinetics, temperature, partial pressures, and electrode microstructure. The dependency of i_0 on so many parameters makes it difficult to define it without the use of semi-empirical relations. In most of the SOFC literature, the exchange current density is expressed as the product of temperature-dependent terms, written in Arrhenius form, and pressure-dependent terms (Costamagna and Honegger 1998; Hosoi et al. 2015):

$$i_0 = \gamma \cdot \exp\left(-\frac{E_{\mathrm{act}}}{RT}\right) \prod_k \left(\frac{p_k}{p_{k,\mathrm{ref}}}\right)^{e_k} \tag{72}$$

where p_k is the partial pressure of the k species involved in the electrochemical reaction as reactant or product and $p_{k,\mathrm{ref}}$ is a reference pressure for the k species. E_{act} is the activation energy of the electrode reaction, which depends on reaction and materials, e_k is a dimensionless exponent and γ is a pre-exponential parameter dependent on electrode materials and microstructure, and in some cases also on the temperature (Leonide 2010). The values of γ and e_k are widely scattered in the literature.

Equation (72) is a semi-empirical relation in which the parameter γ is usually determined by fitting experimental data. Theoretical expressions of the exchange current density have been derived by Hosoi et al. (2015) and Zhu et al. (2005) from the study of reaction mechanisms. In these works, the current density is expressed as a function of rates and equilibrium constants of the elementary reactions that are assumed to compose the entire reaction mechanism. The theoretical formulation of the exchange current density allows to describe its dependency on the partial pressures in a physically based way; however, the high number of constant dependent on reactions and materials makes it necessary to use empirical data for a quantitative evaluation of i_0.

It is worth noting that the current density evaluated by Eq. (70) is expressed per unit of electrochemically active area of the electrode; thus, in order to obtain the volumetric current generated in the electrode, namely i_{v}. The current density has to be multiplied by the active electrode area per unit volume (Costamagna et al. 1998):

$$i_{\mathrm{v}} = A_{\mathrm{v}} i \tag{73}$$

The use of Butler–Volmer equation is consistent when the charge transfer is the rate-determining step of the electrochemical reaction, and its application requires the definition of an equilibrium potential within the electrode; however, if the gas mixture is not in equilibrium, it is not possible to define properly the equilibrium potential and the Butler–Volmer is not applicable. In this case, the current density generated by the electrochemical reaction can be calculated by using an elementary mass-action formulation of the rates of the elementary charge-transfer reactions (Goodwin et al. 2009).

In general, a charge-transfer reaction can be written as:

$$A^{a} + B^{b} \leftrightharpoons C^{c} + D^{d} + \upsilon_{e}e^{-}. \tag{74}$$

where *A*, *B*, *C*, and *D* are the species having charge *a*, *b*, *c*, and *d* that are involved in the transfer of υ_e electrons. For an elementary charge-transfer reaction, the coefficient υ_e assumes the value of +1 for the forward (i.e., anodic) reaction—which "produces" electrons—and −1 for the backward (i.e., cathodic) reaction. The net rate of the reaction is given by the difference in the forward and backward rates of the charge-transfer reaction, which can be written as follows:

$$q_{i,\mathrm{f}} = k_{\mathrm{f}}(T)\prod_{i,r} a_i \exp\left(\frac{\beta_{\mathrm{f}}\mathrm{F}E_{\mathrm{el}}}{RT}\right) \tag{75}$$

$$q_{i,\mathrm{b}} = k_{\mathrm{b}}(T)\prod_{i,p} a_i \exp\left(-\frac{\beta_{\mathrm{b}}\mathrm{F}E_{\mathrm{el}}}{RT}\right) \tag{76}$$

where the rate constants *k* can be expressed in the Arrhenius form, a_i are the activities either of the reactants in the case of the forward reaction or of the products for the backward one. The coefficient β_{f} and β_{b} are the symmetry coefficients, which range between 0 and 1 and are constrained to have sum equal to one. E_{el} is the electrode potential, which is given by:

$$E_{\mathrm{el}} = \phi_e - \phi_i \tag{77}$$

The current generated for unit of TPB length is given by:

$$i_{\mathrm{TPB}} = F\sum_{i}\left(q_{i,\mathrm{f}} - q_{i,\mathrm{b}}\right) \tag{78}$$

where the summation includes all the charge-transfer reactions. The volumetric current density is related to i_{TPB} by the volume-specific TPB length (Janardhanan et al. 2008):

$$i_{\mathrm{v}} = \lambda_{\mathrm{TPB}} i_{\mathrm{TPB}} \tag{79}$$

With this approach, it is possible to calculate the current generated by the electrochemical reactions by avoiding the Butler–Volmer formulation.

Charge transport and conservation In an SOFC stack, the charge transport takes place in the solid phases—ionic and electronic—of cells and components.

Both the ionic and the electronic conductive materials exert a resistance to the charge flow and the movement of charges is driven by the potential difference existing between the electrodes of the cells. The charge flux is referred to as the current density, which is given by the Ohm's Law:

$$i_{i/e} = -\sigma_{i/e} \nabla \phi_{i/e} \tag{80}$$

where $\nabla \phi_{i/e}$ is the gradient of the ionic/electronic potential that drives the charge flow and $\sigma_{i/e}$ is the conductivity of the material. In the case of the impervious solids (i.e., electrolyte, interconnects, and current collectors), the conductivity is that of the pure material, while in the porous electrodes, an effective conductivity must be calculated in order to take into account the presence of pores and electron-conductive phase. The effective conductivity can be calculated by using a statistical approach in which the porous electrode is assumed as system of packed spherical particles (Nam and Jeon 2006). The effective conductivity is given by:

$$\sigma_{i/e}^{\text{eff}} = \sigma_{i/e} \left[(1 - \varepsilon) f_{i/e} P_{i/e} \right]^k \tag{81}$$

where $f_{i/e}$ is the volume fraction of the ionic/electronic phase in the electrode and $P_{i/e}$ is the percolation probability. The exponent k, generally larger than 1, depends on the distribution of the conductive phase in the electrode.

The conservation equation must be applied in the model to enforce the conservation of charge:

$$\frac{\delta \chi_{e/i}}{\delta t} + \nabla \cdot i_{e/i} = \psi_{i/e} \tag{82}$$

where $\chi_{e/i}$ is the volumetric charge density and $\psi_{i/e}$ is the volumetric charge source. The term $\psi_{i/e}$ is different from zero only when the charge-transfer reactions are assumed to take place in the volume of the electrode. In this case, the charge is transferred from the ionic to the electronic phase of the electrodes in the TPB volume of the electrodes; the variation of ionic and electronic currents is given by:

$$\psi_{i/e} = \pm i_{\text{v}} \tag{83}$$

where i_{v} is the current that is transferred at the TPB, which is calculated from Eqs. (73) or (79).

When the electrochemical reactions are imposed at the interface between electrode and electrolyte, the volumetric charge source is null, and the continuity between the ionic and electronic current is imposed on the boundary.

In SOFC models, the charge transport in the electronic conductive materials is usually neglected by assuming them as ideal conductors (i.e., infinite conductivity) and only the transport of ions in the electrolyte is taken into account. Also, the ions transport is frequently assumed to be one-dimensional and normal to the electrolyte/electrode interfaces. This assumption is valid for thin electrolytes and allows to express the potential drop due to ion transport as the difference between the potentials at electrode/electrolyte interfaces:

$$\eta_{\text{ohm}} = \phi_{i,\text{an}} - \phi_{i,\text{cat}} \tag{84}$$

Equation (71) involves the electric/ionic potentials as dependent variables, which are put in relation by Butler–Volmer (70) or elementary rate Eqs. (75) and (76) imposed on the TPB boundary or volume. On the external surface of the electrodes or on the current-collecting plates in the case of a stack, where only the electrical potential is defined because the ionic phase is not present, a boundary condition is needed in order to solve the charge-conservation equation. The cell (or stack) voltage is usually imposed on one of the electrode (or current collector) surfaces, while on the other the ground (i.e., zero) potential is fixed.

Another possible approach is to impose the ground potential on one of the electric boundaries and fix the total current value on the other by imposing a constraint to the integral of the current density:

$$\int_{\partial\Omega} i \cdot \vec{n} \mathrm{d}S = I_{\text{tot}} \tag{85}$$

With this approach, the input of the electrical model is the current, while the voltage of the SOFC cell/stack can be calculated from the difference between the potentials on the two electric boundaries.

5 Modeling the Heterogeneous Chemistry in SOFCs

Chemical reactions can occur within the fuel stream at the typical operating conditions of SOFCs. In particular, when fuels other than hydrogen are fed to the cells, the operating temperatures are sufficient to promote both homogeneous and heterogeneous reactions between the fuel components.

If we consider the typical SOFC mixtures—which contain H_2, H_2O, CO, CO_2 and CH_4—and operating conditions, the reactions that occur within the gas phase are very slow when compared to the heterogeneous ones; thus, the homogeneous chemistry can be safely neglected in the anode domains (Zhu et al. 2005). In particular, when the fuel mixture comes in contact with the porous structure of the anode, typically made by a Ni/YSZ cermet, the heterogeneous reactions are promoted by the presence of the nickel that acts as a catalyst for the reactions, such as methane reforming and gas shifting. However, homogeneous reactions can play a non-negligible role in non-catalytic SOFC regions when dry natural gas (Walters et al. 2003) or higher hydrocarbons are fed to the cells—especially at very high temperatures ($T > 800$ °C)—or when partial oxidation conditions are reached (Gupta et al. 2006) due to the presence of oxygen or air into the fuel stream.

This chapter will focus on the modeling of the heterogeneous reactions in SOFC anodes. Modeling the reactions means to find a suitable mathematical description of the physics that allows to calculate the rates of the reactions. In particular, rates are necessary to interface the chemical model with the thermal and fluidic ones through the source terms of Eqs. (6) and (56).

The problem of describing the heterogeneous reaction rates has been addressed in the SOFC literature in two different ways: by using global expressions for the calculation of an overall reaction rate or through detailed kinetic models that include intermediate reaction steps. Both the approaches are based on the mean-field approximation, which describes the surface state with average quantities and neglects the non-uniformity of the catalytic surfaces.

Global reaction mechanism Two different approaches can be adopted when modeling the chemistry by using global rate expressions: one is based on the assumption that the reaction is controlled by kinetics and the other assumes that the reaction rate is limited by the equilibrium.

The first approach is based on modeling each reaction in a single step whose rate can be generally expressed by a kinetic power law expression:

$$r_{\text{react}} = \gamma_r \prod_i p_i^m \exp\left(-\frac{E_a}{RT}\right) \tag{86}$$

where γ_r and reaction orders m of the i species participating to the reaction are derived from the fitting of experimental data. Alternatively, to power law models, Langmuir–Hinshelwood type models are used to describe the kinetics.

With the second approach, the reaction velocity is expressed through an equilibrium-limited rate expression defined by:

$$r_{\text{react}} = r_{\text{f}}\left(1 - \frac{\prod p^{v}_{\text{products}}}{\prod p^{v}_{\text{reactants}}}\frac{1}{K_{\text{r}}}\right) \tag{87}$$

where v is the stoichiometric coefficient of the gaseous species, K_{r} is the equilibrium constant of the reaction, and r_{f} is the rate of the forward reaction, usually given by a power law expression. It is worth noting that the rate expressed by Eq. (87) goes to zero when the equilibrium composition is reached.

The applicability of the equilibrium rather than the kinetic approach strictly depends on the reaction and on the complexity of the model. In general, if a lumped 0-D model is used for the cell, the equilibrium of the reactions can be safely assumed, while if a multidimensional model is adopted, then it is more appropriate to apply a kinetic description of the reaction for the calculation of local rates.

The approaches described by Eqs. (86) and (87) are valid for any reaction; however, their application requires to fit the kinetic expressions to experimental data measured under reaction conditions that are relevant for the model. Experimental data are available in the literature for the most common reactions and cell materials; however if the model has to include particular reactions, conditions, or materials, it is necessary to build specific experimental data sets in order to characterize the reactions and describe the catalytic activity of materials with respect to the investigated reactions.

The most common reactions included in SOFC models are the heterogeneous methane steam reforming (STR) and water–gas shift (WGS) reactions within Ni/YSZ anodes:

$$CH_4 + H_2O \leftrightharpoons CO + 3H_2 \tag{88}$$

$$CO + H_2O \leftrightharpoons CO_2 + H_2 \tag{89}$$

When these reactions are modeled using a global-mechanism approximation, the kinetic approach is usually adopted for the steam reforming reaction (88), while the water–gas shift (89) is frequently described under the equilibrium assumption. The kinetic expressions used for the steam reforming reactions are commonly derived from experimental studies either over commercial nickel-based catalysts (Xu and Froment 1989; Hou and Hughes 2001) or directly on Ni/YSZ anodes (Drescher et al. 1998; Achenbach and Riensche 1994; Ahmed and Foger 2000; Lee et al. 1990; Belyaev et al. 1995; Dicks et al. 2000). In the study of Nagel et al. (2008) different steam reforming models given by power law; Langmuir–Hinshelwood and equilibrium expressions have been compared showing the effect of the STR kinetics on the temperature distribution in the cell.

In the work of Sanchez et al. (2008), the equilibrium and kinetics approaches have been compared for both reactions (88) and (89). The work highlights that the rates of reactions are controlled either by kinetics or equilibrium depending on the local conditions of the cell, thus the choice between one approach and the other is strictly connected to the peculiarities of the modeled system and cannot be assumed a priori.

Besides the STR and WGS, other reactions frequently included in SOFC models by using global kinetics expressions are as follows: dry reforming ($CH_4 + CO_2 \leftrightharpoons 2CO + 2H_2$), methanation ($CH_4 + 2H_2O \leftrightharpoons CO_2 + 4H_2$), Boudouard ($2CO \leftrightharpoons CO_2 + C$), and methane cracking ($CH_4 \leftrightharpoons C + 2H_2$) (Wang et al. 2011b; Ni 2013).

It is worth noting that the equations introduced in this section for the calculation of an overall reaction rate do not include neither the concentrations of intermediate surface species on the anodic structure nor explicit information on the microstructure of the electrode, even if they are describing heterogeneous kinetics.

Detailed surface reaction kinetics Alternatively to the global rate expressions, the problem of modeling the heterogeneous chemistry can be addressed by using a mass-action formulation of the kinetics of the elementary reaction steps.

The principle is analogous to that showed when modeling the charge transfer with the mass-action formula instead of approximating the reaction with a global mechanism; a multi-step mechanism is developed and a rate is calculated for each step of the reaction.

The total molar rate of the *i*-th reaction step is given by the difference between forward and backward rates of reaction:

$$r_{\text{react},i} = k_{\text{f}i} \prod_{k=1}^{K} [X_k]^{v'_{ki}} - k_{\text{r}i} \prod_{k=1}^{K} [X_k]^{v''_{ki}} \tag{90}$$

where $k_{\text{f}i}$ and $k_{\text{r}i}$ are the rate constants of the reaction, K is the total number of species—gaseous and adsorbed on the surface—involved in the reaction step, $[X_k]$ is the concentration of the k species and the exponents v'_{ki} and v''_{ki} are the stoichiometric coefficients of reactants and products. The concentration of the k species is expressed either as molar volumetric (mol/m^3) for the gas-phase species or as molar superficial (mol/m^2) for the surface species.

If Eq. (90) is applied to all the reaction steps involving the k species, the resulting net molar rate is given by:

$$\dot{s}_k = \sum_{i=1}^{K_\text{r}} \left[\left(v''_{ki} - v'_{ki} \right) k_i \prod_{k=1}^{K_\text{g}+K_\text{s}} [X]_k^{v'_{ki}} \right] \tag{91}$$

In Eq. (91), K_r is the total number of reaction steps, K_g is the number of gas-phase species in the i-th step, and K_s is that of the surface species. The surface molar concentration can be expressed as a function of the surface coverage of the species:

$$[X_k] = \frac{\theta_k \Gamma}{\sigma_k} \tag{92}$$

where θ_k is the surface coverage of the k species, σ_k is the coordination number (i.e., the number of source sites that are occupied by species k), and Γ is the total surface site density.

The rate constants are expressed in Arrhenius form and can be also dependent on the surface coverage of adsorbed species:

$$k_i = A_i T^{\beta_i} exp\left(-\frac{E_{ai}}{RT} \right) \prod_{k=1}^{K_\text{s}} \theta_k^{\mu_{ki}} exp\left(-\frac{\in_{ki}\, \theta_k}{RT} \right) \tag{93}$$

where μ_{ki} and $\in_{ki}$ are parameters for modeling the coverage dependence. When the elementary step is an adsorption reaction of a gas-phase species on the catalyst surface, the rate constant is given by:

$$k_i = \frac{\gamma_{0i}}{(\Gamma)^m} \sqrt{\frac{RT}{2\pi M_\text{g}}} \tag{94}$$

where ς_i is the sticking coefficient of the reaction (i.e., a measure of the probability that the adsorption reaction takes place when the molecule collide with the surface, its value lies between 0 and 1), m is the sum of the stoichiometric coefficients of the reactants, and M_g is the molecular weight of the gas-phase species adsorbed.

When Eq. (91) is solved for all the k species in combination with the fluidic, thermal, and electrochemical equations, the surface coverages are included in the dependent variables and must be determined as a part of the solutions.

The surface coverage of a species depends on the position, because the local temperature and gas species concentrations vary within the electrode. However, the mean-field approximation ensures that the surface species do not interact laterally; thus, the surface coverage in a point of the surface is not influenced by the coverages in the neighboring positions of the computational domains and the time-dependent variation of θ_k can be written as follows:

$$\frac{\partial \theta_k}{\partial t} = \frac{\sigma_k \dot{s}_k}{\Gamma} \tag{95}$$

Equation (95) has to be imposed for all the surface coverages, and the solution of the resulting system of differential equations provides the values of θ_k. However, the times scales of the surface reactions are several order magnitudes lower than those of the variation of temperature and gas-species concentrations. Therefore, the steady state approximation can be applied and the system of Eq. (95) reduces to a set of algebraic equations in which the net molar rates of the surface species are imposed to be equal to zero.

The approach based on elementary reaction mechanisms has a broader validity with respect global mechanism approaches since it can include all the possible chemical reactions occurring within the porous anode. Moreover, the mass-action formulation has a general validity can be applied also to homogeneous chemistry.

A multi-step reaction mechanism for the internal reforming of CH_4–CO–CO_2–H_2–H_2O–O_2 mixtures has been developed and validated over Ni/YSZ cermets by Hecht et al. (2005). The mechanism, which consists of 42-reaction steps that involve 6 gas-phase species and 12 surface species, has been recently applied in the modeling and validation of the heterogeneous chemistry in tubular SOFCs fed by biogas (Santarelli et al. 2013), and its comparison with a global kinetic approach is reported by Hoffman et al. (2009). The multi-step mechanism predicts slower methane conversion with respect to the global kinetic approximation (Hoffman et al. 2009) and shows that thermodynamic equilibrium conditions are not fully achieved inside the anode of a tubular fuel cell, consistently with the experimental observations (Santarelli et al. 2013).

6 SOFC Modeling: Examples of a Multidimensional Approach

The modeling approach presented in this chapter provides a mathematical description of the physical phenomena occurring in SOFCs from a macroscopic point of view.

In the literature, macroscopic SOFC models have been developed from 0-D to 3-D depending on the model objectives. The following sections provide some representative examples of the multidimensional approaches followed in the physically based SOFC modeling.

0-D models Zero-dimensional models are box models that allow the calculation of scalar variables, returning results independent from a spatial description of the physics. These models are typically used to simulate the polarization of an SOFC and analyze the cell performance with the variation of operating conditions (e.g., mean cell temperature and pressure) or geometrical parameters (e.g., electrode thickness).

An example is the zero-dimensional SOFC electrochemical model presented by Chan et al. (2001), which describes the polarization characteristic of a cell operating with H_2/H_2O mixtures by calculating the cell voltage using Eq. (68). The model takes into account the diffusion in porous electrodes by integrating diffusive Fick's flows along the thickness of the electrodes and describes the activation overpotential by the Butler–Volmer Eq. (70). Sensitivity analyses are performed with the model to investigate the effect of electrodes' thickness on polarization curve and cell overpotentials (Fig. 4).

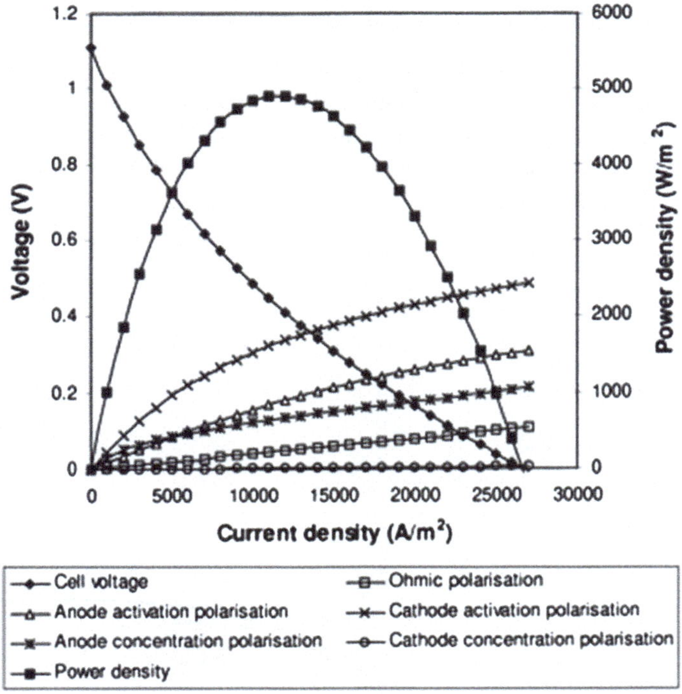

Fig. 4 Cell voltage, polarization losses, and power density for an anode-supported SOFC (anode thickness 750 μm) operating at 800 °C. [Reprinted with permission from Chan et al. (2001)]

1-D models One-dimensional models give a spatial-dependent description of the phenomena. Typically, 1-D models are used to describe the evolution of thermo-electrochemical variables within the cell by considering one dominant geometrical dimension—usually the gas flow direction—and simplifying the equations in the other directions.

A one-dimensional model for the simulation of biogas reforming in a tubular, anode-supported SOFC with Ni/YSZ anode has been presented by Santarelli et al. (2013). The model divides the SOFC in series of elements along the axial direction of the cell and each of them is solved unidimensionally along the radial coordinate by assuming that diffusive transport is dominant over convection. The reforming process is modeled following an elementary kinetics approach by including in the model the multi-step reaction mechanisms developed by Hecht et al. (2005). The anode volume is approximated by a surface located at the channel/electrolyte interface, and the model resolves the species conservation Eq. (3) with chemical source terms calculated from the kinetic model and electrochemical sources evaluated by imposing the current density on the anode boundary and evaluating the species molar fluxes by Faraday's law (7). The model evaluates the evolution of mole fractions of chemical species along the fuel channel for a given temperature and current (Fig. 5). Model results are close to gas compositions obtained from experimental measurements and indicate that thermodynamic equilibrium is not completely achieved within the anode channel, as at the outlet composition is different from that predicted by chemical equilibrium.

2-D models Two-dimensional models are generally used for the simulation of cells and SRUs. Cell models of tubular SOFCs can be implemented by taking advantage of the axis-symmetric geometry, while planar SOFC models require the selection of a 2-D section representative of the entire cell/SRU, which is typically a cross section parallel to gas flow.

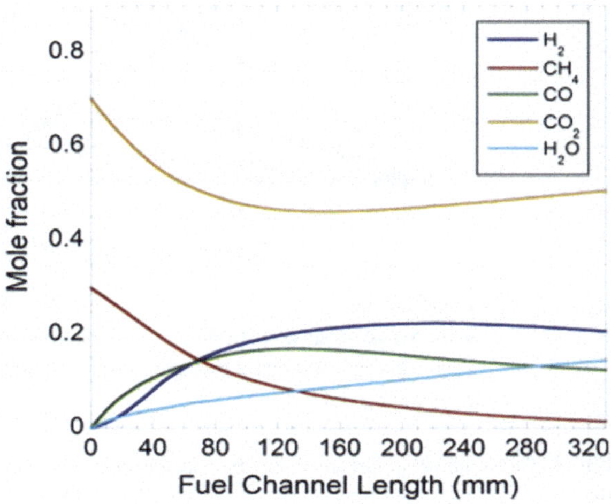

Fig. 5 Simulated mole-fraction profiles of chemical species along the fuel channel with CO_2/biogas ratio of 1, at 800 °C and 50 % of fuel utilization. [Reprinted with permission from Santarelli et al. (2013)]

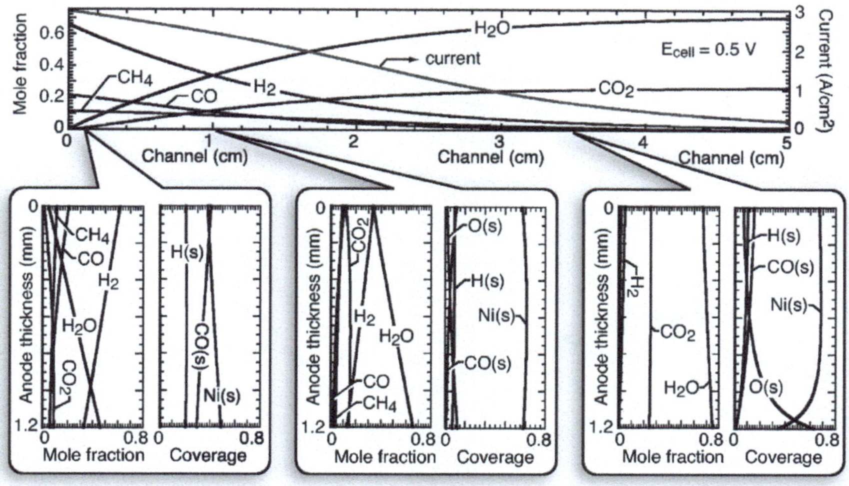

Fig. 6 Simulated distribution of species and current density for a fuel mixture of 66 % H_2, 22 % CO, and 12 % CH_4 entering the anode channel. Cell is operating at constant temperature (800 °C) and atmospheric pressure. The *upper panel* shows mole fractions and current density as functions of distance along the channel length. *Lower panels* show mole fractions and surface coverages through the thickness of the porous anode at different positions along the channel. [Reprinted with permission from Zhu et al. (2005)]

An example of 2-D model that encompasses both the electrochemical and chemical phenomena occurring in an SOFC is that developed by Zhu et al. (2005). The model describes a planar, anode-supported SOFC operating with carbon/hydrogen mixtures with co-flowing channels. A steady plug-flow model is applied to describe the channel flows, and the DGM model (29) is used to describe the diffusion in porous electrodes. Electrochemistry is implemented by considering Nernst (64) and Butler–Volmer Eqs. (70) for the hydrogen oxidation reaction, while the heterogeneous chemistry is described by implementing the elementary kinetics of the reaction mechanism assumed for hydrogen/carbon mixtures over Ni/YSZ. The model predicts the current–voltage performance of the cell and the species distribution in the channels and within the anode structure (Fig. 6).

3-D models Three-dimensional models are generally used for the simulation at full-stack or SRU level to accurately describe fluidic and thermal fields on complex geometries.

An example is the 3-D model developed by Qu et al. (2011). The model describes an anode-supported planar SOFC with corrugated bipolar plates acting both as gas channels and current collectors (Fig. 7).

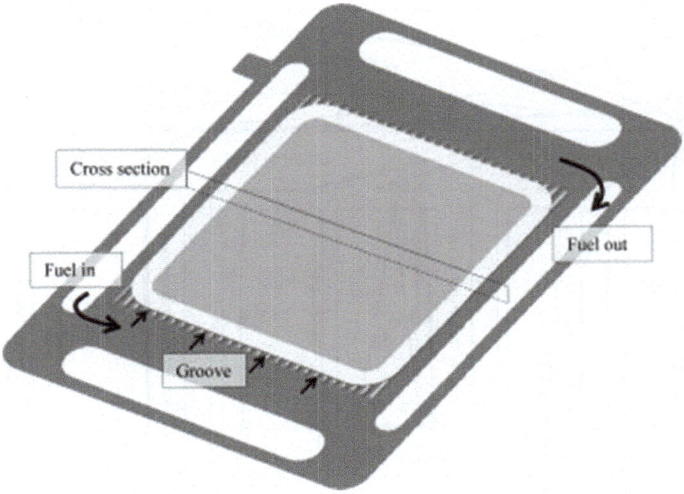

Fig. 7 Schematic drawing of the fuel flow and configuration of the modeled SOFC. [Reprinted with permission from Qu et al. (2011)]

The model applies the Navier–Stokes equations in the fluidic domains and describes the diffusion by using the SMM model (15). Heat transfer by conduction and convection is considered in the model, and the electrochemistry is described by applying the polarization Eq. (68) in which Nernst (64) and Butler–Volmer equations (70) are used to describe reversible voltage and activation overpotentials. The conservation of mass, momentum, energy, and species are solved together with the electrochemical equations, and the distributions of temperature, flow velocity, pressure, and species concentrations through the cell structure and gas channels are obtained.

The model is applied to study the current distribution in the SRU and results indicate that quite uniform distributions of current density over the active cell area can be achieved with the investigated geometry. The results also indicate that the geometry of cathode gas channel has a non-negligible effect on the oxygen distribution and thus on the overall cell performance (Fig. 8).

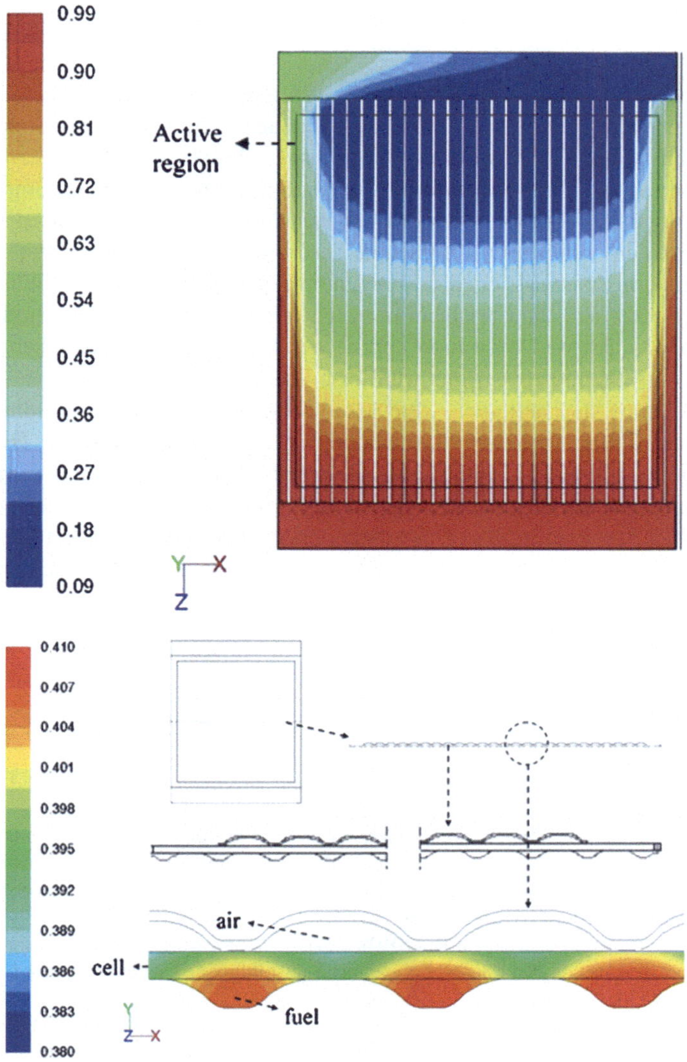

Fig. 8 Mole fraction of H_2 in the anode. *Upper panel top view* of anode surface; *lower panel* SRU cross section (half-cell length). Simulations performed at 973 K, with 5000 A/m^2 average current density and 95 % H_2–5 % H_2O inlet fuel and air on the cathode side. [Reprinted with permission from Qu et al. (2011)]

7 Summary and Conclusions

A SOFC is a complex system consisting of different components, each one composed of peculiar materials in which interconnected physical phenomena occur simultaneously involving gas and solid phases. Modeling an SOFC can be performed following different approaches, depending on the particular component on which the model is focused. This chapter introduced a physically based modeling approach, which entails the macroscopic description of the phenomena.

The equations of mass, energy, momentum, charge transport, and conservation were introduced, and their application in the different domains (i.e., fluid, porous, and solid) was assessed. Common approaches for modeling electrochemistry and heterogeneous chemical reactions were discussed.

Results from the literature were also presented to show multidimensional applications of the physically based modeling approach described. The mathematical models introduced and the examples given show that a universal formulation suitable for solving all SOFC modeling problems does not exist. Indeed, the way in which the description of the interconnected phenomena is established in a model framework strongly depends on the level of approximation desired and on the final objective of the model.

References

Achenbach, E., & Riensche, E. (1994). Methane/steam reforming kinetics for solid oxide fuel cells. *Journal of Power Sources, 52*(2), 283–288.

Ahmed, K., & Foger, K. (2000). Kinetics of internal steam reforming of methane on Ni/YSZ-based anodes for solid oxide fuel cells. *Catalysis Today, 63*(2), 479–487.

Amhalhel, G., & Furmański, P. (1997). Problems of modeling flow and heat transfer in porous media. *Journal of Power Technologies, 85*, 55–88.

Andersson, M., Yuan, J., & Sundén, B. (2010). Review on modeling development for multiscale chemical reactions coupled transport phenomena in solid oxide fuel cells. *Applied Energy, 87* (5), 1461–1476.

Andersson, M., Yuan, J., & Sundén, B. (2012). SOFC modeling considering electrochemical reactions at the active three phase boundaries. *International Journal of Heat and Mass Transfer, 55*(4), 773–788.

Andersson, M., Yuan, J., & Sundén, B. (2013). SOFC modeling considering hydrogen and carbon monoxide as electrochemical reactants. *Journal of Power Sources, 232*, 42–54.

Bagotsky, V. S. (Ed.). (2005). *Fundamentals of electrochemistry* (2nd ed., Vol. 44). New York: John Wiley & Sons.

Bear, J. (1972). *Dynamics of fluids in porous media*. New York: American Elsevier.

Belyaev, V. D., Politova, T. I., Mar'ina, O. A., & Sobyanin, V. A. (1995). Internal steam reforming of methane over Ni-based electrode in solid oxide fuel cells. *Applied Catalysis, A: General, 133*(1), 47–57.

Bertei, A., & Nicolella, C. (2015). Common inconsistencies in modeling gas transport in porous electrodes: The dusty-gas model and the Fick law. *Journal of Power Sources, 279*, 133–137.

Bird, R., Steward, W., & Lightfoot, E. (2006). *Transport phenomena*. Amsterdam: John Wiley & Sons. (Revised 2nd edn).

Brinkman, H. C. (1949a). A calculation of the viscous force exerted by a flowing fluid on a dense swarm of particles. *Applied Scientific Research, 1*(1), 27–34.

Brinkman, H. C. (1949b). On the permeability of media consisting of closely packed porous particles. *Applied Scientific Research, 1*(1), 81–86.

Bruggeman, D. A. G. (1935). Calculation of the various physical constants of heterogeneous substances. Dielectric constants and conductivities of mixtures of isotropic substances. *Annual Physik, 24*, 636–664.

Brus, G., Miyawaki, K., Iwai, H., Saito, M., & Yoshida, H. (2014). Tortuosity of an SOFC anode estimated from saturation currents and a mass transport model in comparison with a real micro-structure. *Solid State Ionics, 265*, 13–21.

Cai, Q., Adjiman, C. S., & Brandon, N. P. (2011). Investigation of the active thickness of solid oxide fuel cell electrodes using a 3D microstructure model. *Electrochimica Acta, 56*(28), 10809–10819.

Carman, P. C. (1956). *Flow of gases through porous media*. Waltham: Academic Press.

Cayan, F. N., Pakalapati, S. R., Elizalde-Blancas, F., & Celik, I. (2009). On modeling multi-component diffusion inside the porous anode of solid oxide fuel cells using Fick's model. *Journal of Power Sources, 192*(2), 467–474.

Chan, S. H., Khor, K. A., & Xia, Z. T. (2001). *Journal of Power Sources, 93*(1), 130–140.

Costamagna, P., Costa, P., & Antonucci, V. (1998). Micro-modelling of solid oxide fuel cell electrodes. *Electrochimica Acta, 43*(3), 375–394.

Costamagna, P., & Honegger, K. (1998). Modeling of solid oxide heat exchanger integrated stacks and simulation at high fuel utilization. *Journal of the Electrochemical Society, 145*(11), 3995–4007.

Costamagna, P., Selimovic, A., Del Borghi, M., & Agnew, G. (2004). Electrochemical model of the integrated planar solid oxide fuel cell (IP-SOFC). *Chemical Engineering Journal, 102*(1), 61–69.

Curtiss, C. F., & Bird, R. B. (1999). Multicomponent diffusion. *Industrial and Engineering Chemistry Research, 38*(7), 2515–2522.

Damm, D. L., & Fedorov, A. G. (2004). Spectral radiative heat transfer analysis of the planar SOFC. In *Proceedings of the ASME IMECE*, Anaheim, CA, November 13–19, 2004. Paper No. IMECE2004-60142.

Damm, D. L., & Fedorov, A. G. (2005). Radiation heat transfer in SOFC materials and components. *Journal of Power Sources, 143*(1), 158–165.

Damm, D. L., & Fedorov, A. G. (2006). Local thermal non-equilibrium effects in porous electrodes of the hydrogen-fueled SOFC. *Journal of Power Sources, 159*(2), 1153–1157.

Désilets, M., Proulx, P., & Soucy, G. (1997). Modeling of multicomponent diffusion in high temperature flows. *International Journal of Heat and Mass Transfer, 40*(18), 4273–4278.

Dicks, A. L., Pointon, K. D., & Siddle, A. (2000). Intrinsic reaction kinetics of methane steam reforming on a nickel/zirconia anode. *Journal of Power Sources, 86*(1), 523–530.

Drescher, I., Lehnert, W., & Meusinger, J. (1998). Structural properties of SOFC anodes and reactivity. *Electrochimica Acta, 43*(19), 3059–3068.

Elizalde-Blancas, F., Celik, I. B., Rangel-Hernandez, V., Hernandez-Guerrero, A., & Riesco-Avila, J. M. (2013). Numerical modeling of SOFCs operating on biogas from biodigesters. *International Journal of Hydrogen Energy, 38*(1), 377–384.

Ferguson, J. R., Fiard, J. M., & Herbin, R. (1996). Three-dimensional numerical simulation for various geometries of solid oxide fuel cells. *Journal of Power Sources, 58*(2), 109–122.

Ferrero, D., Lanzini, A., Leone, P., & Santarelli, M. (2015). Reversible operation of solid oxide cells under electrolysis and fuel cell modes: Experimental study and model validation. *Chemical Engineering Journal, 274*, 143–155.

Froment, G. F., Bischoff, K. B., & De Wilde, J. (1990). *Chemical reactor analysis and design* (Vol. 2). New York: Wiley.

Fuller, E. N., Schettler, P. D., & Giddings, J. C. (1966). New method for prediction of binary gas-phase diffusion coefficients. *Industrial and Engineering Chemistry, 58*(5), 18–27.

Funahashi, Y., Shimamori, T., Suzuki, T., Fujishiro, Y., & Awano, M. (2007). Fabrication and characterization of components for cube shaped micro tubular SOFC bundle. *Journal of Power Sources, 163*(2), 731–736.

García-Camprubí, M. (2011). *Multiphysics models for the simulation of solid oxide fuel cells*. (Ph. D., dissertation). University of Zaragoza.

García-Camprubí, M., Sánchez-Insa, A., & Fueyo, N. (2010). Multimodal mass transfer in solid-oxide fuel-cells. *Chemical Engineering Science, 65*(5), 1668–1677.

Geisler, H., Kromp, A., Weber, A., & Ivers-Tiffée, E. (2014). Stationary FEM model for performance evaluation of planar solid oxide fuel cells connected by metal interconnectors I. Model framework and validation. *Journal of the Electrochemical Society, 161*(6), F778–F788.

Goldin, G. M., Zhu, H., Kee, R. J., Bierschenk, D., & Barnett, S. A. (2009). Multidimensional flow, thermal, and chemical behavior in solid-oxide fuel cell button cells. *Journal of Power Sources, 187*(1), 123–135.

Goodwin, D. G., Zhu, H., Colclasure, A. M., & Kee, R. J. (2009). Modeling electrochemical oxidation of hydrogen on Ni–YSZ pattern anodes. *Journal of the Electrochemical Society, 156* (9), B1004–B1021.

Greene, E. S., Chiu, W. K., & Medeiros, M. G. (2006). Mass transfer in graded microstructure solid oxide fuel cell electrodes. *Journal of Power Sources, 161*(1), 225–231.

Grew, K. N., & Chiu, W. K. (2012). A review of modeling and simulation techniques across the length scales for the solid oxide fuel cell. *Journal of Power Sources, 199*, 1–13.

Gupta, G. K., Hecht, E. S., Zhu, H., Dean, A. M., & Kee, R. J. (2006). Gas-phase reactions of methane and natural-gas with air and steam in non-catalytic regions of a solid-oxide fuel cell. *Journal of Power Sources, 156*(2), 434–447.

Haberman, B. A., & Young, J. B. (2004). Three-dimensional simulation of chemically reacting gas flows in the porous support structure of an integrated-planar solid oxide fuel cell. *International Journal of Heat and Mass Transfer, 47*(17), 3617–3629.

Hajimolana, S. A., Hussain, M. A., Daud, W. A. W., Soroush, M., & Shamiri, A. (2011). Mathematical modeling of solid oxide fuel cells: A review. *Renewable and Sustainable Energy Reviews, 15*(4), 1893–1917.

Hanna, J., Lee, W. Y., Shi, Y., & Ghoniem, A. F. (2014). Fundamentals of electro-and thermochemistry in the anode of solid-oxide fuel cells with hydrocarbon and syngas fuels. *Progress in Energy and Combustion Science, 40*, 74–111.

Hao, Y., & Goodwin, D. G. (2008). Numerical study of heterogeneous reactions in an SOFC anode with oxygen addition. *Journal of the Electrochemical Society, 155*(7), B666–B674.

He, W., Lu, W., & Dickerson, J. H. (2014a). *Chapter 2: Gas diffusion in porous media gas transport in solid oxide fuel cells* (1st ed., pp. 9–17). New York: Springer.

He, W., Lu, W., & Dickerson, J. H. (2014b). *Gas transport in solid oxide fuel cells*. New York: Springer.

Hecht, E. S., Gupta, G. K., Zhu, H., Dean, A. M., Kee, R. J., Maier, L., et al. (2005). Methane reforming kinetics within a Ni–YSZ SOFC anode support. *Applied Catalysis, A: General, 295*(1), 40–51.

Hernández-Pacheco, E., Singh, D., Hutton, P. N., Patel, N., & Mann, M. D. (2004). A macro-level model for determining the performance characteristics of solid oxide fuel cells. *Journal of Power Sources, 138*(1), 174–186.

Hirschfelder, J. O., Curtiss, C. F., Bird, R. B., & Mayer, M. G. (1954). *Molecular theory of gases and liquids* (Vol. 26, p. 10). New York: Wiley.

Ho, T. X., Kosinski, P., Hoffmann, A. C., & Vik, A. (2008). Numerical modeling of solid oxide fuel cells. *Chemical Engineering Science, 63*(21), 5356–5365.

Ho, T. X., Kosinski, P., Hoffmann, A. C., & Vik, A. (2009). Modeling of transport, chemical and electrochemical phenomena in a cathode-supported SOFC. *Chemical Engineering Science, 64*(12), 3000–3009.

Hofmann, P., Panopoulos, K. D., Fryda, L. E., & Kakaras, E. (2009). Comparison between two methane reforming models applied to a quasi-two-dimensional planar solid oxide fuel cell model. *Energy, 34*(12), 2151–2157.

Hosoi, T., Yonekura, T., Sunada, K., & Sasaki, K. (2015). Exchange current density of SOFC electrodes: Theoretical relations and partial pressure dependencies rate-determined by electrochemical reactions. *Journal of the Electrochemical Society, 162*(1), F136–F152.

Hou, K., & Hughes, R. (2001). The kinetics of methane steam reforming over a Ni/α-Al_2O catalyst. *Chemical Engineering Journal, 82*(1), 311–328.

Hsu, C. T., & Cheng, P. (1990). Thermal dispersion in a porous medium. *International Journal of Heat and Mass Transfer, 33*(8), 1587–1597.

Hussain, M. M., Li, X., & Dincer, I. (2005). Multi-component mathematical model of solid oxide fuel cell anode. *International Journal of Energy Research, 29*(12), 1083–1101.

Iwai, H., Shikazono, N., Matsui, T., Teshima, H., Kishimoto, M., Kishida, R., et al. (2010). Quantification of SOFC anode microstructure based on dual beam FIB-SEM technique. *Journal of Power Sources, 195*(4), 955–961.

Janardhanan, V. M., & Deutschmann, O. (2006). CFD analysis of a solid oxide fuel cell with internal reforming: Coupled interactions of transport, heterogeneous catalysis and electrochemical processes. *Journal of Power Sources, 162*(2), 1192–1202.

Janardhanan, V. M., & Deutschmann, O. (2007). Numerical study of mass and heat transport in solid-oxide fuel cells running on humidified methane. *Chemical Engineering Science, 62*(18), 5473–5486.

Janardhanan, V. M., Heuveline, V., & Deutschmann, O. (2008). Three-phase boundary length in solid-oxide fuel cells: A mathematical model. *Journal of Power Sources, 178*(1), 368–372.

Jiang, Y., & Virkar, A. V. (2003). Fuel composition and diluent effect on gas transport and performance of anode-supported SOFCs. *Journal of the Electrochemical Society, 150*(7), A942–A951.

Joos, J., Carraro, T., Weber, A., & Ivers-Tiffée, E. (2011). Reconstruction of porous electrodes by FIB/SEM for detailed microstructure modeling. *Journal of Power Sources, 196*(17), 7302–7307.

Jung, H. Y., Kim, W. S., Choi, S. H., Kim, H. C., Kim, J., Lee, H. W., et al. (2006). Effect of cathode current-collecting layer on unit-cell performance of anode-supported solid oxide fuel cells. *Journal of Power Sources, 155*(2), 145–151.

Kast, W., & Hohenthanner, C. R. (2000). Mass transfer within the gas-phase of porous media. *International Journal of Heat and Mass Transfer, 43*(5), 807–823.

Kerkhof, P. J. (1996). A modified Maxwell–Stefan model for transport through inert membranes: The binary friction model. *The Chemical Engineering Journal and the Biochemical Engineering Journal, 64*(3), 319–343.

Kim, J. H., Liu, W. K., & Lee, C. (2009). Multi-scale solid oxide fuel cell materials modeling. *Computational Mechanics, 44*(5), 683–703.

Kishimoto, M., Iwai, H., Saito, M., & Yoshida, H. (2011). Quantitative evaluation of solid oxide fuel cell porous anode microstructure based on focused ion beam and scanning electron microscope technique and prediction of anode overpotentials. *Journal of Power Sources, 196*(10), 4555–4563.

Klein, J. M., Bultel, Y., Georges, S., & Pons, M. (2007). Modeling of a SOFC fuelled by methane: From direct internal reforming to gradual internal reforming. *Chemical Engineering Science, 62*(6), 1636–1649.

Kong, W., Zhu, H., Fei, Z., & Lin, Z. (2012). A modified dusty gas model in the form of a Fick's model for the prediction of multicomponent mass transport in a solid oxide fuel cell anode. *Journal of Power Sources, 206*, 171–178.

Krishna, R., & Wesselingh, J. A. (1997). The Maxwell–Stefan approach to mass transfer. *Chemical Engineering Science, 52*(6), 861–911.

Lage, J. L. (1993). Natural convection within a porous medium cavity: Predicting tools for flow regime and heat transfer. *International Communications in Heat and Mass Transfer, 20*(4), 501–513.

Lanzini, A., Leone, P., & Asinari, P. (2009). Microstructural characterization of solid oxide fuel cell electrodes by image analysis technique. *Journal of Power Sources, 194*(1), 408–422.

Laurencin, J., Kane, D., Delette, G., Deseure, J., & Lefebvre-Joud, F. (2011). Modelling of solid oxide steam electrolyser: Impact of the operating conditions on hydrogen production. *Journal of Power Sources, 196*(4), 2080–2093.

Lee, A. L., Zabransky, R. F., & Huber, W. J. (1990). Internal reforming development for solid oxide fuel cells. *Industrial and Engineering Chemistry Research, 29*(5), 766–773.

Lee, K. T., Vito, N. J., & Wachsman, E. D. (2013). Comprehensive quantification of Ni–$Gd_{0.1}Ce_{0.9}O_{1.95}$ anode functional layer microstructures by three-dimensional reconstruction using a FIB/SEM dual beam system. *Journal of Power Sources, 228*, 220–228.

Lehnert, W., Meusinger, J., & Thom, F. (2000). Modelling of gas transport phenomena in SOFC anodes. *Journal of Power Sources, 87*(1), 57–63.

Leonide, A. (2010). *SOFC modelling and parameter identification by means of impedance spectroscopy* (Ph.D. dissertation). Karlsruher Institut für Technologie.

Li, Y., Gemmen, R., & Liu, X. (2010). Oxygen reduction and transportation mechanisms in solid oxide fuel cell cathodes. *Journal of Power Sources, 195*(11), 3345–3358.

Mason, E. A., & Malinauskas, A. P. (1983). *Gas transport in porous media: The dusty-gas model* (pp. 1–202). New York: Elsevier.

Matsuzaki, K., Shikazono, N., & Kasagi, N. (2011). Three-dimensional numerical analysis of mixed ionic and electronic conducting cathode reconstructed by focused ion beam scanning electron microscope. *Journal of Power Sources, 196*(6), 3073–3082.

Menon, V., Janardhanan, V. M., Tischer, S., & Deutschmann, O. (2013). Internal multi-physics phenomena of SOFC with direct internal reforming. *ECS Transactions, 57*(1), 2475–2484.

Modest, M. F. (2013). *Radiative heat transfer* (3rd ed.). New York: Academic Press.

Moon, H., Kim, S. D., Park, E. W., Hyun, S. H., & Kim, H. S. (2008). Characteristics of SOFC single cells with anode active layer via tape casting and co-firing. *International Journal of Hydrogen Energy, 33*(11), 2826–2833.

Murthy, S., & Fedorov, A. G. (2003). Radiation heat transfer analysis of the monolith type solid oxide fuel cell. *Journal of Power Sources, 124*(2), 453–458.

Nagel, F. P., Schildhauer, T. J., Biollaz, S. M., & Stucki, S. (2008). Charge, mass and heat transfer interactions in solid oxide fuel cells operated with different fuel gases—a sensitivity analysis. *Journal of Power Sources, 184*(1), 129–142.

Nam, J. H., & Jeon, D. H. (2006). A comprehensive micro-scale model for transport and reaction in intermediate temperature solid oxide fuel cells. *Electrochimica Acta, 51*(17), 3446–3460.

Ni, M. (2009). Computational fluid dynamics modeling of a solid oxide electrolyzer cell for hydrogen production. *International Journal of Hydrogen Energy, 34*(18), 7795–7806.

Ni, M. (2013). Modeling and parametric simulations of solid oxide fuel cells with methane carbon dioxide reforming. *Energy Conversion and Management, 70*, 116–129.

Ni, M., Leung, M. K., & Leung, D. Y. (2007). Parametric study of solid oxide fuel cell performance. *Energy Conversion and Management, 48*(5), 1525–1535.

Nield, D. A., & Bejan, A. (2006). *Chapter 1: Mechanics of fluid flow through a porous medium—convection in porous media* (3rd ed., pp. 1–26). New York: Springer.

Noren, D. A., & Hoffman, M. A. (2005). Clarifying the Butler–Volmer equation and related approximations for calculating activation losses in solid oxide fuel cell models. *Journal of Power Sources, 152*, 175–181.

Novaresio, V., García-Camprubí, M., Izquierdo, S., Asinari, P., & Fueyo, N. (2012). An open-source library for the numerical modeling of mass-transfer in solid oxide fuel cells. *Computer Physics Communications, 183*(1), 125–146.

Park, E. W., Moon, H., Park, M. S., & Hyun, S. H. (2009). Fabrication and characterization of Cu–Ni–YSZ SOFC anodes for direct use of methane via Cu-electroplating. *International Journal of Hydrogen Energy, 34*(13), 5537–5545.

Poling, B. E., Prausnitz, J. M., & O'connell, J. P. (2001). *The properties of gases and liquids* (Vol. 5, pp. 11.5–11.9). New York: McGraw-Hill.

Pollard, W. G., & Present, R. D. (1948). On gaseous self-diffusion in long capillary tubes. *Physical Review, 73*(7), 762.

Qu, Z., Aravind, P. V., Boksteen, S. Z., Dekker, N. J. J., Janssen, A. H. H., Woudstra, N., et al. (2011). Three-dimensional computational fluid dynamics modeling of anode-supported planar SOFC. *International Journal of Hydrogen Energy, 36*(16), 10209–10220.

Ramshaw, J. D. (1990). Self-consistent effective binary diffusion in multicomponent gas mixtures. *Journal of Non-Equilibrium Thermodynamics, 15*(3), 295–300.

Sanchez, D., Chacartegui, R., Munoz, A., & Sanchez, T. (2008). On the effect of methane internal reforming modelling in solid oxide fuel cells. *International Journal of Hydrogen Energy, 33*(7), 1834–1844.

Sánchez, D., Munoz, A., & Sánchez, T. (2007). An assessment on convective and radiative heat transfer modelling in tubular solid oxide fuel cells. *Journal of Power Sources, 169*(1), 25–34.

Santarelli, M., Quesito, F., Novaresio, V., Guerra, C., Lanzini, A., & Beretta, D. (2013). Direct reforming of biogas on Ni-based SOFC anodes: Modelling of heterogeneous reactions and validation with experiments. *Journal of Power Sources, 242*, 405–414.

Shi, Y., Cai, N., & Li, C. (2007). Numerical modeling of an anode-supported SOFC button cell considering anodic surface diffusion. *Journal of Power Sources, 164*(2), 639–648.

Shi, Y., Li, C., & Cai, N. (2011). Experimental characterization and mechanistic modeling of carbon monoxide fueled solid oxide fuel cell. *Journal of Power Sources, 196*(13), 5526–5537.

Suwanwarangkul, R., Croiset, E., Fowler, M. W., Douglas, P. L., Entchev, E., & Douglas, M. A. (2003). Performance comparison of Fick's, dusty-gas and Stefan–Maxwell models to predict the concentration overpotential of a SOFC anode. *Journal of Power Sources, 122*(1), 9–18.

Tseronis, K., Kookos, I. K., & Theodoropoulos, C. (2008). Modelling mass transport in solid oxide fuel cell anodes: A case for a multidimensional dusty gas-based model. *Chemical Engineering Science, 63*(23), 5626–5638.

Veldsink, J. W., Van Damme, R. M. J., Versteeg, G. F., & Van Swaaij, W. P. M. (1995). The use of the dusty-gas model for the description of mass transport with chemical reaction in porous media. *The Chemical Engineering Journal and the Biochemical Engineering Journal, 57*(2), 115–125.

Walters, K. M., Dean, A. M., Zhu, H., & Kee, R. J. (2003). Homogeneous kinetics and equilibrium predictions of coking propensity in the anode channels of direct oxidation solid-oxide fuel cells using dry natural gas. *Journal of Power Sources, 123*(2), 182–189.

Wang, K., Hissel, D., Péra, M. C., Steiner, N., Marra, D., Sorrentino, M., et al. (2011a). A review on solid oxide fuel cell models. *International Journal of Hydrogen Energy, 36*(12), 7212–7228.

Wang, S., Worek, W. M., & Minkowycz, W. J. (2012). Performance comparison of the mass transfer models with internal reforming for solid oxide fuel cell anodes. *International Journal of Heat and Mass Transfer, 55*(15), 3933–3945.

Wang, Y., Weng, S., & Weng, Y. (2011b). Numerical investigation of the chemical and electrochemical characteristics of planar solid oxide fuel cell with direct internal reforming. *Frontiers in Energy, 5*(2), 195–206.

Webb, S. W., & Pruess, K. (2003). The use of Fick's law for modeling trace gas diffusion in porous media. *Transport in Porous Media, 51*(3), 327–341.

Welty, J. R., Wicks, C. E., Rorrer, G., & Wilson, R. E. (2001). *Fundamentals of momentum, heat, and mass transfer* (4th ed.). New York: John Wiley & Sons.

Wilke, C. R. (1950a). A viscosity equation for gas mixtures. *The Journal of Chemical Physics, 18*(4), 517–519.

Wilke, C. R. (1950b). Diffusional properties of multicomponent gases. *Chemical Engineering Progress, 46*, 95–104.

Wilson, J. R., Cronin, J. S., & Barnett, S. A. (2011). Linking the microstructure, performance and durability of Ni-yttria-stabilized zirconia solid oxide fuel cell anodes using three-dimensional focused ion beam–scanning electron microscopy imaging. *Scripta Materialia, 65*(2), 67–72.

Wu, W., Wang, G. L., Guan, W. B., Zhen, Y. F., & Wang, W. G. (2013). Effect of contact method between interconnects and electrodes on area specific resistance in planar solid oxide fuel cells. *Fuel Cells, 13*(5), 743–750.

Xu, J., & Froment, G. F. (1989). Methane steam reforming, methanation and water–gas shift: I. Intrinsic kinetics. *AIChE Journal, 35*(1), 88–96.

Yakabe, H., Hishinuma, M., Uratani, M., Matsuzaki, Y., & Yasuda, I. (2000). Evaluation and modeling of performance of anode-supported solid oxide fuel cell. *Journal of Power Sources, 86*(1), 423–431.

Yakabe, H., Ogiwara, T., Hishinuma, M., & Yasuda, I. (2001). 3-D model calculation for planar SOFC. *Journal of Power Sources, 102*(1), 144–154.

Zheng, K., Sun, Q., & Ni, M. (2013). Local non-equilibrium thermal effects in solid oxide fuel cells with various fuels. *Energy Technology, 1*(1), 35–41.

Zhu, H., & Kee, R. J. (2003). A general mathematical model for analyzing the performance of fuel-cell membrane-electrode assemblies. *Journal of Power Sources, 117*(1), 61–74.

Zhu, H., & Kee, R. J. (2008). Modeling distributed charge-transfer processes in SOFC membrane electrode assemblies. *Journal of the Electrochemical Society, 155*(7), B715–B729.

Zhu, H., Kee, R. J., Janardhanan, V. M., Deutschmann, O., & Goodwin, D. G. (2005). Modeling elementary heterogeneous chemistry and electrochemistry in solid-oxide fuel cells. *Journal of the Electrochemical Society, 152*(12), A2427–A2440.

The manufacturer's authorised representative in the EU is Springer Nature Customer Service Centre GmbH, Europaplatz 3, 69115 Heidelberg, Germany. If you have any concerns regarding our products, please contact ProductSafety@springernature.com

Printed and bound by CPI Group (UK) Ltd, Croydon, CR0 4YY
30/09/2026
02225572-0014